PREVENTION AND COMPENSATION OF MARINE POLLUTION DAMAGE

RECENT DEVELOPMENTS IN EUROPE, CHINA AND THE US

Comparative Environmental Law & Policy Series

VOLUME 9

Editors
Eric W. Orts
Professor of Legal Studies and Director, Environmental Management Program
The Wharton School, University of Pennsylvania

Kurt Deketelaere
Professor of Law and Director, Institute for Environmental and Energy Law
University of Leuven

Editorial Board
Professor Ben Boer, *University of Sydney*
Professor Michael Faure, *University of Maastricht*
Professor Alberto José Blanco-Uribe Quintero, *Central University of Venezuela*
Professor Richard Macrory, *University College London*
Professor Ari Ekroos, *Helsinki University of Technology*
Professor Jerzy Sommer, *Polish Academy of Sciences*
Professor Willemien du Plessis, *Potchefstroom University*
Professor Richard Revesz, *New York University*
Professor Daniel Esty, *Yale University*
Professor René Seerden, *University of Maastricht*
Professor Cary Coglianese, *Harvard University*

The aim of the Editors and the Editorial Board of this series is to publish works of excellent quality that focus on the comparative study of environmental law and policy of countries of groups of countries.

Through the comparative study of environmental law and policy, the Editors and the Editorial Board hope:

- to contribute to the improvement of the quality of environmental law and policy in general and environmental quality in particular;
- to increase the access to environmental information and environmental justice for students, academics, non-governmental organizations, government institutions, and business;
- to facilitate cooperation between academic and non-academic communities in the field of environmental law and policy throughout the world.

PREVENTION AND COMPENSATION OF MARINE POLLUTION DAMAGE

RECENT DEVELOPMENTS IN EUROPE, CHINA AND THE US

MICHAEL G. FAURE AND JAMES HU (EDS.)

A C.I.P. catalogue record for this book is available from the Library of Congress.

ISBN 90 411 2338 5

Published by:
Kluwer Law International
P.O. Box 316
2400 AH Alphen aan den Rijn
The Netherlands

Sold and distributed in North, Central and South America by:
Aspen Publishers, Inc.
7201 McKinney Circle
Frederick, MD 21704
United States of America

Sold and distributed in all other countries by:
Turpin Distribution Services Ltd
Stratton Business Park
Pegasus Drive
Biggleswade
Bedfordshire SG18 8TQ
United Kingdom

Printed on acid-free paper

© 2006 Kluwer Law International

All rights reserved. No part of this publication may be reproduced, stored in a retrieval system, or transmitted in any form or by any means, mechanical, photocopying, recording or otherwise, without prior written permission of the publishers.

Permission to use this content must be obtained from the copyright owner. Please apply to Permissions Department, Wolters Kluwer Legal, 111 Eighth Avenue, 7th Floor, New York, NY 10011-5201, United States of America, E-mail: permissions@kluwerlaw.com, Website: www.kluwerlaw.com.

PART I: PREVENTION AND SANCTIONING OF MARINE POLLUTION DAMAGE

FOREWORD

1 DEFINING THE PROBLEM

It hardly needs any explanation that oil pollution damage causes a lot of turmoil in Europe and China as well as in the US. Europe has already been awakened by the dangers of oil pollution damage, first with the Torrey Canyon incident, then with the Amoco Cadiz and more recently with the Erica and the Prestige. With the well known Exxon Valdez case that caused a huge ecological disaster off the coast of Alaska, the consequences of oil pollution became very apparent to the US as well. China has also recently become aware of the risks of oil pollution damage and is all the more interested in taking adequate measures to prevent and compensate for oil pollution damage, given its huge coastline and therefore its increased exposure to the risk of oil pollution damage.

The many incidents with oil tankers leaking enormous quantities of crude oil have shown the huge and devastating consequences of oil pollution damage. The world has probably been exposed most of all to these horrible consequences during Exxon Valdez incident in Alaska, when pictures of birds, but also of mammals like seals, all covered with oil, were shown all over the world. In addition to the ecological disaster such an incident causes, many also point to the huge economic consequences of these (but also smaller) incidents. Many people working in the coastal zones are dependent upon fish, clean water and beaches for their source of income, not only because of fishery activities, but sometimes also merely because of tourism. It often takes many years before the damaged coastlines are fully recovered, if at all.

It therefore hardly needs to be demonstrated that all pollution incidents are of high social relevance. The same is true, moreover, for the role the law could play in fighting these incidents. Indeed, already since the 1960s, conventions have been drafted with precisely the aim, on the one hand, to prevent the likelihood of oil pollution damage and, on the other hand, to provide adequate compensation in case such an incident would occur. However, many of the above mentioned large-scale incidents have quite clearly shown that these conventions were apparently not able to provide adequate compensation. For that reason, serious amendments to the international conventions were made, but the question still arises whether the legal framework currently in place is adequate for dealing with oil pollution damage.

More specifically, the question arises whether today's international framework is indeed sufficient to, on the one hand, prevent oil pollution damage and, on the other hand, to provide compensation for it. Serious questions can still be posed, for example, concerning the financial caps (limitations) that traditionally exist in maritime law. Up to now, financial limitations on liability have been defended in maritime law since the usual victims in maritime law are the cargo interests who can seek protection themselves in case of the loss of cargo via first party insurance. However, this justification for financial caps is of course not applicable when it is not a contracting party but a third party who suffers the loss. Hence, the question arises whether the current financial caps, even after the recent revisions, are adequate to provide compensation for oil pollution damage. These serious doubts persist, as can be seen in the fact that the European Union took an initiative to set up its own Fund precisely because the amounts covered under the conventions have proven to be insufficient to cover the losses caused in the Erica incident.[1] Recent incidents, as were mentioned above, have shown that even after the implementation of amendments to the international conventions, one can question whether the still limited amounts available are sufficient to provide adequate compensation.

In addition, the question of course arises whether compensation for oil pollution damage should merely be provided via liability rules and traditional insurance. Already in the past, fund solutions have been implemented to deal (partially) with oil pollution damage as well. An open question is therefore how an optimal combination of liability rules, insurance and fund solutions can be put into place in order to provide adequate compensation.

Oil pollution damage is a highly important domain from a legal perspective since it clearly shows that sometimes a wide gap may exist between the 'law in the books' and the 'law in practice'. Indeed, many countries have (usually as a result of international conventions) implemented legislation with the aim of preventing (intentional or accidental) oil pollution damage. The major problem is, however, that many of these incidents occur at high seas where various questions arise with regard to enforcement. This issue not only raises questions concerning the competence of authorities, but also raises questions with respect to the enforcement strategies to be followed and the potential sanctions to be applied. Many have argued that an effective enforcement of oil pollution prevention legislation is often lacking.

Research into legal aspects of oil pollution damage is therefore of great importance both for Europe and for Asia; even more important is the way the contributions in this book propose to engage in comparative research. Indeed, many European countries have (usually as a result of international conventions) implemented oil pollution legislation, usually aiming at the prevention of oil pollution damage and providing for compensation via liability rules and (usually mandatory) insurance. The same is true in the case of Asian countries like China, for example, which has, with its 18.000 km coastline, quite an important incentive to prevent marine oil pollution. China is at the point of redrafting its maritime code (a process which has already been partially concluded). Within this redrafting process China is now also considering whether to introduce a separate chapter in its maritime code, dealing more specifically with the regulation of pollution damage to the marine environment. Thus, it is clearly in the mutual interests of Asian countries like China and European countries to learn from each other's experiences with legal remedies against oil pollution damage. This can be particularly of interest to China within this redrafting process. However, as was indicated above, recent incidents in Europe have also cast doubt with respect to the adequacy of the current legal framework. Therefore, reforms aiming at a more adequate protection against oil pollution damage are also taking place in Europe as well.

The aim of this book is therefore to address the legal remedies against oil pollution damage from a comparative perspective, that is, from both a Chinese and a European

---

[1] See in this respect the proposals for a Regulation on the establishment of a fund for the compensation for oil pollution damage in European waters (*OJ* C227, 24 September 2002).

angle. The basic goal of this comparative approach is that, via this process of mutual learning, new theoretical and empirical insights can be reached that may lead to an improvement of current regulatory structures in Europe and China.

Moreover, this book aims, on the one hand, to gain new theoretical insights which may indicate how the law should be shaped in an optimal way to provide adequate protection against oil pollution damage. It is believed that through this comparative research new theoretical insights can be gained.

2 PROJECT HISTORY

This book is the result of a longstanding cooperation between various Chinese and European institutions that have worked together to make possible the research project on which this book is based. The Chinese editor of this book, James Hu, has had a longstanding relationship with the Catholic University of Leuven (more particularly with Prof. Marc Huybrechts), as a result of which they engaged in joint research and an exchange of students in the area of maritime law. The European editor of this book, Michael Faure, cooperated closely (through the Research Institute Metro) with the Institute of Energy and Environmental Law of the Catholic University of Leuven (more particularly with Prof. Kurt Deketelaere) on issues of environmental law and with the Max Planck Institute for Foreign and International Criminal Law in Freiburg as far as the issue of environmental criminal law was concerned (more particularly, with Prof. Günter Heine and Dr. Thomas Richter), the latter having written a doctoral dissertation on environmental criminal law in China. Hence, through Leuven, the environmental law professor Michael Faure met with the maritime law professor James Hu in Dalian, and there the plan emerged to set up an international conference on marine pollution by analyzing the legal remedies in China and Europe. However, given the fact that the organizers of the project were well aware that especially in the US a large body of literature had emerged on both prevention and compensation of marine pollution, two well-known American scholars in this field were invited to join the project (James Boyd and Mark Cohen).

Originally, the conference resulting from this research project was to take place in July 2003 in cooperation with Dalian Maritime University. However, the outbreak of SARS in China in March 2003 made the organization of the conference in the summer of 2003 impossible. As a result, it was postponed for one year. The benefit of this postponement was that, in the meantime, the Chinese organizer, James Hu, moved from Dalian Maritime University to Shanghai Maritime University, and as a result of this move, scholars from both universities became engaged in the project. Moreover, the Jiangsu Maritime Safety Administration (with important competences especially in the Yangtze River Delta) became involved in the project, thanks to which many practical issues concerning the application of legal remedies in specific cases could also be analyzed.

The conference resulting from this collaborative project was held (with substantial support from the Jiangsu Maritime Safety Administration) in Nanjing from 2-5 July 2004. More than thirty papers were presented by various practitioners and scholars in the field and even more papers were presented in the documentation material for the conference.

Pursuant to an editorial review process, the papers were rewritten after the conference and eighteen of them were selected to appear in this book.

3 METHODOLOGY

As was already shown above, an assortment of different approaches have been taken by the various papers presented in this book.

3.1 LEGAL MULTIDISCIPLINARY

First of all, a legal multidisciplinary approach was used by many of the authors, since it is believed that the problem of oil pollution damage is so important that various legal disciplines should be used to provide a full picture of the optimal ways in which to remedy it. The following disciplines will therefore be drawn upon and combined:

- maritime law: it is obvious that one should start with maritime law, given the fact that marine oil pollution has its main source in ships, and more particularly in tankers.
- environmental law: in environmental law as well, oil pollution damage has been analysed in detail. Many scholars writing about environmental liability and transboundary pollution have analysed the legal schemes regarding oil pollution damage. Thus the insights from environmental law should be used as well.
- liability law: so far, the schemes chosen to compensate for oil pollution damage have originated in liability law. Indeed, the conventions usually rely upon a strict liability rule, even though the amount of compensation due to the victim is limited (financial caps). Thus, liability law should be used as well, in order to investigate, among other reasons, the functions and goals of tort law. It needs to be examined whether these goals can be achieved within the framework of oil pollution damage.
- insurance law: as was indicated above, most conventions have introduced not only strict liability, but also mandatory insurance. Mandatory insurance always raises important questions, from insurance, legal and economic perspectives. For one, it is not clear what type of insurance market should (or will) provide the insurance. Furthermore, it must be settled whether the insurance contract itself should be regulated as well and whether exceptions between the insurance company and the insured can be involved against the third party (the victim).
- criminal law: it was already indicated that effective prevention of oil pollution damage will only be possible if an adequate system of enforcement can also be organized. However, in that case, important questions arise, among others, as to the precise territorial scope of jurisdiction or the particular states that have the powers to police tankers. There is also the question concerning the extent of the States' power to prosecute for oil pollution damage committed within (or outside) its territories. In addition, there lies the question whether enforcement will take place merely on the basis of administrative law or whether criminal law enforcement is possible as well. Finally, criminal law is also needed in order to address the optimal sanctions in cases of oil pollution damage. This is all the more important since it has been indicated in

the literature that the current insurers (the P&I clubs) in practice provide full insurance for the fines imposed in case of oil pollution damage. The insurance of fines may well jeopardize effective deterrence of oil pollution damage. In that case, the question arises whether sanctions other than fines could be imposed and how they could be executed.

3.2 COMPARATIVE APPROACH

This book does not only engage in a legal multidisciplinary approach, but will also rely upon a comparative approach. Thus, the relevant international conventions will be examined, as well as their implementation in Asia and Europe. The importance of comparative research in this domain has already been indicated above.

3.3 LAW AND ECONOMICS

Above, we already pointed to the enormous economic importance of oil pollution damage to the coastal zones involved. Oil pollution damage imposes huge costs and already for this reason research into the optimal legal remedies to fight against oil pollution damage is warranted. Since oil pollution damage can also be seen as an economic problem, we also believe an economic analysis of the law can be used to provide insights into the optimal way of shaping these legal remedies. Indeed, the economic analysis of law, which has largely been developed in US since the 1970s, has paid a great deal of attention to the optimal way in which legal instruments should be shaped in order to prevent accidents. Additionally, optimal compensation mechanisms have been compared. Hence, it is believed that these economic insights can certainly add significant value to the research on the optimal remedies for fighting against oil pollution damage.

For instance, the precise function of liability rules, safety regulations and insurance has been examined in an extensive range of literature within the 'law and economics' tradition. Law and economics scholars have addressed the potential dangers (and benefits) of financial caps, and economists have also addressed optimal enforcement strategies in cases of oil spills. Therefore, the economic analysis of law will equally be used as a methodology within some of the contributions contained in this book (see for instance the contributions by James Boyd and by Michael Faure and Wang Hui).

3.4 CASE STUDIES

Finally, within the set-up of the research project from which this book resulted, it was also considered important to provide good case studies of major marine pollution incidents. Case studies can indeed indicate why in specific instances prevention may have failed (and the marine pollution was therefore caused). In addition, a clear look at various case studies can also provide important indications of the instruments used for compensation. Given the fact that the project was set up just after the Prestige case occurred off the coast of Galicia in Spain, a Spanish expert was for example asked to report on the legal regime applicable to this recent case (see the contribution by Maria Paz Garcia Rubio).

4 FRAMEWORK OF THE PROJECT

The project at the origin of this book was originally (as indicated above) initiated by the Maastricht European Institute for Transnational Legal Research (METRO) in collaboration with the Dalian Maritime University (Dalian, China), equally enjoying the support of the Ius Commune Research School. Subsequently, the Jiangsu Maritime Safety Administration and Shanghai Maritime University joined the project as well, by participating both as far as the contents are concerned and with financial support.

5 TOPICS

Although the issues of prevention and compensation of marine pollution damage are relatively broad, the organizers of the conference and the subsequent book focused on a few specific topics of high academic and societal interest.

A first issue that was obviously to be undertaken is simply to analyze the state of legislation in the international conventions as far as prevention of and compensation for oil pollution damage is concerned. Thus, the conventions are to be described both according to the legal history and more recent reforms. In addition, the implementation of the conventions (if this is the case) in Asia and Europe is to be described and particular attention shall be paid to the relative differences between the two regimes. Moreover, attention will also be paid to the reasons for separate legal regimes in addition to the international conventions. An example of this is the European initiative for a European Fund in addition to the international conventions.[2]

A second topic aims at critically reviewing the current structure of prevention of and compensation for oil pollution damage, whereas the first topic provides a positive analysis of the contents of the conventions the second topic chooses a normative approach. It is clear that particularly in this second topic approaches that allow for a critical analysis, like 'law and economics,' can be used. It can be analysed, for example, whether the current structure of liability rules, combined with insurance and regulation, provides for optimal incentives to prevent oil pollution damage. Also, the above-mentioned financial caps on liability will be critically analysed.

A third topic focuses on how optimal compensation for oil pollution damage could be given, given the shortcomings one can discover today. Indeed, as was mentioned earlier, all agree that the financial limitations of liability, even after the recent revisions, do not guarantee full compensation for oil pollution damage in cases of serious accidents. Therefore, within this third topic it shall be critically analysed why the current insurance markets (more particularly the P&I clubs) are not able to generate higher amounts of compensation. In addition, it will be examined whether a further restructuring of the conventions (or national legislation) is possible and who should bear the costs of increased compensation. Moreover, alternative routes to compensation (other than liability

---

[2] See in this respect e.g. the above mentioned initiative for a 'European' oil pollution fund, but also Directive 2002/84/EC amending the Directives on maritime safety and the prevention of pollution from ships (*OJ* L324/53, 29 November 2002).

and liability insurance) shall be examined as well. Thus, the question of first party insurance by victims (coastal zones) can be examined, as well as the possibility of (supplementary) compensation funds and maybe even the intervention of the state.

A fourth topic will clearly focus on how the optimal enforcement of oil pollution legislation can be achieved. Within this topic, questions of criminal law will play an important role. Hence, the issue of the appropriate jurisdiction in cases of oil pollution crimes has to be examined, but also the optimal control techniques and sanctions. Within this topic, empirical material with respect to optimal legal techniques to prevent oil spills will play an important role. So too will the economic analysis of criminal law, which may indicate how public resources can be optimally allocated to prevent oil pollution crimes.

6 STRUCTURE OF THE PRESENTATION

The book is divided in three parts. Part I deals with the prevention and sanctioning of marine pollution damage, thereby largely focusing on the way in which the law deals with the prevention of marine pollution and the way in which infringements are sanctioned (through administrative or criminal law).

Part II deals with specific case studies concerning compensation for marine pollution damage by focusing on the situation in Indonesia, Belgium, the US and Spain.

Part III deals particularly with compensation for marine pollution damage in China. It of course addresses the way in which the international conventions have been implemented in China (Han Lixing and Guan Zhengyi), but also deals with the application of law in civil liability for oil pollution damage (James Hu and Yang Bo). This part of the book hence clearly shows the extent to which China is struggling with the implementation of the international conventions and with restructuring its own legal system in order to provide a more adequate compensation for oil pollution damage. Part IV contains a set of comparative conclusions made by the editors.

7 CONTRIBUTORS

The contributors to this book come from Europe (Michael Faure, Günter Heine, Marc Huybrechts, Maria Paz Garcia Rubio, Thomas Richter and Katrien Van Damme), from the US (James Boyd and Mark Cohen) from Indonesia (Etty Agoes) and from China (too many to name here). The Chinese contributors are largely connected either to the Shanghai or the Dalian Maritime University or to the Jiangsu Maritime Safety Administration. In addition, various judges (e.g. Cheng Pingping) have contributed to the book as well. A complete list of the contributors and their affiliation is provided after the table of contents.

8 ACKNOWLEDGEMENTS

As editors of this book, we are grateful to the many people who made this project possible. First of all, we would like to thank METRO, the Ius Commune Research School, the Science Committee of the Law Faculty of Maastricht University and especially the Jiangsu Maritime Safety Administration for the generous financial support that made this project and this book possible. In this respect, we would especially like to mention Zhang Tongbin, executive director of Jiangsu Maritime Safety Administration and Professor Yu Shicheng, president of Shanghai Maritime University. In addition, we are grateful to everyone else who made possible both the preceding conference and the publication of this book. In this respect, we owe a special thanks to Martin McGann who revised the English of the non-native English speakers and who reviewed the texts, footnotes and referencing. The Chinese texts have also been reviewed by Wang Hui. Finally, we also owe thanks to the administrative centre of the Maastricht European Institute for Transnational Legal Research (METRO) and especially to Marina Jodogne and Yleen Simonis for their editorial assistance in the preparation of the publication of this book.

The texts were finalized in September 2004, and for that reason developments after that date have not been treated.

Michael Faure and James Hu

Maastricht/Shanghai, March 2005

TABLE OF CONTENTS

Michael Faure and James Hu
FOREWORD I
1   Defining the Problem i
2   Project History iii
3   Methodology iv
3.1   Legal Multidisciplinary iv
3.2   Comparative Approach v
3.3   Law and Economics v
3.4   Case Studies v
4   Framework of the project vi
5   topics vi
6   Structure of the Presentation vii
7   Contributors vii
8   Acknowledgements vii

TABLE OF CONTENTS IX

LIST OF CONTRIBUTORS XXIII

LIST OF ABBREVIATIONS XXV

Part I: Prevention and Sanctioning of Marine Pollution Damage

Wang Hui
RECENT DEVELOPMENTS IN THE EU MARINE OIL POLLUTION REGIME 1
1   Introduction 1
2   Basic framework of EU marine oil pollution legislation 2
2.1   1978-1992: First Set of Actions 3
2.1.1 The 1978 Resolution: Start of Community Action 3
2.1.2 The 1990 Resolution 4
2.2   1993-2000: Second Set of Actions 4
2.2.1 The 1993 Communication 5
2.2.2 1993 Directive 93/75/EEC 5
2.2.3 1994 Council Regulation 2978/94 5
2.2.4 The 2000 Decision: Legal Basis for Community Action 5
2.3   Post Erika & Prestige 5
2.3.1 Erika I and II Packages 6
2.3.2 Post-Prestige 6
2.4   New Challenges for the Enlarged EU 7
3   Specific issues 7

3.1	Port State Control	7
3.1.1	Background – Paris MOU on Port State Control	8
3.1.2	1995 Directive 95/21/EC	8
3.1.3	2000 Erika I-Directive 2001/106/EC	9
3.1.4	Prestige	10
3.2	Notification/Report System	11
3.2.1	Directive 93/75/EEC	11
3.2.2	Erika II	11
3.3	Classification Society	12
3.3.1	Council Directive 94/57/EC	12
3.3.2	Commission Directive 97/58/EC	13
3.3.3	Erika I	13
3.3.4	Prestige	14
3.4	Single Hull Tanker	14
3.4.1	International v. US Regimes on Single Hull/Double Hull Tankers	14
3.4.2	Council Regulation 2978/94	15
3.4.3	Erika I – Regulation 417/2002	15
3.4.4	Prestige	16
3.5	Liability and Compensation	17
3.5.1	Post-Erika	17
3.5.1.1	Erika I Proposals	17
3.5.1.2	Erika II Proposals	18
3.5.2	Post-Prestige: The 2002 Communication & 2003 Proposal	18
3.5.3	Limitation of Liability	18
3.5.3.1	Erika	19
3.5.3.2	The 2002 Communication	20
3.5.3.3	The 2003 Proposal	20
3.5.4	Definition of Pollution Damage	21
3.5.5	Balance of Responsibilities	21
4	Concluding remarks	22

List of References ... 24

Mark A. Cohen
OIL POLLUTION PREVENTION AND ENFORCEMENT MEASURES AND THEIR
EFFECTIVENESS: A SURVEY OF EMPIRICAL RESEARCH FROM THE US ... 25

1	Introduction	25
2	Oil Spill Prevention and Enforcement Mechanisms in the US	25
3	Economic Theory of Oil Spill Prevention and Enforcement	27
4	Empirical Evidence on Oil Spill Prevention and Enforcement	31
5	Concluding Remarks	36

List of References ... 38

Günter Heine
MARINE (OIL) POLLUTION: PREVENTION AND PROTECTION BY CRIMINAL
LAW – INTERNATIONAL PERSPECTIVES, CORPORATE AND/OR
INDIVIDUAL CRIMINAL LIABILITY 41
1   Introduction 41
2   Problems and Shortcomings in International Laws of Marine Pollution 42
2.1   General Questions 42
2.1.1   Jurisdiction 42
2.1.2   Technical Safety, Control Management, Compensation 43
2.1.3   Compensation Shortcomings 44
2.2   The Need for Reform 45
3   Combating and Preventing Marine Pollution Through Criminal Law 46
3.1   International Frameworks: Opportunities and Limitations 46
3.2   Criminal Liability: Shortcomings of Individual Responsibility 49
3.3   Corporate Criminal Liability 51
3.3.1   Approaches by an International View 52
3.3.2   Sanctions 53
4   Summary and Concluding Remarks 54

List of References 57

Thomas Richter
INTERDEPENDENCIES BETWEEN CRIMINAL LAW AND OIL POLLUTION
REGULATION IN CHINA 61
1   Initital Position in China 61
1.1   Marine Oil Pollution in the Chinese Reality 61
1.2   Economic Development and Energy in China 61
1.3   Marine Oil Pollution and Focus of the Report 65
2   The Legal Milieu 66
2.1   International Law 66
2.2   National Law: Oil Pollution Regulations 66
2.2.1   The Relevant Legislation 66
2.2.2.   Administrative Sanctions and Other Forms of Sanctions 67
3   Approach of the Criminal Law 68
3.1   Preconditions 68
3.2   Development of Crimes against the Environment 68
3.3   Crimes having to do with Oil Pollution 69
3.3.1   Art. 338 PL (Considerable Accident with Environmental Pollution) 69
3.3.2   Art. 408 PL (Misconduct in Office while Supervising and Controlling the
          Environment) 70
3.3.3   Art. 136 PL (Trouble with Dangerous Objects) 71
4   Interdependencies between Environmental Criminal Law and Oil Pollution
          Regulation 72
4.1   General Remarks 72
4.2   Dependence of Criminal Law on Administrative Regulation 73

4.2.1	Function	73
4.2.2	Form	74
4.2.3	Integration in the Structure of Criminal Law	75
4.3	Mens Rea	76
4.4	False Allowances	77
5	Conclusions	78
5.1	Conclusion on the General Interdependencies between Criminal Law and Environmental Law	78
5.2	Conclusion on the Interdependencies between Criminal Law and Oil Pollution Regulation	79

List of References 80

Chen Xuebin
JURIDICTION OF THE PEOPLE'S REPUBLIC OF CHINA OVER POLLUTION
FROM VESSELS IN ITS EXCLUSIVE ECONOMIC ZONE 83

1	Introduction	83
2	Legal Resources for China to exercise its Jurisdiction	84
3	Substantive Requirements for China to exercise its Jurisdiction	86
4	Procedural Requirements for China to exercise its Jurisdiction	89
5	Jurisdiction exercised by China in Civil Proceedings	91
6	Revision of Marine Environment Protection Laws	93
7	Conclusions	95

List of References 96

Part II: Compensation of Marine Pollution Damage: Case Studies

Etty R. Agoes
INDONESIA'S LAWS AND REGULATIONS CONCERNING POLLUTION OF
THE SEA BY OIL: CASE STUDIES ON COMPENSATION FOR OIL POLLUTION
DAMAGES 99

1	Introduction	99
2	Institutional Arrangements	100
3	Non-Governmental Organizations	101
4	Environmental Legislation in Indonesia	101
4.1	The 1982 UNCLOS (as ratified by Law No. 17 of 1985) and the Marine Environment	102
4.2	IMO Conventions, as ratified by Presidential Decrees of 1978, 1986 and 1999	103
4.3	Relevant National Laws and Regulations	104
4.3.1	Law No. 6 of 1996 on Indonesian Territorial Waters	104
4.3.2	Law No. 23 of 1997 concerning the Management of the Living Environment	105

4.3.3	Government Regulation No. 19 of 1999 on Marine Pollution Control	107
4.3.4	Law No. 21 of 1992 on Shipping	107
4.3.5	Decree of the Minister for Communication No. 86 of 1990 on the Prevention of Pollution of Oil from Vessels	107
4.3.6	Law No. 22 of 2001 on Oil and Gas	108
4.3.7	Law No. 22 of 1999 on Regional Government	108
4.3.8	Other Laws and Regulations	108
5	Problems in implementing Environmental Legislation	109
5.1	Government Agencies	110
5.2	Department of Marine Affairs and Fisheries	110
6	Regional Initiatives	111
7	Draft National Contingency Plan	111
8	Cases involving Pollution of the Sea by Oil	113
8.1	The Nagasaki Spirit Marine Pollution Casualty	113
8.2	The MT Natuna Sea Oil Spill	114
8.3	The MT King Fisher Incident	115
9	Conclusion	116

List of References 117

Marc A. Huybrechts and Katrien N. Van Damme
PROTECTION OF THE MARINE ENVIRONMENT UNDER BELGIAN LAW 119

1	Introduction	119
2	Contents of the Belgian Statute Marine Environment	120
3	Importance of the North Sea and the Risk of Pollution	121
4	The Belgian Statue Marine Environment	122
4.1	General Approach	122
4.1.1	Field of Application	122
4.1.2	Aim and Basic Principles	123
4.2	Prevention and Avoidance of Damage to and Disruption of the Marine Environment	124
4.2.1	Damages to and Disruption of the Marine Environment	124
4.2.1.1	Damage to the Marine Environment	124
4.2.1.2	Disruption of the Marine Environment	125
4.2.2	Compensation	125
4.2.2.1	Who has the Right to claim for Compensation?	126
4.2.2.2	Limitation of Liability	126
4.2.2.3	Compensation: Restoration in Kind?	127
4.2.2.4	Bail or Guarantee Declaration	128
5	Pollution in Practice	128
5.1	Air Surveillance by the Management Unit of the North Sea Mathematical Models (MUMM)	129
5.2	Reception Facilities and Port State Control	130
5.3	North Sea Disaster Plan and Places of Refuge	130
5.4	Legal Aspects: Judicial Proceedings	131

5.5	Future Developments: Zero Tolerance!	133
5.5.1	Maximize the Chances of Apprehension: the Introduction of Unmanned Cameras	133
5.5.2	Referee Magistrate North Sea	133
5.5.3	Active International Cooperation	133
5.5.4	Equipment	133
5.5.5	Compensation for Disruption of the Environment	134
6	Conclusion	134

List of References 135

James Boyd
COMPENSATION FOR OIL POLLUTION DAMAGES: THE AMERICAN OIL POLLUTION ACT AS AN EXAMPLE FOR GLOBAL SOLUTIONS? 137

1	Introduction	137
2	OPA and US Natural Resource Damages Law	139
2.1	What is a Natural Resource Damage?	140
2.2	Common Law Foundations	141
2.3	Administrative Issues	142
3	Calculating Damages	144
3.1	The History of NRD Assessment	145
3.2	The Challenge of Ecosystem Service Assessment	147
3.3	The Politics of NRD	150
3.4	Damages: Lost Benefits Versus Replacement Cost	151
4	NRD Claims – Scale and Frequency	154
5	Financial Assurance's Role in Compensation and Deterrence	157
6	Conclusion	160

List of References 161

María Paz García Rubio
THE PRESTIGE CASE. INTERNATIONAL AND SPANISH LEGAL REGIME FOR COMPENSATING DAMAGE 165

1	Raising the problem	165
2	General Overview of the System of Civil Liability as a Mechanism for Compensating Environmental Damage	166
3	International Framework. International Convention on Civil Liability for Oil Pollution Damage and International Convention on the Establishment of an International Fund for Compensation for Oil Pollution Damage	167
4	The Spanish System of Civil Liability and its Application in the *Prestige* case	170
5	The Royal Decree-Law 4/2003, of 23 June 2003	172
5.1	Legal Framework of the Agreement between State and Victims regulated in the Royal Decree-Law	173

5.2	Subjective Framework. The Victims	173
5.3	Waiving to the Claims by the Victims	175
6	Final Note	176

List of References 178

Part III: Compensating Marine Pollution Damage in China

Han Lixin and Guan Zhengyi
THE ENFORCEMENT OF INTERNATIONAL CONVENTIONS FOR THE
PREVENTION OF POLLUTION FROM SHIPS AND COMPENSATION FOR
POLLUTION DAMAGE IN CHINA 181

1	Introduction	181
2	Brief Introduction to the International Conventions for the Prevention of Pollution from Ships and Compensation for Pollution Damage to which China has acceded	181
3	The Main Measures of enforcing International Conventions	183
3.1	Adoption	183
3.2	Transformation	183
3.3	Mixture of Adoption and Transformation	184
4	The Current Status Quo for the Application of International Conventions on the Prevention of Pollution from Ships and for Compensation for Pollution Damage in China	184
4.1	The Existing Regulations for the Application of Conventions in China	184
4.2	China's Practice in Implementing the Conventions	185
4.3	The Current Status Quo for the Application of International Conventions on the Prevention of Pollution from Ships and for Compensation for Pollution Damage in China in Practice	186
5	Suggestions on the Application of International Conventions	189

List of References 191

James Hu and Yang Bo
APPLICATION OF LAW IN CIVIL LIABILITY FOR OIL POLLUTION DAMAGE
CAUSED BY COASTAL VESSELS IN CHINA 193

1	Introduction	193
2	Diversity of Views regarding Application of Law	194
2.1	Application of CLC	194
2.2	Application of Civil Law	194
2.3	Application of the Maritime Code	195
3	Application of the CLC	195
3.1	The Application of International Conventions in China Generally	195

3.2	Non-applicability of CLC to Cases without Foreign Elements	197
3.3	Application of the CLC to Coastal Vessels involving Foreign Elements	198
3.4	Analysis on Views in Favour of the Application of the CLC to Cases without Foreign Elements	199
3.4.1	By Virtue of *the Administrative Regulations on Preventing Marine Pollution from Ships*	199
3.4.2	By Virtue of Governmental Notices for Implementing the 1969 CLC	200
3.5	Interest-orientated Consideration	200
4	Applicable domestic laws	201
4.1	Common Features of Application	201
4.2	Priority Issues	202
5	The Necessity of a Specific Legal Regime	203
6	Conclusions	204

List of References 205

Zhao Yuelin, James Hu and Wang Hua
A STUDY OF THE LEGAL NATURE OF COMPULSORY CLEAN-UP COSTS UNDER CHINESE LAW 207

1	Introduction	207
2	Compulsory Clean-up Acts as Compulsory Administrative Measures	208
3	Legal Nature of the Compulsory Clean-up Costs borne by Responsible Parties	210
3.1	Compulsory Clean-up Costs as Costs of Substitution of Performance under Administrative Law	211
3.1.1	Compulsory Clean-up Acts as Indirect Compulsory Acts	211
3.1.2	Compulsory Clean-up Costs as those of Substituted Performance of Carrying out Compulsory Administrative Measures	211
3.2	Tortious Liability under the Civil Law	213
3.2.1	Compulsory Clean-up Costs within Tortious Liability	213
3.2.2	Compulsory Clean-up Costs within the Scope of Civil Compensation Liability	215
3.3	Contribution to the Realisation of Uniformity of Fairness and Efficiency	216
3.4	The International Tendency to treat Payment of Compulsory Clean-up Costs as Civil Liability	219
4	Conclusions	220

List of References 221

Ma Jing-jing and Du Jiang
DISCUSSION ON THE NATIONAL CLAIM SYSTEM FOR OIL POLLUTION DAMAGE FROM SHIPS 223

1	Introduction	223

2	The Need to set up the National Claim System for Oil Pollution Damage caused by Ships	223
3	Establishment of the Subject of the National Claim for Oil Pollution Damage from Ships	225
3.1	Qualification of the Country to propose the Oil Pollution Claim	225
3.2	Exercise by the Administrative Authority of a Right to file Claims on behalf of the Country	228
3.2.1	Administrative Authority Entitlement to bring an Action as Trustee of National Property	228
3.2.2	Equality of the Administrative Authority as a Civil Subject in Claims for Damages from Oil Pollution	229
4	Practice in National Claims of Oil Pollution Damage from Ships	230
5	The Scope of a National Claim System for Oil Pollution Damage from Ships	233
5.1	Compensation for the Clean-up Charges	233
5.2	Compensation for Pure Economic Loss	234
5.3	Punitive Damage	236
6	Conclusion	238

List of References 239

Chen Pingping
A STUDY ON THE TYPES OF LIABILITY OF THE INSURER FOR OIL POLLUTION AND THAT OF THE PARTY LIABLE 241

1	Introduction	241
2	Presentation of the Problem	241
3	Relative Concepts of Liability Insurance	242
3.1	Origin and Development of Liability Insurance	242
3.2	Substitute Liability for Compensation of the Liability Insurer	244
3.3	Victim's Right of Direct Action	245
3.3.1	Doctrine of Original Acquisition (Doctrine of Statutory Rights)	245
3.3.2	The Doctrine of Payment without Evasion of Liability	245
3.4	Restriction on the Insurer's Right to call on Defenses	246
4	Issues Related to Compensation Liability for Oil Pollution Damage	247
4.1	The Current Status of Legislation in Respect of Oil Pollution Damage in China	247
4.2	The Principle of Imputation of Compensation Liability for Oil Pollution Damage	249
4.3	The Promotion of Compulsory Liability Insurance for Oil Pollution	250
4.4	The System of Direct Action	251
5	Non-typical Joint and Several Liability between the Insurer for Oil Pollution and the Party Liable for Oil Pollution	253
5.1	Non-typical Joint and Several Liability in the Framework of the Current Law	253
5.2	The Legislative Trend of the Joint and Several Liability	256

6	Perfection of Legislation Regarding Compensation for Damages of Ship Oil Pollution	257
6.1	The Substantive Content of the Direct Action	257
6.2	Oil Tanker to be covered by Compulsory Insurance	258
6.3	Establishment of a Scheme of Payment on Account	259
7	Concluding Remarks	260

List of References ... 261

Wu Lijing and Zhang Min
JUDICIAL AUTHENTICATION OF MARINE POLLUTION DAMAGES ... 263

1	Introduction	263
2	The Current System of Judicial Authentication of Marine Pollution Damages in China	264
2.1	The Current Judicial Authentication System in China	264
2.1.1	Authentication Determination System	264
2.1.2	Authentication Subject System	265
2.1.3	Expert Witness Selecting and Appointing System	265
2.2	System of Judicial Authentication of Marine Pollution Damages	266
3	Analysis of Objectivity and Authenticity of Marine Pollution Damages Authentication Reports	267
3.1	Features of Judicial Authentication	267
3.2	Essence and Features of the Authentication Report	268
3.2.1	Essence of the Authentication Report	268
3.2.2	Features of the Authentication Report	268
3.2.2.1	Dualism of Subjectivity and Objectivity	268
3.2.2.2	The Dual Tendency of Authenticity and Distortion	268
3.3	Possibilities for the Distortion of the Authentication Report and their Removal	269
3.3.1	Analysis of the Causes of Distortion	269
3.3.1.1	Inauthentic Materials	269
3.3.1.2	Improper Process	269
3.3.1.3	Incorrect Theories used in Authentication	269
3.3.1.4	Subjectivity of the Expert Witness	269
3.3.2	Prevention of Authentication Report Distortion	270
3.3.2.1	Selection and Appointment of Proper Expert Witness	270
3.3.2.2	Strict Conformance to the Regulations and Procedural Rules	270
3.3.2.3	Judicial Authentication Review	271
4	The Role of the Authentication Report in Litigation for Marine Pollution Damages Compensation	271
4.1	Evidential Effect of the Authentication Report	271
4.2	Status of the Expert Witness in Litigation	272
5	Conclusion	273

List of References ... 274

Li Zhiwen and An Shouzhi
THE NECESSITY OF INCORPORATING MARINE ENVIRONMENTAL TORTS
INTO THE MARITIME LEGAL SYSTEM 275
1 Introduction 275
2 Definition and Types of Marine Environmental Torts 275
2.1 Characteristics of Marine Environmental Torts 276
2.2 Categorization of Marine Environmental Torts 277
3 Analysis of the Deficiency of the Status in the Civil Remedies for Marine
 Environmental Torts in China 277
3.1 The Status of Civil Remedies for Marine Environmental Torts regulated by
 Law in China 277
3.2 Analysis of China's current Civil Remedy System for Marine Environmental
 Torts, and its Deficiencies 278
4 The Necessity to incorporate Marine Environmental Torts into the Marine
 Tort Law 279
4.1 The Trend of expanding the Scope of Modern Marine Tort Law 279
4.2 The Avoidance of Contradiction in the System 280
4.3 Protection of the Legal Rights and Interests of the Parties concerned 281
4.3.1 The Marine Tort Legal System is useful for the Development and Utilisation
 of Economic Activities at Sea 281
4.3.2 The Special System of Compensation in Marine Tort Law contributes to
 Sufficient Compensation, and to the Development of the Ocean 281
4.4 Facilitation of Judicial Remedies of Marine Environmental Torts 282
5 Conclusions 283

List of References 284

Guo Ping and Zhao Lujun
THE LEGAL PROTECTION OF THE MARINE ENVIRONMENT IN CHINA:
CURRENT SITUATION AND CHALLENGES 285
1 Introduction 285
2 Chinese Legislation on the Marine Environmental Protection 285
2.1 Law of Environmental Protection (LEP) 285
2.2 Law of Marine Environment Protection (LMEP) 286
3 Related Administrative Rules and Regulation concerning the Protection of
 the Marine Environment 288
3.1 Administrative Regulations on Preventing Ships from Polluting Sea Areas
 (ARPSP) 288
3.2 Administrative Regulation of Preventing Pollution from Dismantling and
 Repair of Ships 288
3.3 Administrative Regulations on Environmental Protection on Ocean Oil
 Exploration And Exploitation (REPOE) 289
3.4 Administrative Regulations on Dumping Wastes at Sea (RDW) 290
3.5 Administrative Regulations on Preventing Ocean Pollution from
 Land-Source Pollutants 290

3.6	Administrative Regulations on Preventing and Curing Ocean Pollution from Coastal Construction Projects	291
3.7	Implementing Rules on Ocean Administrative Penalty	292
4	Relations and Existing Problems between Laws, Administrative Regulations and Rules on Marine Environment Protection	293
4.1	Relations between Laws, Administrative Regulations and Rules on Marine Environment Protection of China	293
4.2	Problems existing in Laws, Administrative Regulations and Rules on Marine Environment Protection	293
5	Preliminary Considerations and Methods to solve the Problem	297
5.1	Perfecting the Relevant Legislation	297
5.2	Organizations and Forms	297
5.3	Promoting the Legal Effect Level of the Department's Regulations and Rules	298
6	Conclusions	298
List of References		299

Wei Zhije and Song Ruqing
RISK ANALYSIS OF NOXIOUS LIQUID SUBSTANCES TRANSPORTATION BY WATER IN THE YANGTZE DELTA AND THE RESEARCH OF ITS COUNTERMEASURES 301

1	Introduction	301
2	Analysis of the Risk in NLS Transportation by Water in the Yangtze Delta	301
2.1	Existing Conditions in NLS Transportation in the Yangtze Delta	302
2.2	Status of NLS Pollution Accidents in the Yangtze Delta	303
2.3	Development Trend in NLS Carriage in the Yangtze Delta	303
2.4	Risk Analysis of Pollution Accident Hazards in the Yangtze Delta	304
3	Countermeasures (Prevention, Salvage and Compensation)	306
3.1	Means to Control the Occurrence of Pollution Accidents	307
3.1.1	Control of the Factor of Personnel	307
3.1.2	Control over the Factor of Vessels	308
3.1.3	Control over the Factor of Loading and Unloading Terminals (Facilities)	309
3.1.4	Control over the Factor of Environment for Navigation	309
3.1.5	Control over the Factor of Administration	310
3.2	Countermeasures of Emergency Response to Pollution Accidents	310
3.2.1	Emergency Plans	311
3.2.2	Information System for Emergency Response	312
3.2.3	Guarantee System for Emergency Response	312
3.2.4	System of Technical Support for Emergency Response	313
3.3	Compensation Regime for Pollution Accidents	314
4	Conclusions	315
List of References		316

Michael Faure and Wang Hui
FINANCIAL CAPS FOR OIL POLLUTION DAMAGE: CHINA AND THE
INTERNATIONAL CONVENTIONS 317
1   Introduction 317
2   International Regimes 318
2.1   Pre-1969 Regime 318
2.2   1969/1971 Regime 319
2.2.1   CLC Convention 1969 & Fund Convention 1971 319
2.2.1.1 CLC Convention 1969 319
2.2.1.2 Fund Convention 1971 321
2.2.2   Voluntary Schemes 321
2.2.2.1 TOVALOP 322
2.2.2.2 CRISTAL 322
2.3   1984/1992 Regime 322
2.3.1   1984 Protocols 322
2.3.2   1992 Protocols 323
2.3.3   Disappearance of TOVALOP & CRISTAL 324
2.3.4   Winding up of the 1969/1971 Regime 324
2.4   The 2000/2003 Regime 325
2.4.1   The 2000 Amendments 325
2.4.1.1 Background 325
2.4.1.2 2000 Amendments 326
2.4.2   2003 Supplementary Fund Protocol 326
3   Civil Liability for Oil Pollution Damage in China 328
3.1   International Conventions Effective in China 328
3.1.1   CLC Convention 328
3.1.2   Fund Convention 328
3.2   Domestic Legislations in China 329
3.2.1   China Maritime Code 330
3.2.2   Marine Environment Protection Law 330
3.3   Practical Issues 331
3.3.1   Negative Effect on Clean-up 331
3.3.2   No Compulsory Insurance for Cabotage Tankers & Low Financial Limits
        under Chinese Law 331
3.3.3   No Compensation Fund in China 332
4   Analysis 332
4.1   Introduction 332
4.2   Financial Caps on Tort Liability: An Economic Perspective 333
4.2.1   Financial Caps Cause Underdeterrence 333
4.2.2   Strict Liability versus Negligence 333
4.2.3   Subsidization Effect 334
4.2.4   In sum 335
4.3   Financial Caps to Increase Insurability? 335
4.3.1   Capacity as Insurability Problem 335
4.3.2   Limitation of the Duty to Insure 336
4.3.3   Contractual Limitations 337

4.3.4	Unlimited Liability to Control Moral Hazard	337
4.3.5	Financial Caps to increase the Insurability of Oil Pollution?	337
4.4	A Duty to seek Insurance Coverage	338
4.5	China	339
5	Concluding Remarks	340

List of References 344

Part IV: Comparative Conclusions

Michael Faure and James Hu
COMPARATIVE CONCLUSIONS 351
1	Introduction	351
2	Prevention	351
3	Monitoring and Sanctions	352
4	The International Compensation Regime and the National Application	353
5	Compensating Marine Pollution Damage in China	354
6	Damage Assessment, Claims Handling and Insurance	355
7	Results	356
8	Policy	357
9	Look Forward	358

LIST OF CONTRIBUTORS*

Agoes Etty R., Faculty of Law, Padjadjaran University, Bandung, Indonesia

An Shouzhi, Jinghai and Co, Law firm, Guangdong, China

Boyd James, Resources for the future, Washington D.C., USA

Chen Ping Ping, Xiamen Maritime University, Xiamen, China

Chen Xuebin, Asia-Pacific Great Wall (Shanghai) Law firm, Shanghai, China

Cohen Mark A., Vanderbilt Centre for Environmental Management Studies, Owen Graduate School of Management, Vanderbilt University, Nashville, US

Du Jiang, Shanghai Maritime University, Shanghai, China

Faure Michael, Metro Institute, Maastricht University, The Netherlands

Guo Ping, Dalian Maritime University, Dalian, China

Guan Zhengyi, Dalian Maritime Court, Dalian, China

Han Lixin, Dalian Maritime University, Dalian, China

Heine Günter, University of Berne, Switzerland

Hu James, Shanghai Maritime University, Shanghai, China

Huybrechts Marc A., Attorney at the Antwerp and Brussels Bar, University of Antwerp and Catholic University of Leuven.

Li Zhiwen, Dalian Maritime University, Dalian, China

Ma Jing-jing, Shanghai Maritime University, Shanghai, China

Paz Garcia Rubio Maria, Universidad de Santiago de Compostella, Spain

---

* We have, for reasons of uniformity followed the Chinese tradition of putting the family name first, also for the non-Chinese contributors.

Richter Thomas, Max Planck Institute for International and Foreign Criminal Law, Freiburg, Germany

Song Ruqing, Maritime Safety Administration, Jiangsu, China

Van Damme Katrien N., Attorney at the Antwerp Bar, Belgium

Wang Hua, Dalian Maritime University, Dalian, China

Wang Hui, Ph.D. student, Institute for Energy and Environmental Law, Faculty of Law, Catholic University of Leuven, Belgium

Wei Zhijie, Maritime Safety Administration, Jiangsu, China

Wu Lijing, Tsinghua University, Beijing, China

Yang Bo, Shanghai Maritime University, Shanghai, China

Zhang Min, Dalian Maritime University, Dalian, China

Zhao Lujun, Dalian Maritime University, Dalian, China

Zhao Yuelin, Dalian Maritime University, Dalian, China

LIST OF ABBREVIATIONS

ABS	American Bureau of Shipping
APA	Administrative Procedures Act (US)
ARPSP	Administrative Rules and Regulation concerning the Protection of the Marine Environment (China)
Art.	Article
Bapedal	Badan Pengendalian Dampak Linglungan (Environmental Impact Management Agency – Indonesia)
B.S.	Belgisch Staatsblad
CERCLA	Comprehensive Environmental Response Compensation and Liability Act (US)
CLC	International Convention on Civil Liability for Oil Pollution Damage
CMC	China Maritime Code
CMI	Comité Maritime International
COD	Chemical Oxygen Demand
COLREG's	International Regulations for Preventing Collisions at Sea
COMPOSNAV	Naval Operation Command
COPE	Fund for Compensation for Oil Pollution in European Waters
CRISTAL	Contract Regarding an Interim Supplement to Tanker Liability for Oil Pollution
CV	Contingent Valuation
CWA	Clean Water Act (US)
DO	Dissolved Oxygen
DOI	Department of Interior (US)
EEC	European Economic Community
EEZ	Exclusive Economic Zone
EIA	Environmental Impact Assessment
EMA	Environmental Management Act (Indonesia)
EPA	Environmental Protection Agency (US)
EU	European Union
FL	Fisheries Law (China)
GAO	General Accounting Office (US)
GPCL	General Principles of the Civil Law of China

HNS Convention	International Convention on Liability and Compensation for Damage in Connection with the Carriage of Hazardous and Noxious Substances by Sea
HSNI	Indonesian Fishermen Association
ICO	Offical Institute of Credit (Spain)
IMO	International Maritime Organization
IOPC-Fund	International Oil Pollution Compensation Fund
ISM	International Safety Management Code
Km	Kilometer
LAPP	Law on Air Pollutio Prevention (China)
LBA	Land Based Activities
LEP	Law on Environmental Protection (China)
LMEP	Law of marine Environment Protection (China)
LPTWP	Law on Prevention and Treatment of Water Pollution (China)
MARPOL	International Convention for the Prevention of Pollution from Ships
MC	Maritime Code (of China)
MEPL	Marine Environment Protection Law (China)
METRO	Maastricht European Institute for Transnational Legal Research
Mio.	Million
MOU	Memorandum of Understanding
MPR	People's Consultative Assembly (Indonesia)
MSA	Maritime Safety Administration (China)
MUMM	Management Unit of the North Sea Mathematical Models
NLS	Noxious Liquid Substances
NOAA	National Oceanic and Atmospheric Administration (US)
NOCOP	National Operation Centre for Combating Oil Pollution (Indonesian)
NRD	Natural Resource Damage
OECD	Organization for Economic Cooperation and Development
OILPOL	International Convention for the Prevention of Pollution of the Sea by Oil
OJ	Official Journal
OPA	Oil Pollution Act
OPL	Oil Pollution Law (China)

OSRAP	Oil Spill Response Action Plan (Asean)
P&I-club	Protection and Indemnity Club
PL	Penal Law (China)
POLREP	Report on Oil Pollution (Belgian)
PRC	People's Republic of China
RDW	Administrative Regluations on Dumping Wastes at Sea (China)
REPOE	Administrative Regulations on Environmental Protection on Ocean Oil Exploration and Exploitation (China)
RFC	Revolving Fund Committee
RMB	Ren Min Bi (Chinese currency)
SAR	Special Administrative Region (China)
SBT	Segregated Ballast Tank
SDR	Special Drawing Rights
SEPA	State Environment Protection Administration (China)
SMPL	Special Maritime Procedure Law (China)
SOA	State Oceanic Administration (China)
SOLAS	International Convention for the Safety of Life at Sea
Sq.	Square
STCW	International Convention on Standards of Training, Certification and Watchkeeping for Seafarers
TOVALOP	Tanker Owner's Voluntary Agreement Concerning Liability for Oil Pollution
UAV	Unmanned Air Vehicle
UK	United Kingdom
UNCLOS	United Nations Convention on the Law of the Sea
US	United States
USCG	United States Coast Guard
VLCC	Very Large Crude Carrier

RECENT DEVELOPMENTS IN THE EU MARINE OIL POLLUTION REGIME

Wang Hui

1 INTRODUCTION

The European Union, due to its geography (a vast coastline along the Atlantic and the Mediterranean sea) and its history (many EU Member States have a long merchant marine tradition, in particular, the UK, Denmark, Greece and the Netherlands) is very much dependent on maritime business. The EU oil trade is the largest in the world, with its crude oil imports representing about 27% of the total world trade.[1] However, while 90% of the total oil trade with the European Union is carried out by sea transportation, on the other hand, the coasts in Europe are extremely vulnerable to the pollution risks.[2] Hence, it is in the interest of the European Union to protect European waters from the risks of pollution, and especially oil pollution. As specified in the well-known Art. 175 of the EU Treaty, environmental protection is one of the basic aims of the European Union.[3]

As marine oil pollution cannot be limited by the borders between countries, so many of the pollution problems cannot be tackled effectively without joint action by the Member States at the Community level. It is even questionable whether this can be done without cooperation at the international level, among countries outside the EU. The European Union thus not only aims at the prevention of and compensation for oil pollution damage under the Community framework, but also aims at the protection of its interests at the international level.[4]

The competent body in this respect at the international level is the International Maritime Organization (IMO). Over the years, the IMO has developed a series of conventions addressing marine oil pollution (as well as pollution from other sources) caused by accidents and other operations at sea. International conventions with respect to oil pollution can be distinguished between, on the one hand, measures aiming at prevention *ex ante* and, on the other, measures aiming at compensation *ex post*. The

---

[1] 'EUROSTAT and OECD/IEA statistics' (2000), *Journal de la Marine Marchande*; and COM(2000) 142 final, Communication from the Commission to the European Parliament and the Council on the Safety of the Seaborne Oil Trade, 21 March 2000, 8.
[2] Some 70% of the European Union's oil imports are transported along the Brittany coast and through the English Channel. This is why the same regions are hit again and again. For a more detailed analysis, please refer to COM(2000) 142 final, 21 March 2000, 4.
[3] Environmental policy was built into the Treaty by the Single European Act of 1987 and its scope was extended by the Treaty on European Union 1992.
[4] For an overview of the protection of the marine environment in the EU see M. Hedemann-Robinson, 'Protection of the Marine Environment and the European Union – Drifting or making headway? Some critical reflections on internal and external dimensions to law, policy and practice' (2004) 12 *Environmental Liability* 3-18.

Michael G. Faure and James Hu (eds), Prevention and Compensation of Marine Pollution Damage.
Recent Developments in Europe, China and the US, 1-24.
© *2006 Kluwer Law International. Printed in the Netherlands.*

prevention of marine pollution is the subject of, *inter alia*, the MARPOL,[5] SOLAS,[6] STCW[7] and ISM[8] Conventions.[9] Compensation is mainly regulated under the regime built up by the CLC[10] and the Fund[11] Conventions.

The EU has built its own legislation, mainly on the basis of IMO Conventions and Resolutions. The difference, and benefit, of a legislative basis separate from the international regime is that the Community has a greater chance of adopting more detailed, effective and quick legislation. Furthermore, while the IMO is unable to impose rules on states, and can only rely on states to sign and ratify the international conventions, the EU may impose binding rules on its Member States through its various institutions.[12] There has been a reciprocal effect as well though, and Europe's activism in this domain has triggered several important changes in the international regime.

This paper will specifically focus on the EU-approach towards the prevention and compensation of oil pollution damage. It will be shown that originally Europe relied largely on the existence of the international conventions and hence had no initiatives in the domain of oil pollution that were particularly its own. It will be shown that later, especially after Europe was confronted with the major oil spill from the Amoco Cadiz in France in 1978, the Commission started to take several initiatives aiming, in the first place, at the prevention of oil spills. Then, in a second phase, Europe also realized that the prevention and compensation mechanism designed by the international conventions might be ineffective and insufficient in some cases. Finally it will be demonstrated that, more particularly after the recent Erika (1999) and Prestige (2002) incidents, the Commission has developed various initiatives, not only to increase maritime safety, but also to facilitate better compensation for victims of oil pollution.

2 BASIC FRAMEWORK OF EU MARINE OIL POLLUTION LEGISLATION

When the European (Economic) Community was founded in 1957, there was no Community-wide maritime safety legislation. The necessity for such a Community

---

[5] The International Convention for the Prevention of Pollution from Ships, 1973, as amended by the 1978 Protocol, often referred to as MARPOL 73/78.
[6] The International Convention for the Safety of Life at Sea, often referred to as the SOLAS Convention, 1974, amended in 1978.
[7] The International Convention on Standards of Training, Certification and Watchkeeping for Seafarers, 1978, often referred to as the STCW Convention. The revised STCW Convention came into force in 1997.
[8] The International Safety Management Code is often referred to as ISM Code. This has become mandatory since 1998.
[9] For a detailed analysis of these international conventions, please refer to P. Sands, *Principles of International Environmental Law*, (2nd edn, Cambridge University Press, Cambridge, 2003), pp. 391-459.
[10] International Convention on Civil Liability for Oil Pollution Damage, often referred to as CLC Convention. It was adopted in 1969, and modified in 1976, 1985, 1992, 2000 and most recently in 2003.
[11] International Convention on the Establishment of an International Fund for Compensation for Oil Pollution Damage, 1971 (Fund Convention).
[12] V. Power, *EC Shipping Law,* (2nd edn, Lloyd's of London Press, London, UK, 1998), p. 14. Take, for instance, the fact that the EC Commission is empowered to pursue the Member States for failure to notify, adequately transpose into national law or implement directives.

measure was not recognized until 1973 when some influential shipping powers joined the Community (Denmark and the UK particularly). Especially after witnessing some major oil disasters in the 1970s, many of which occurred off the European coast (e.g. Amoco Cadiz sank off Brittany on 16 March 1978), the Community began to act in 1978, setting up ambitious action programs that resulted in some formal declarations and resolutions encouraging member states to ratify the international conventions concerning maritime safety.

2.1 1978-1992: FIRST SET OF ACTIONS

2.1.1 The 1978 Resolution: Start of Community Action

Community action to combat marine pollution started with the Council Resolution of 26ᵗ June 1978,[13] which was adopted in response to the oil spill caused by Amoco Cadiz in 1978. The Resolution aimed to set up an action program for the control and reduction of pollution. It has resulted in several studies and proposals from the Commission, but only a few concrete decisions[14] have been taken at Community level. However, cooperation between the Member States has progressively developed since then.[15]

The Council also made a Recommendation to the effect that the Member States should ratify the SOLAS Convention 1978 and the MARPOL Convention 73/78, which, the Council stated, 'make a substantial contribution towards protecting the marine environment against pollution by ships, particularly oil tankers'.[16]

On 21 December 1978, the Council of Ministers adopted three legislative decisions concerning maritime safety:

- Council Recommendation 79/114/EEC on the ratification of the 1978 STCW Convention.[17] It recommended that the Member States sign the 1978 STCW Convention by 1 April 1979 and ratify it as soon as possible, but no later than 31 December 1980.

---

[13] Council Resolution of 26 June 1978 setting up an action program of the European Communities on the control and reduction of pollution caused by hydrocarbons discharged at sea, *OJ* C162, 8 July 1978, 1-4. This Resolution was later supplemented to further deal with other harmful substances.

[14] E.g. Commission Decision 80/686/EEC, of 25 June 1980, setting up an Advisory Committee on the control and reduction of pollution caused by hydrocarbons discharged at sea, which was amended later by 85/208/EEC (Commission Decision of 25 March 1985) and 87/144/EEC (Commission Decision of 13 February 1987); Council Decision 81/971/EEC, of 3 December 1981, establishing a Community information system for the control and reduction of pollution caused by hydrocarbons discharged at sea; Council Decision 86/85/EEC of 6 March 1986, establishing a Community information system for the control and reduction of pollution caused by the spillage of hydrocarbons and other harmful substances at sea, which was later amended by 88/346/EEC (Commission Decision of 16 June 1988).

[15] Decision 2850/2000, of 28 December 2000.

[16] Council Recommendation 78/584/EEC, of 26 June 1978, on the ratification of Conventions on Safety in Shipping.

[17] Council Recommendation of 21 December 1978 on the ratification of the 1978 International Convention on Standards of Training, Certification and Watchkeeping for Seafarers, *OJ* L194/17, 19 April 1979.

- Council Directive 79/115/EEC on deep-sea pilots.[18] It regulated the compulsory pilotage of vessels by deep-sea pilots in sensitive maritime areas like the North Sea and the English Channel, intending to ensure that ships wishing to use pilotageservice in these sensitive areas would have available to them adequately qualified deep-sea pilots for such service.
- Council Directive 79/116/EEC concerned the minimum communication requirements for large tankers entering or leaving Community ports, such that safety could be adequately monitored.[19]

2.1.2 The 1990 Resolution

In 1990, another Resolution[20] was adopted to urge the Member States to ratify the international conventions, and in particular the SOLAS and the MARPOL Conventions.

On 4 March 1991, the Council adopted a Regulation on the transfer of ships from one register to another within the Community.[21] This regulation aimed to facilitate the change of flags between Member States on the basis of mutual recognition of safety and pollution prevention certificates as laid down in the main IMO Conventions.

As we can see from above, during this first period the EU mainly recommended that the Member States ratify various kinds of international conventions aiming at the prevention of marine pollution, but did not take particular initiatives of its own and only a few legislative decisions were taken. The EU approach changed after 1993.

2.2 1993-2000: SECOND SET OF ACTIONS

After the Exxon Valdez disaster in 1989, the US formed its own regime regarding oil pollution; the Oil Pollution Act 1990, which in turn led to some important changes in the international regime. Faced with challenges to its low intervention approach posed by the US and at the international level, the Community adopted its 'common policy on safe seas' in 1993, which is considered to be the real start of EU maritime safety measures. Following this Communication, a wider range of directives were adopted in which the Commission took up its responsibility for the promotion of maritime safety. Moreover, the Community further examined many different elements to identify influences on maritime safety (including, *inter alia*, port State control, ships structures, classification societies, crew quality and compensation for oil pollution damage) and adopted several decisions during the Nineties to reinforce maritime safety.

---

[18] Council Directive 79/115/EEC, of 21 December 1978, concerning pilotage of vessels by deep-sea pilots in the North Sea and English Channel, *OJ* L33/32, 8 February 1979, 32.

[19] Council Directive 79/116/EEC, of 21 December 1978, concerning minimum requirements for certain tankers entering or leaving Community ports, *OJ* L33, 8 February 1979, 33.

[20] Council Resolution 90/818/EEC, of 19 June 1990, on the Prevention of Accidents Causing Marine Pollution, amending Annex II of 86/280/EEC on Limit Values and Quality Objectives for Discharges of Certain Dangerous Substances listed in the Annex to Directive 76/464/EEC, *OJ* L206/01, 18 August 1990.

[21] Council Regulation 613/91/EEC, of 4 March 1991, on the transfer of ships from one register to another within the Community, *OJ* L68 of 15 March 1991, 1, amended by Commission Regulation 2158/93/EEC, *OJ* L194 of 3 August 1993, 5.

2.2.1 The 1993 Communication

In 1993, the Commission published the long-awaited Communication on Safe Seas.[22] In the Communication, the Commission analysed the situation of maritime safety in Europe and outlined a framework for a common maritime policy. The basic principles of the policy included the convergent implementation of existing international rules, the uniform enforcement of international rules by the port States and the reinforcement of the EU's role as the driving force for rule making at the international level. Several important legislative acts were proposed and adopted in the following five years as an implementation of the action program attached to the Communication. They still remain the core of the EU's maritime safety policy today.[23]

2.2.2 1993 Directive 93/75/EEC

This Directive set up a notification system for ships carrying dangerous or polluting goods bound for, or leaving EU ports. It has been amended a few times, most recently in its repeal, on 5 February 2004, through the adoption of the Erika II package. This will be discussed further below.

2.2.3 1994 Council Regulation 2978/94

In November 1994, the Council adopted Regulation 2978/94 to promote the use of segregated ballast tanks (SBT) and double hull oil tankers through the lowering of their port charges, and to make the IMO Resolution A.747 (18) mandatory in the EU. This will be discussed below together with its further amendments.

2.2.4 The 2000 Decision: Legal Basis for Community Action

At present, the role of Europe in the field of responses to marine pollution finds its legal basis in a Decision of 20 December 2000, in which the European Parliament and the Council set up a Community framework for cooperation in the field of accidental or deliberate marine pollution for the period from 1 January 2000 to 31 December 2006.[24]

2.3 POST ERIKA & PRESTIGE

Two major oil tanker accidents (Erika in 1999 and Prestige in 2002) have, again, cast doubts on the effectiveness of the international regime and EU legislation in this area. The Community has adopted more stringent maritime safety measures and proposed an EU

---

[22] COM(1993) 66 final, 24 February 1993.
[23] The website of DG Transport of the European Union provides more detail on the measures taken: <http://europa.eu.int/comm/transport/maritime/safety/index_en.htm>.
[24] Decision 2850/2000/EC of the European Parliament and the Council, of 20 December 2000, setting up a Community framework for co-operation in the field of accidental or deliberate marine pollution, published in *OJ* L332 of 28 December 2000, 1-6.

wide compensation fund with a higher upper limit in reaction to these oil spills. These were taken into account at the IMO discussions concerning the current international regime, which resulted in some changes at the international level as well.

2.3.1 Erika I and II Packages

After the Erika disaster of 12 December 1999, the EU has considerably reinforced its legislative arsenal in order to better protect European waters from the risks of accidental oil spills. Two sets of legislative proposals were adopted by the Commission, in March and December 2000: the so-called Erika I[25] and Erika II packages[26] respectively.

The Erika I package addressed serious gaps in EU maritime safety legislation that were highlighted by the Erika disaster. It strengthened the existing rules on port State control and classification societies, and also set up an accelerated timetable for the phasing out of single hull oil tankers. The Commission, in the Erika I package, also developed some criteria for examining the existing international compensation system. It held that despite some shortcomings in the system it should still serve as the basis for future measures.

The Erika II package then presented additional measures designed to bring about a radical improvement in maritime safety in EU waters. It proposed the creation of a European Maritime Safety Agency and the establishment of a Community maritime traffic monitoring system, and the setting up of a European Fund for compensation of oil pollution damage (COPE Fund) with an updated ceiling of €1 billion (instead of €200 million under the international conventions).

2.3.2 Post-Prestige

When the Prestige sank off the Spanish coast on 19 November 2002, the Commission again responded very quickly. It adopted a Communication on improving safety at sea as early as 3 December 2002,[27] whereby the measures in the Erika I and II packages were to be further strengthened. This Communication specially focused on updating existing international rules on compensation and civil liability, and the introduction of penal sanctions against those who cause pollution through gross negligence.

In May 2003, after the adoption of a Supplementary Fund Protocol to the Fund Convention, the Commission urged the Member States to join the updated international regime. This was supported by the Parliament in January 2004. This will be discussed in further detail below.

---

[25] COM(2000) 142 final, Communication from the Commission to the European Parliament and the Council on the Safety of the Seaborne Oil Trading, 21 March 2000.

[26] COM(2000) 802 final, Communication from the Commission to the European Parliament and the Council on a Second Set of Community Measures on Maritime Safety following the Sinking of the Oil Tanker Erika, 6 December 2000.

[27] COM(2002) 681 final, Communication on improving safety at sea in response to the Prestige accident, 3 December 2002.

2.4 NEW CHALLENGES FOR THE ENLARGED EU

Since May 2004, with ten new Member States joining the EU, the enlarged Community is confronted with more exciting challenges centring on how to enforce the stringent maritime safety measures in those new Member States that have a lower standard of maritime safety infrastructure. More particularly, for countries with more lenient registry policies like Cyprus, Latvia, and Malta, it will be challenging but crucial to monitor their functions as flag States, in order to control their registered fleets and thereby improve maritime safety. In a report[28] in November 2003 that analysed the progress of the (then) future Member States in their preparation for accession to the EU, Cyprus and Malta were found to lack effective safety control for sea-going ships, and Lithuania and Poland were found to fail to meet obligations concerning the inspection and control of the fleet.

These developments, particularly concerning the most recent evolution of the issues on prevention and liability for oil pollution, merit some further discussion.

3 SPECIFIC ISSUES

This paper is mainly concerned with the most recent developments of the EU regime on marine oil pollution. Hence, the earlier legislation in these specific areas cannot be discussed in detail.

3.1 PORT STATE CONTROL

Flag States have a primary role to play in maritime safety as a result of IMO conventions,[29] but the power of flag State control is often weakened by the practice of displaying flags of convenience.[30] In such cases, effective controls being exercised by port States can be very useful for the aim of marine environment protection, as specified in the UNCLOS Convention.[31] The EU is in favour of promoting the approach of port State control, as its vast coastline, which is a busy marine oil transport route, is heavily exposed to the risk of marine oil pollution. Hence, inspection of vessels by the EU Member States where ports are located is considered by the Community as an effective tool to reduce

---

[28] COM(2003) 675 final, Comprehensive monitoring report of the European Commission of 5 November 2003 on the state of preparedness for EU membership of the Czech Republic, Estonia, Cyprus, Latvia, Lithuania, Hungary, Malta, Poland, Slovenia and Slovakia.

[29] P. Birnie and A. Boyle, *International Law and the Environment*, (2nd edn, Oxford University Press, Oxford, 2002), pp. 347-403.

[30] There are certain 'open registry' countries like the Bahamas, Bermuda, Cyprus, Liberia and Panama. These open registries often have low or no tax requirements, which attracted some shipping companies that were originally registered in EU countries to 'flag out' to these open registries. Although major shipping countries like France and Greece have adopted policies to encourage ship registry, there is still quite a large amount of fleet registered in other countries. For example, by 1990 some 75% of the UK (another major shipping country) owned fleet was flagged out. For a more detailed discussion on this issue, refer to *supra* note 12, 25-53.

[31] Seminar on Flag State Implementation and Port State Control, Alexandria, Egypt, 7-10 May 2002, Arab Academy of Science & Technology and Maritime Transport.

substandard shipping in its waters and to defend its waters from potential oil pollution.

The EU legislation concerning port State control is built on IMO Resolutions and the Paris Memorandum of Understanding (Paris MOU). Directive 95/21/EC was adopted to establish common criteria and harmonized procedures for port State control. This Directive has been amended several times, and the latest amendment is part of the Erika I package.

3.1.1 Background – Paris MOU on Port State Control

In reaction to the Amoco Cadiz incident in March 1978, some 14 states signed the Paris Memorandum of Understanding on 26 January 1982.[32] It was agreed that the elimination of substandard shipping would be best achieved through coordination of port States and should be based on the provisions of relevant international conventions that were widely accepted. The Paris MOU thus provided a basis for cooperation among the Member States and established harmonized inspection procedures.

The Paris MOU provides a framework for the signatory countries to carry out their inspection duties and to cooperate with each other. Under the Paris MOU, the port State may check visiting foreign ships to see if they comply with the relevant international conventions on safety, pollution prevention and seafarers living and working conditions. It is a means of enforcing compliance where the ship-owner and the flag State fail to do so.[33] It gives the port State a legal basis for refusing entry to substandard ship.

3.1.2 1995 Directive 95/21/EC

In 1989 the Commission, in its Communication to the Council, recommended improvement of the effectiveness of port State control.[34] The Member States were hence recommended to ratify the international conventions mentioned in the Paris MOU[35] and to carry out the inspections required in the Paris MOU. However, this was only a recommendation and there was no Community legislation on port State control until 1995,

---

[32] There are 19 signatory countries to the Paris MOU to date: Belgium, Canada, Croatia, Denmark, Finland, France, Germany, Greece, Iceland, Ireland, Italy, the Netherlands, Norway, Poland, Portugal, the Russian Federation, Spain, Sweden and the United Kingdom.

[33] Z. Oya Ozcayr, *The Role of Port State Control*, The Impact on Caspian Oil and Gas Development on Turkey and Challenges facing the Turkish Straits Conference, Istanbul, 9 November 2001.

[34] COM(89) 266 final, Commission Recommendation on Improving the Effectiveness of Port State Control in the Community, submitted by the Commission to the Council on 2 August 1989.

[35] Paris MOU is not an international convention, but an administrative agreement, and it did not extend the scope of port state control beyond the requirements in the relevant international conventions. Section 2 of Paris MOU refers to the following international conventions as relevant instruments: International Convention on Load Lines 1966, as amended and its 1988 Protocol (LOADLINES 66/88); International Convention for the Safety of Life at Sea 1974, 1978 and its 1988 Protocol (SOLAS 74/78/88); International Convention for the Prevention of Pollution from Ships 1973 and its 1978 Protocol (MARPOL 73/78); International Convention on Standards of Training, Certification and Watchkeeping for Seafarers 1978, as amended (STCW 78); Convention on the International Regulations for the Prevention Collisions at Sea 1972, as amended (COLREG 72); International Convention on Tonnage Measurement of Ships 1969 (TONNAGE 1969); Merchant Shipping (Minimum Standards) Convention 1976 (ILO Convention no.147).

when the Council adopted a Directive regarding port State control.[36] This Directive made the Paris MOU mandatory and significantly strengthened its provisions by setting up harmonized inspection rules and criteria. The principal objective of the Directive is to reduce the operation of substandard ships in Community waters, thereby improving safety as well as marine environmental protection by inspecting vessels at EU ports.[37]

However, problems exist with the enforcement of the port State control measures. There are Member States failing to inspect the required number of ships to the required standards. In 2002, the Commission brought France and Ireland to the Court of Justice for not observing the annual minimum threshold of 25% of ships to be inspected by the port State as provided for by the Port State Control Directive.[38] The legal action may ultimately lead to fines being imposed on Member States if they consistently fail to implement these maritime safety rules.

3.1.3 2000 Erika I-Directive 2001/106/EC

The Erika incident in 1999 has again shown the weakness of the existing regime. The Commission therefore decided to tighten up the Community regulatory framework (including port State control) in its first set of proposals in reaction to the Erika accident.[39]

In its Erika I Proposal, the Commission promoted tougher measures against manifestly substandard ships. It required that ships of old age with bad detention records to be banned. Furthermore, oil tankers more than 15 years old (instead of 20 or 25 years in Directive 95/21/EC) are to be subjected to enhanced surveys. Also, instead of being optional, the provisions in Directive 95/21/EC concerning potentially hazardous ships (according to the Annex to the Directive, this includes oil tankers, gas and chemical tankers, etc.) were to become mandatory.

The Erika I legislation on port State control finally entered into force on 22 July 2003,[40] which is the deadline for the Member States to transpose the Erika I legislation into national law. The Commission initiated legal proceedings against eight Member

---

[36] Council Directive 95/21/EC, of 19 June 1995, concerning the enforcement, in respect of Shipping Using Community Ports and sailing in the waters under the jurisdiction of the Member States, of international standards for ship safety, pollution prevention and shipboard living and working conditions (port state control), *OJ* L157, 7 July 1995.

[37] *Ibid.*, preamble.

[38] IP/02/931.

[39] Proposal for a Directive of the European Parliament and of the Council amending Council Directive 95/21/EC concerning the enforcement, in respect of shipping using Community ports and sailing in the waters under the jurisdiction of the Member States, of international standards for ship safety, pollution prevention and shipboard living and working conditions (port state control), COM(2000) 142 final. Communication from the Commission to the European Parliament and the Council on the Safety of the Seaborne Oil Trade, Brussels, 21 March 2000.

[40] Directive 2001/106/EC of the European Parliament and of the Council, of 19 December 2001, amending Council Directive 95/21/EC concerning the enforcement, in respect of shipping using Community ports and sailing in the waters under the jurisdiction of the Member States, of international standards for ship safety, pollution prevention and shipboard living and working conditions (port state control), *OJ* L19, 22 January 2002, 17-31.

States for failing to notify transposition of the EU maritime safety legislation on port State control into national law.[41]

3.1.4 Prestige

In reaction to the Prestige accident in November 2002, the Commission submitted a Communication to the Parliament and the Council,[42] in which *inter alia* an earlier implementation of Erika measures including port State control was proposed. For this reason, an indicative list of ships that may be banned in accordance with Article 7b of Directive 95/21/EC on port State control was published on 3 December 2002.[43] The Commission hoped that the ship-owners and flag States concerned in the list would thus take measures to improve the safety standards.

In the Communication, the Commission also stressed that in practice inspections on vessels only calling at European ports for refuelling were insufficient compared with those making full commercial calls. As it might be the only chance to exercise some control over such refuelling call vessels, inspections, it was stated, should be improved in order to create an even standard of inspection in all the EU ports.

As we can see from above, port State control may be an effective measure of protecting coastal States from the risk of substandard shipping. However, it only applies when the ships enter and call at the port, and only a certain number of ships may be checked. There are still chances that substandard ships might escape the inspections. As such, the control of flag States should remain the primary safety measure, and cooperation between the flag State and the port State should be enhanced.

Among the ten new accession States, Malta and Cyprus have large oil tanker fleets that are among the world's top fifteen tanker fleets.[44] At the same time, they have long tradition of open registries. This gives the Community an opportunity to reassert itself as a major player on a global scale.[45] The strict implementation of International and Community safety standards on these countries as flag States is crucial for the maritime safety and marine environment protection of the enlarged European Union.

---

[41] By 22 July 2003, only Denmark, France, Germany, Spain and the UK had transposed the whole Erika I measures into national law. Therefore, the Commission has initiated the infringement procedure against 10 Member States, namely Austria, Belgium, Finland, Greece, Ireland, Italy, Luxembourg, the Netherlands, Portugal and Sweden for failing to communicate national measures implementing key maritime safety Directives. Belgium and Ireland have so far only transposed the Port State Control Directive and the Netherlands, the Ship Inspection and Survey Organisation Directive.

[42] COM(2002) 681 final, Communication from the Commission to the European Parliament to the European Parliament and to the Council on improving safety at sea in response to the Prestige accident, Brussels, 3 December 2002.

[43] This indicative black list was followed up on 25 July 2003 with a second list of ships likely to be banned from European waters, and on 1 November 2003 a list of ships that should definitively be banned from EU ports.

[44] COM(2000) 142 final, 10.

[45] M. Frendo, 'The future of open registers in the European Union', (2000) 3 *Lloyd's Maritime and Commercial Law Quarterly* 383-393.

3.2 NOTIFICATION/REPORT SYSTEM

The coastal States are entitled to impose specific requirements as conditions for the entry of foreign vessels into their ports.[46] This is crucial for the EU as most of the European oil trade is sea born, and seas and coasts with dense traffic routes are exposed to serious danger. For this reason, the EU has adopted such provisions in its maritime safety measures in order to prevent substandard ships from entering its waters.

3.2.1 Directive 93/75/EEC

As early as 1978, in reaction to the Amoco Cadiz accident, the Council adopted a Directive concerning the minimum requirements for certain tankers entering or leaving Community ports.[47] It required that tankers, before entering the Community ports, notify the competent authority of the port State of the general details of the tanker and the nature and quantity of the cargo. The fulfilment of such an obligation was left to the Member State of the registered flag.

This Directive was modified several times to take into account the development of international conventions. One important modification was in September 1993 (the so-called Hazmat Directive),[48] in order to include all cargo vessels carrying dangerous or pollutant goods (in addition to tankers). Under the Hazmat Directive, port authorities may take precautionary actions in regard of vessels using EU ports so as to prevent and minimize accidents. Also, the required information to be provided to the port authorities must include detailed information on the cargo carried, and any other information contributing to prevention and minimization of accidents. Failure to provide such information may lead to the detention of ships.

3.2.2 Erika II

The Directive 93/75/EEC entered into force in 1995. The Commission evaluated the application of this Directive in its Erika II proposals and considered it 'inadequate'.[49] The Commission argued that the established system was not clear and hence not correctly applied, and that there was no procedure or standard format for transmission or exchange of data, a lack of appropriate means of communication and no clear definition of competent authorities. However, the Commission intended to implement much broader goals that could not be covered by simply amending the Directive. Therefore, the Commission proposed a new instrument. The Commission hopes that with the assistance of modern technology, by means of monitoring maritime traffic, closer surveillance of

---

[46] Para. 3, art. 211 UNCLOS Convention.
[47] Council Directive 79/116/EEC, of 21 December 1978, concerning minimum requirements for certain tankers entering or leaving Community ports, *OJ* L33, 8 February 1979, 33.
[48] Council Directive 93/75/EEC, of 13 September 1993, concerning vessels bound for or leaving Community ports and carrying dangerous or polluting goods.
[49] COM(2000) 802 final, Communication from the Commission to the European Parliament and the Council on a second set of Community Measures on Maritime Safety following the sinking of the oil tanker Erika, Brussels, 6 December 2000, 10.

ships posing higher risks and increasing intervention at sea in cases of threat to the environment and shipping safety, the new instrument will achieve prevention of pollution at sea.[50]

As part of the Erika II packages, Directive 2002/59/EC[51] was adopted in June 2002, as a result of which the previous Directive was repealed on 5 February 2004. The new Directive was aimed at extending the reporting obligations by making use of modern technology like EDI (electronic data interchange) and AIS (automatic identification of ships), and establishing a Community vessel traffic monitoring and information system. The Member States were required to apply the provisions laid down in the Directive by 5 February 2004 the latest.

These modifications to the Community legislation coincide with the most recent development at the international level. Take, for instance, the new Regulation 11 of the SOLAS convention concerning ship reporting systems, requiring ships in transit in particular areas to notify the coastal authorities responsible and give them certain types of information. Further, the new regulation 12 of the SOLAS Convention on vessel traffic services provides that ships travelling through sea areas where traffic density is high or where there are serious risks of accidents should receive assistance regarding shipping and weather information, and, where appropriate, with traffic assistance and routing services.

3.3 CLASSIFICATION SOCIETY

3.3.1 Council Directive 94/57/EC

The classification society plays an important role in maritime safety, as it has the competence to inspect the quality of the ships. However, due to economic pressure and even such organizations operating without sufficient expertise and professionalism, the confidence of the maritime industry and the public in the classification societies has recently declined.[52]

As early as 1993, the role of classification societies in enhancing maritime safety was recognized by the Community. As specified in the preamble of Common Policy on safe seas of 8 June 1993, 'the Council has set the objective of removing all substandard vessels from Community waters and has given priority to Community action to secure the effective and uniform implementation of international rules by elaborating common standards for classification societies'.

However, Community regulation for the monitoring of classification societies did not begin until 1994 with the adoption of Council Directive 94/57/EC.[53] This Directive laid down strict criteria for Classification Societies in order for them to be recognized

---

[50] *Ibid.*, 11.
[51] Directive 2002/59/EC of the European Parliament and of the Council of 27 June 2002 establishing a Community vessel traffic monitoring and information system and repealing Council Directive 93/75/EC, *OJ* L208, 10-27.
[52] COM(2000) 142 final, 23.
[53] Council Directive 94/57/EC, of 21 November 1994, on common rules and standards for ship inspection and survey organisations and for the relevant activities of maritime administrations, *OJ* L319, 12 December 1994, 20-27.

Community wide. Only highly reliable and professionally competent bodies can be recognized by the EU to carry out statutory surveys and certification on behalf of EU Member States. This Directive has been amended several times, and has been further reinforced in the Erika I package and post-Prestige measures.

3.3.2 Commission Directive 97/58/EC

Subsequent to the IMO Assembly, at its 19th session, adopting Resolution A.789(19), on specifications on the survey and certification functions of recognized organizations acting on behalf of the administration,[54] the Commission felt it appropriate to incorporate such an amendment in EC Law. Hence, the Commission adopted Directive 97/58/EC[55] in order to incorporate the provision of this Resolution.

3.3.3 Erika I

The transposition of Directive 94/57/EC into national law of the Member States was not completed until 1998 and the Commission had to launch twelve infringement procedures. In addition, the Directive was criticized for lacking harmonized and centralized controls of the recognized organizations.

As such, in Erika I package, harmonized Community arrangements on Classification Societies were proposed. Directive 2001/105/EC,[56] based on the Erika I proposal, was thus adopted to tighten and harmonize the Community legislation on Classification Societies. Classification societies, in order to be recognized, must comply with more stringent qualitative criteria, the failure of which might lead to the suspension or even withdrawal of the recognition. In the case of transfer of class, both gaining and losing organizations are subject to stringent control. This was aimed at preventing the practice of class hopping (changing class to avoid carrying out requested repairs), which constitutes a potential danger to the marine environment. Moreover, the financial liability of the recognized classification societies is also harmonized at Community level (limited to different levels according to the seriousness of the negligence, or even potentially unlimited).

When the Erika I package entered into force on 22 July 2003, the Commission initiated the infringement procedure against nine Member States (Austria, Belgium, Finland, Greece, Ireland, Italy, Luxembourg, Portugal and Sweden) for failing to communicate to the Commission any national measures implementing Directive 2001/105/EC.

---

[54] Annex to Directive 94/57/EC is based on IMO Resolution A.739(18) on guidelines for the authorization of organisations acting on behalf of the administration.
[55] Commission Directive 97/58/EC, of 26 September 1997, amending Council Directive 94/57/EC on common rules and standards for ship inspection and survey organisations and for the relevant activities of maritime administrations, *OJ* L274, 7 October 1997, 8.
[56] Directive 2001/105/EC of the European Parliament and of the Council, of 19 December 2001, amending Council Directive 94/57/EC on common rules and standards for ship inspection and survey organisations and for the relevant activities of maritime administrations, *OJ* L019, 22 January 2002, 9-16.

3.3.4 Prestige

In its Communication after the Prestige accident,[57] the Commission emphasized the importance of closer monitoring of the performance of the classification societies. The Commission made its determination clear that they would 'not hesitate to launch procedures to suspend or even withdraw the recognition of those organizations which do not give enough guarantees in safety terms'.[58] It was also concerned with the practice of classification societies issuing seaworthiness certificates for vessels on behalf of the shipowner on a commercial basis, and at the same time on behalf of the flag State that is supposed to monitor the vessel. The Commission considered it necessary to change these practices and keep these checks separate. However, there was no specific measure regarding how to carry out such checks more separately per se.

3.4 SINGLE HULL TANKER

Design defects of the ship's structure might cause serious oil spills. Among these, the one that arouses most public concern today is that of the single hull design.[59]

3.4.1 International v. US Regimes on Single Hull/Double Hull Tankers

Following the Exxon Valdez accident in 1989, dissatisfied with the ineffectiveness of the international regime concerning oil pollution, the US in 1990 began its own regime with the adoption of the Oil Pollution Act (OPA 90). Under OPA 90, double hull requirements were unilaterally imposed on both new and existing oil tankers in the form of age limits (between 23 and 30 years from 2005) and deadlines (2010 and 2015) for the phasing out of single hull tankers. Hence, under the US regime, the single hull tanker will be phased out earlier than in the international regime. Faced with the challenge of the separate US regime, the IMO introduced double hull standards into the MARPOL Convention in 1992, but with a more lenient age limits and phasing out schedule for the single hull tankers. It required that all oil tankers delivered after July 1996 should have a double hull. Hence, there should be no single hull oil tankers built after that date. Single hull oil tankers above 20,000 DWT, delivered before 6 July 1996, are required to comply with double hull requirements by the age of 25 or 30, depending on whether or not they have segregated ballast tanks. The gradual phasing out of single hull tankers will take until 2026.

There is, then, a problem because of the difference between the US and the international system. Single hull tankers, banned from US waters from 2005 onwards due to the age limit, might start to operate in other regions of the world. On the other hand, the

---

[57] COM(2002) 681 final, Communication from the Commission to the European Parliament and to the Council on improving safety at sea in response to the Prestige accident, Brussels, 3 December 2002.
[58] *Ibid.*, 7.
[59] In single hull tankers, the oil is separated from the seawater only by the bottom and side plating. In case of collision or grounding, when the hull is damaged, the risk that the cargo oil in the tanks be discharged into the sea and cause pollution to the environment is very high. Compare this to double hull designs, which protect the cargo tank with a second inner plate at a sufficient distance form the outer shell, thus reducing the pollution risk.

EU, as the largest oil trade area, will be very likely exposed to the risk of increasing pollution danger caused by old single hull tankers.

3.4.2 Council Regulation 2978/94

In November 1994, the Council adopted a Regulation to promote the use of segregated ballast tanks (SBT)[60] and double hull oil tankers by lowering their port charges.[61] This was a step in the right direction by the EU, but whether it went far enough to prevent environmental oil pollution was questionable, especially considering the running aground of the 25-year-old single hull tanker Erika in 1999.

3.4.3 Erika I – Regulation 417/2002

Realising the potential danger posed by old age single hull tankers banned from the US waters shifting their trade to the EU waters, the Commission in the Erika I package proposed that the EU should take measures to phase out single hull tankers in European waters[62] and that the proposed system should be aligned to the age limits and end-dates provided for in the US OPA 90. Hence, the appropriate measures should take effect before 2005, the time from which single hull tankers will be banned from US waters.

In order to encourage the use of double hull tankers, the Commission proposed as an accompanying measure a financial incentive mechanism in the form of reduction of port and pilotage dues. This mechanism was based on the principles established in Council Regulation 2978/94. As mentioned before, the difference lay in the distinction of double hull and single hull oil tankers with regard to the levels of reduction of port and pilotage dues. Another accompanying measure is to notify the IMO about the Community legislation on the accelerated phasing in of double hull or equivalent design standards for single hull oil tankers, once it is adopted.

There has been concern in the industry that the accelerated phasing out of single hull tankers might have an impact on the price of oil products. However, the Commission, in its proposal, referred to a study on OPA 90 carried out by the US National Research Council in 1998, which concluded that such an influence would be insignificant.[63] Compared with the huge amount of clean up costs and even the irreversible damage to the environment *per se*, the Commission judged that the additional cost does not outweigh the

---

[60] SBT is designed to reduce the risks of operational pollution by ensuring that ballast water never comes into contact with oil. However, this is only one way to minimise the pollution risks. For more discussion, please see M. Faure and R. Van den Bergh, *Objectieve Aansprakelijkheid, Verplichte Verzekering en Veiligheidsregulering* (Maklu, Antwerp, 1989), p. 278.

[61] Council Regulation 2978/94/EC, of 21 November 1994 on the implementation of IMO Resolution A.747(18), on the application of tonnage measurement of ballast spaces in segregated ballast oil tankers, *OJ* L319, 12 December 1994, 1-6.

[62] Proposal for a Regulation of the European Parliament and of the Council on the accelerated phasing-in of double hull or equivalent design requirement for single hull oil tankers, 20 December 2002.

[63] The Assessment study on the Oil Pollution Act published by the USA National Research Council in 1998 concluded that the impact of OPA 90 on the cost of the oil products was estimated to be approximately 10 US cents per barrel, or about one tenth of the cost of transportation, which in itself only represent between 5 to 10 % of the total product cost. The final impact on the price of the oil products will thus be less than 1%.

benefits, if, with the proposed measures, the re-occurrence of such accidents in Community waters can be prevented. As such, the Commission considered it a reasonable price to pay to ensure an effective reduction in the risks of pollution.[64] Moreover, the Commission wanted to signal to the industry that it is necessary and unavoidable to invest in new double hull tankers without delay.

There was also discussion in the IMO on the phasing out of single hull tankers. The IMO was aware that without reaching concrete and acceptable results in a short time period, there was a chance that the EU might adopt its own regional rules. Under pressure from the EU, on 27 April 2001 the IMO adopted an amendment of the MARPOL Convention to introduce a new global timetable for accelerating the decommissioning of single hull oil tankers (this amendment entered into force on 1 September 2002). Based on the amended international regime, the EU adopted Regulation 417/2002 on 18 February 2002,[65] and the ambitious schedule to phase out single hull tankers was slowed down.

3.4.4 Prestige

The Commission regretted the slow pace of implementing the Erika measures, as otherwise the Prestige accident could have been prevented.[66] After the Prestige accident, the Commission proposed a regulation[67] to prohibit the transport of heavy fuel oil in single hull tankers either bound for or leaving EU ports, and to speed up the decommissioning of single hull tankers. The proposed schedule falls back on the one originally proposed in Erika I package.

The Commission had observed the cases of the Baltic Carrie and Pindar incidents in the Baltic Sea, in which dramatic damage was avoided thanks to their double bottom or double hull construction.[68] By contrast, the Prestige accident, concerning a 26-year-old single hull tanker carrying heavy fuel oil, had disastrous consequences. The practice of transporting heavy fuel oil in old age tankers (due to the low commercial value of heavy fuel oil) has posed great dangers to the maritime safety and environment. Thus, the Commission decided to forbid the transport of heavy fuel oil in single hull tankers.

This proposed Regulation was adopted in July 2003 and entered into force in October 2003.[69] At the same time, the EU Member States worked actively at the IMO level to promote the adoption of such enhanced measures at the international level. Fortunately,

---

[64] COM(2000) 142 final, 27.
[65] Regulation 417/2002 of 18 February 2002 on the accelerated phasing in of double hull or equivalent design requirements for single hull oil tankers, *OJ* L64, 7 March 2002, 1.
[66] Commission Memo of 21 October 2003, Safer seas: the fight goes on, 2. If the Erika I proposal had been accepted, the Prestige would have been phased out in 2002 at the age of 23, however, under the finally adopted Regulation, the Prestige fell to be phased out in March 2005.
[67] Proposal to the European Parliament and the Council for a Regulation of the European Parliament and of the Council amending Regulation 417/2002/EC on the accelerated phasing in of double hull or equivalent requirements for single hull oil tankers and repealing Council Regulation 2978/9420/EC, Brussels, December 2002.
[68] COM(2002) 681 final, 10.
[69] Regulation 1726/2003/EC of the European Parliament and of the Council, of 22 July 2003, amending Regulation 417/2002/EC on the accelerated phasing in of double hull or equivalent requirements for single hull oil tankers, *OJ* L249, 1 October 2003, 1-4.

agreement on the same arrangement was achieved at the IMO, closing the gap between the international and EU regimes.

3.5 LIABILITY AND COMPENSATION

The issue of civil liability and compensation for oil pollution is mainly regulated by international conventions, and particularly the CLC Convention of 1969 and the Fund Convention of 1971 (which was revised in 1992, and more recently in 2000 and 2003). The 1992 Protocols entered into force in 1996 and all EU Member States with a coastline (not including the 10 new accession Member States) are parties to the 1992 Protocols.[70] In contrast to the prevention legislation on oil pollution, there is no particular Community legislation as such to address the civil liability for oil pollution.

That said, the most important change in the EU approach to compensation for oil pollution damage took place after the Erika oil spill, and so we will mainly discuss changes since 2000.

3.5.1 Post-Erika

3.5.1.1 Erika I Proposals

In the Erika I package, the Commission developed three criteria to examine the extent to which the existing liability and compensation system was fully satisfactory. It then critically reviewed the international regime according to these criteria:[71]

The *first* criterion was prompt compensation. The CLC and the Fund regime, according to the Commission, had been able to handle most of the cases promptly and without extensive or lengthy judicial procedure.

The *second* criterion provided was whether there is a sufficiently high compensation limit. The limit set by the CLC and the Fund Convention was considered insufficient to cover any potential disaster. Hence, victims of serious oil spills may remain incompletely compensated.

The *third* criterion was the dissuasive effect of the regime. In this respect, the CLC and the Fund regime were also considered unsatisfactory due to the fact that the fiscal limitation of ship-owners' general strict liability was almost unbreakable while, additionally, cargo owners held no individual responsibility at all. Consequently, no parties in oil transportation activity were subject to sufficient incentives to give up the practice of deliberately employing vessels in appalling conditions.

On the basis of these considerations, the adequacy of the international regime for oil pollution compensation, as laid down in the CLC and the Fund Convention of 1992, was

---

[70] COM(2000) 142 final, 21 March 2000, 33. According to the statistics from the IOPC Fund, as of 1 March 2004, all the current EU Member States, with the exceptions of Austria and Luxemburg (which are the only two EU countries with no coastline), are contracting states of the 1992 CLC and the Fund Conventions. Luxemburg is a contracting state to the 1969 CLC Convention, while Austria has not joined any of the international conventions.
[71] COM(2000) 142 final, 21 March 2000, 34-37.

consequently questioned by the Commission in Erika II package from an EU perspective, especially as to the limitation of liability and the definition of pollution damage.

3.5.1.2 *Erika II Proposals*

As was noted in the Erika I package, the compensation limits set up by the international conventions were not sufficient in cases of compensation for major oil spills.[72] The Commission therefore decided to create a European mechanism to ensure adequate and prompt compensation without delay. Moreover, in the Erika II package, the Commission raised the issue of the extension of the scope of application of the international conventions in the area of pollution damage covered, and, further, that the balance of responsibilities of all parties involved should be re-examined.

3.5.2 Post-Prestige: The 2002 Communication & 2003 Proposal

In its Prestige Communication, which specially focused on the update of existing international rules on compensation and civil liability, the Commission decided that the measures in the Erika I and II packages should be strengthened.

Thanks to the efforts of the EU, a new protocol to the Fund Convention was adopted in May 2003, which established a supplementary, higher value, compensation fund. This will substantially improve the capacity to compensate pollution victims compared to the previous situation. The Commission then proposed that the Member States ratify the Supplementary Fund Convention as soon as possible.[73]

The findings of the Commission corresponded with criticisms in the literature where it had equally been held that there are three major shortcomings in the current scope of the international regime in the sense of (1) too low amounts of limitation, (2) exclusive channelling of liability to the tanker owner (excluding liability of others) and (3) a too narrow definition of environmental damage.[74]

3.5.3 Limitation of Liability

A first area of debate and divergence between the EU and the international regime is the limitation of liability. The Commission felt that in the cases of recent high-profile European oil spills, where the compensation claims came close to the limitation, there

---

[72] For an overview of the evolution in the limitations see M. Faure and H. Wang, 'The international regimes for the compensation of oil pollution damage: are they effective?', (2003) 12 *Review of European Community and International Environmental Law* 242-253.

[73] COM(2003) 534 final, Proposal for a Council Decision authorising the Member States to sign, ratify or accede to, in the interest of the European Community, the Protocol of 2003 to the International Convention on the Establishment of an International Fund for Compensation for Oil Pollution Damage, 1992, and authorising Austria and Luxemburg, in the interest of the European Community, to accede to the underlying instruments, 8 September 2003.

[74] For a summary of this criticism on the international conventions see E.H.P. Brans, *Liability for damage to public natural resources. Standing, damage and damage assessments* (Kluwer Law International, The Hague, 2001), pp. 344-360 and M. Faure and H. Wang, 'Liability for Oil Pollution – the EU approach', (2004) 12 *Environmental Liability* 55-67.

were unacceptably long delays in payment.[75] According to the Commission, the main reason for this was the uncertainty as to the final costs of an oil spill being caused by the restrictive limits to compensation.

Under the CLC Convention, the amendment procedure would not allow for more than a 50% increase of the existing limit.[76] The first decisions to approve this increase were taken by the IMO in October 2000 and the Amendments took effect on 1 November 2003.

The new amounts of compensation under the 2000 Amendments, which raised the limitation of liability in the 1992 Protocols, are as follows:

- for a ship not exceeding a Gross Tonnage of 5000, the liability is limited to 4.51 million SDR[77] (under 1992 Protocols, the limit was 3 million SDR);
- for a ship with a Gross Tonnage of 5000 to 14000, liability is limited to 4.51 million SDR plus 631 SDR for each additional gross tonne over 5000 (under the 1992 Protocols, it was 3 million SDR plus 420 SDR for each additional gross tonne over 5000);
- for a ship over 14000 tonnage, liability is limited to 89.77 million SDR (under the 1992 Protocols, it was 59.7 million SDR).

Such an increase might seem impressive. However, the Erika incident in 1999 had already raised doubts on whether this amount would be sufficient to cover such large losses, and even before the 2000 protocols had entered into force (in 2003) the Prestige incident in 2002 virtually proved that the changes did not provide a satisfactory solution in case of disastrous oil spills. Hence, the European Commission judged that in addition to these 2000 protocols, further new steps were necessary.

3.5.3.1 Erika

The Commission felt that a 50% increase of the existing limits was insufficient to guarantee adequate protection for victims in the event of a major oil spill in Europe today and in the future.[78] The Commission held the view that all oil spills must be adequately and promptly compensated. However, statistics showed that some 10 out of 100 oil spills dealt by the Fund raised serious doubts as to the effect of the limits on the efficacy of compensation. Moreover, the international procedures to achieve concrete results in the

---

[75] The examples given by the Commission to justify its argument were, *inter alia*, Aegean Sea in Spain 1992, Braer in the UK 1993, Sea Empress in the UK 1996. The claimants in these oil spills have indeed encountered a delay in compensation without knowing if they would receive full compensation.

[76] Art. 15 of the 1992 Protocol to the CLC Convention and art. 33 of the 1992 Protocol to the Fund Convention provide the procedures for amendments. According to art. 15 of the CLC Convention 1992, under the title Amendments of Limitation Amounts, there at least one quarter of the Contracting States must propose amendment of the limits of liability, and the amendments shall be adopted by a two-thirds majority of the contracting States present and voting in the Legal Committee, subject to procedural requirements. Any amendment made under these articles enters into force, at the earliest, 36 months after its adoption by the IMO Legal Committee.

[77] Special Drawing Rights. An international reverse asset and unit of account of the IMF and other international organisations.

[78] The explanatory memorandum of the Erika II proposals.

sense of increasing the limit would be lengthy. Therefore, in the Erika II package the Commission proposed to complement the then existing international regime by creating a European supplementary fund, the so-called COPE Fund,[79] in order better to compensate oil pollution victims in Europe. The compensation under the COPE Fund would be based on the same principles as the current international fund system, but the financial limits would be raised to EUR 1 billion, which was considered by the Commission to be a sufficient amount.

Although the Commission criticized the international regime of CLC and the Fund for not being able to provide sufficient compensation to the victims of the Erika spill,[80] the efficiency of the proposed regional fund was doubtful. The COPE Fund was designed by the Commission to speed up compensation. However, it was to be applied only to those not fully compensated under the international regime. As such, it was questionable whether this proposal was optimal since in this model the victim first has to wait for a decision handed down by the international authorities. The proposed COPE Fund also received strong objections from the oil industry. According to them, insofar as they would have to be contributors to such a fund, their liability would be increased to an unreasonable extent. Hence, the Council preferred to refer the discussion to the competent international body, that is, the IMO, in order to obtain a similar agreement that would also be able to be applied worldwide.[81]

3.5.3.2 The 2002 Communication

In the Communication, the Commission asked the Member States to work determinedly within the IMO to promote the rapid implementation, by the IOPC Fund, of an additional compensation scheme for the victims of oil spills with an increased limit of up to EUR 1 billion. It was very likely at that moment that if the proposal failed at international level, the EU would have to address the question within the EU framework, similar to the model of the US system.[82]

3.5.3.3 The 2003 Proposal

It is likely that the efforts of the EU were responsible for the Supplementary Fund Protocol to the 1992 Fund Convention being adopted in May 2003. According to this protocol, a supplementary fund of 750 SDR (at the time of adoption this corresponded to approximately EUR 920 million or USD 1,000 million) have to be established. The creation of this new fund will of course substantially improve the capacity to compensate

---

[79] The Fund for Compensation for Oil Pollution in European waters, referred to as the COPE Fund.
[80] In Commission Memo Memo/01/387 of 2000, Erika, two years on, the Commission found that the people worst affected by the Erika spill working in maritime and other related sectors (e.g. Tourism) had not been fully compensated for their losses. The Commission considered that the international scheme was to be blamed for not covering full costs of an incident of this scale.
[81] The Transport Council in December 2000 adopted conclusions on the necessity to achieve ameliorations to the existing international regime, including 'a substantial increase in liability and compensation ceilings'.
[82] In the context of the 1990 Oil Pollution Act (OPA 1990), the US decided not to join the international arrangement, but set up its own system with a compensation fund of USD 1 billion.

pollution victims as compared to the previous situation. Therefore, there no longer seems to be a need to develop a regional European Fund for oil pollution.

Art. 21 of the Supplementary Fund Convention stipulates that it will enter into force three months after eight States, representing a total of at least 450 million tonnes of contributing oil, have become contracting parties to it. Hence, the actions by EU Member States alone will be sufficient to bring the Supplementary Fund Protocol into operation. The Commission addressed the Council on 8 September 2003 in order to urge the Member States to ratify the Supplementary Fund Convention as soon as possible with the hope of bringing the updated regime into operation before the end of 2003.[83]

However, this proposed schedule was postponed to 30 June 2004 in the Recommendation of the Committee on Legal Affairs and the Internal Market to the Parliament adopted in January 2004.[84]

3.5.4 Definition of Pollution Damage

The sufficiency of the compensation regime is not only to be evaluated in terms of the amount of compensation. Rather, as the Commission has suggested, it is also to be evaluated in terms of the types of damage that are covered by the regime. The Commission held the view that if the range of damage types were to be extended, the amounts available for compensation should be raised accordingly. Hence, the substantial increase of financial limits is further justified by the expanding definition of the damage to be covered.

The international regime established under the CLC and Fund Conventions covered pollution damage, including preventative measures and, to a limited extent, environmental damage per se for accidents occurring in the coastal waters (up to 200 miles) of the States. The Commission considered that the introduction of rules at the Community level in this respect would enhance the implementation of the 'polluter pays' principle and would thus be in line with the White Paper, which aims to extend the definition of pollution damage.[85] However, in the newly adopted environmental liability directive, marine oil pollution damage is explicitly excluded.

3.5.5 Balance of Responsibilities

The Commission stated that an adequate liability and compensation system should reflect a fair balance between the parties involved in the activities, whereby incentives are given to the stakeholders to take preventive measures. However, the ship-owner's right to limit their liability was almost unbreakable, and furthermore his liability was solely calculated

---

[83] COM(2003) 534 final, Proposal for a Council Decision authorising the Member States to sign, ratify or accede to, in the interest of the European Community, the Protocol of 2003 to the International Convention on the Establishment of an International Fund for Compensation for Oil Pollution Damage, 1992, and authorising Austria and Luxemburg, in the interest of the European Community, to accede to the underlying instruments, 8 September 2003.

[84] Recommendation A5-0042/2004 final of 29 January 2004.

[85] Remember that the White Paper on Environmental Liability proposed to bring damage to biodiversity and damage in the form of contamination of sites together under the heading 'Environmental Damage'.

on the basis of the size of the ship, without taking into account other relevant factors, like the nature of cargo carried and the amount of oil spilled. The Commission thus suggested that the threshold for loss of limitation rights should be lowered, so that, at the least, proof of gross negligence on behalf of the shipowner should trigger unlimited liability. Such a measure would produce both preventive and punitive effects according to the Commission. This will be further discussed below.

4 CONCLUDING REMARKS

In this paper, the EU legislation concerning prevention and liability for marine oil pollution, and especially the most recent developments therein, have been examined. The European Union is not only working as a driving force in the international arena to promote more stringent marine environment policies, but has moreover recognized the ineffectiveness of previous EU laws. As a result, it has striven to keep the Community legislations updated to keep Community laws in line with the international regimes. However, it will take some time before the adequacy of some of the newly amended legislation can be judged.

Prevention and compensation are two sides of the same coin. Focusing on prevention is important since preventive measures are always better than cures. However, prevention cannot be always successful, and oil spills will nevertheless occur, as has been revealed by recent events. In such cases, the question of how to compensate the victims adequately arises.

The EU originally took the point of view that marine oil pollution was an international problem that could be better solved at the international level. Henceforth, the EU merely counted on its Member States to ratify the various international conventions aiming at the promotion of maritime safety. However, that approach has changed since 1993, when the EU realized that the international conventions had not been effectively implemented throughout the world. Despite more stringent rules being established in the international conventions, the lack of overall international monitoring, sanctions and courts left the IMO without any auditing authority to check whether countries were observing the relevant rules. Hence the EU decided to include those international standards in EU legislation and to check for compliance. Through its special institutional structure, the European Community makes the international legislation binding for the Member States. When the states do not apply the rules, they will have to appear before the Court of Justice and may be convicted, and failure to comply will lead to penalties.

European activism in this domain has especially increased after the Erika and Prestige incidents within European territory. This led to the decision of the Community to phase out single hull tankers rapidly, to enhance oversight of classification societies, and to increase the compensation limit for the oil pollution victims. The decisions at the Community level in turn promoted a modification of the international conventions.

In the last ten years, much has happened in the domain of marine oil pollution, not only in Europe, but also at the international level. In Europe, moreover, rapid development is continuing even on a daily basis. There is obviously more legislation at the Community level that has been adopted or that is to be adopted. While it has not been possible to discuss all of these developments here, one can still hope that all of these increasing regulatory instruments will finally lead to the desired goal of reducing the risks of marine pollution and better protection of pollution victims.

LIST OF REFERENCES

P. Birnie and A. Boyle, *International Law and the Environment,* (2nd edn, Oxford University Press, Oxford, 2002).

E.H.P. Brans, *Liability for Damage to Public Natural Resources. Standing, Damage and Damage Assessments* (Kluwer Law International, The Hague, 2001).

M. Faure and H. Wang, 'The International Regimes for the Compensation of Oil Pollution Damage: are they Effective?', (2003) 12 *Review of European Community and International Environmental Law* 242-253.

M. Faure and H. Wang, 'Liability for Oil Pollution – the EU approach', (2004) 12 *Environmental Liability* 55-67.

M. Faure and R. Van den Bergh, *Objectieve Aansprakelijkheid, Verplichte Verzekering en Veiligheidsregulering* (Maklu, Antwerp, 1989).

M. Frendo, 'The future of open registers in the European Union', (2000) 3 *Lloyd's Maritime and Commercial Law Quarterly* 383-393.

M. Hedemann-Robinson, 'Protection of the Marine Environment and the European Union – Drifting or making headway? Some critical reflections on internal and external dimensions to law, policy and practice' (2004) 12 *Environmental Liability* 3-18.

Z. Oya Ozcayr, *The Role of Port State Control,* The Impact on Caspian Oil and Gas Development on Turkey and Challenges facing the Turkish Straits Conference, Istanbul, 9 November 2001.

V. Power, *EC Shipping Law,* (2nd ed, Lloyd's of London Press, London, UK, 1998).

P. Sands, *Principles of International Environmental Law,* (2nd edn, Cambridge University Press, Cambridge, 2003).

OIL POLLUTION PREVENTION AND ENFORCEMENT MEASURES AND THEIR EFFECTIVENESS: A SURVEY OF EMPIRICAL RESEARCH FROM THE US

Mark A. Cohen

1 INTRODUCTION

This chapter reviews the empirical research on the effectiveness of oil pollution prevention and enforcement measures in the United States. I consider both 'prevention' or 'monitoring' activities like government inspections, and 'enforcement' activities like sanctions, remedial actions, and other mechanisms designed to punish or bring a firm into compliance to reduce the frequency or size of spills. Over the past 20 years, there have been a series of independent studies by academic researchers analyzing Coast Guard and oil spill data with the goal of determining the effectiveness of alternative monitoring, enforcement and penalty policies. While the overwhelming finding from these studies is that enforcement efforts reduce oil spills, questions abound about the proper mix of enforcement techniques and whether the costs of more enforcement would exceed the benefits.

This paper is organized as follows. In Section 2, I briefly review the legal prevention and enforcement mechanisms available to government officials as well as those available to the pubic at large in the US. In Section 3, I provide an overview of the economic theory that relates to oil spill prevention and enforcement. In Section 4, I examine what is known about the relative effectiveness of various government prevention and enforcement mechanisms. Section 5 contains a brief conclusion and thoughts about how this research might inform policymakers in other countries. It also explores what little is known about non-regulatory mechanisms like informal pressure, private tort law, and criminal sanctions.

2 OIL SPILL PREVENTION AND ENFORCEMENT MECHANISMS IN THE US

Under US environmental laws, it is illegal to discharge oil into either inland or coastal waters. Not only is it illegal to discharge oil – whether accidentally or intentionally – but the polluter is required by law to report the spill to the US Coast Guard and to clean up the spill. If the polluter does not clean up the spill, the Coast Guard is authorized to clean up any remaining oil and to charge the polluter for any costs incurred. The Coast Guard is also authorized to impose monetary fines.[1] The Coast Guard by itself is authorized to penalize no more than USD 5,000 per spill. If it believes a higher fine is warranted, it must go to court and convince a judge to impose a more severe penalty. A court may also award damages to the government for publicly owned natural resources that have been harmed as a result of the oil spill. In addition to imposing a monetary sanction for spilling oil, a court may impose an even higher penalty for not voluntarily reporting the spill to the proper authorities. For example, failure to report to the Coast Guard is a criminal offence that may carry with it a prison sentence.

Of course, the Coast Guard's toolkit is not unlimited. It cannot simply impose a penalty – laws limit the amount of a penalty, and in some cases, the amount of cleanup costs that may be imposed upon an oil spiller. Prior to 1970, for example, vessel owners were liable for cleanup only for an amount equal to the remaining value of the vessel. In one extreme case, owners of the completely destroyed super tanker *Torrey Canyon* were liable for only USD 50 in damages (the value of the remaining life raft!) even though the cleanup costs were over USD 16 million.[2] As a result of this incident – and others – Congress increased the liability limits in 1977 and once again in 1990.[3]

Although most oil spill enforcement activities are entrusted to the Coast Guard, the US Environmental Protection Agency and the US Department of Justice might also become involved – especially for very large spills. In fact, the original Refuse Act of 1899 made it a criminal offence to spill oil in navigable waters in the US – even if the oil spill was unintentional. In the US, criminal offences must be brought by a US Attorney (i.e. through the US Department of Justice) – not directly by the Coast Guard or the Environmental Protection Agency. While this 'strict liability' crime has rarely been enforced, the potential exists and has been used in several high profile cases like the Exxon Valdez.

In addition to these government enforcement mechanisms, there are various legal and extra-legal mechanisms that might provide an incentive for firms to reduce the probability and severity of oil spills. Citizens who are harmed by a pollution incident may sue in court to recover their damages. They may also sue on behalf of the 'public' to ensure that the government is adequately enforcing its mandate. For example, if the government does not take legal action to penalize a firm that has spilled oil, citizens may go to court and request that penalties be imposed or that government enforcement actions are otherwise imposed. Aside from that legal option, it is also possible that external market pressures exert some influence on firm behaviour and help prevent oil spills from occurring. Thus, for example, it is possible that the negative reputation effect of being known as an oil polluter will induce firms to take certain precautions. Presumably, this is especially the case for publicly-traded firms as well as firms with brand-name reputations, where consumers and/or shareholder might provide their own form of punishment.

In addition to remedies and 'punishments,' the Coast Guard has several different mechanisms at its disposal to prevent proactively and reduce the harm caused by oil spills. First, it may inspect a tanker or transfer operation to ensure that all required equipment

---

[1] 33 USC §1321, Oil and hazardous substances liability. The maximum penalty is $10,000 per violation or per day, although some fines may be as high as $25,000.
[2] The owners eventually paid about $3 million as the spill occurred off the coast of Brittany and was subject to an international treaty – not US law – which had a higher liability limit.
[3] Water Quality Improvement Act of 1970 and Oil Pollution Control Act of 1990.

has been installed and properly maintained (e.g. backup or containment equipment) and that personnel have been properly trained. These inspections can take place at any time – either when oil is being transferred or when a tanker is simply waiting at port. Secondly, the Coast Guard may actually be on site to monitor an oil transfer operation. Third, the Coast Guard may devote resources to detecting an oil spill. For example, they might patrol a port area by boat or helicopter looking for an oil sheen. Fourth, if they detect a spill, they might conduct an investigation to determine who is culpable for the spill – and what degree of negligence was involved. Finally, as we discussed above, the Coast Guard may require the polluter to clean up or clean up the spill themselves and may impose a penalty on the company or individuals responsible for the spill.

3 ECONOMIC THEORY OF OIL SPILL PREVENTION AND ENFORCEMENT

Although the purpose of this paper is to examine the empirical evidence on the effectiveness of oil spill prevention and enforcement, it is useful to review the underlying theories that explain why government actions are expected to prevent oil spills.[4] Economists who study firm compliance and deterrence invariably start with the 'optimal penalty' model of Becker (1968) that is based on deterring criminal activity.[5] The basic insight of that seminal article is that potential offenders respond to both the probability of detection and the severity of punishment if detected and punished. Two key assumptions underlying the ability of the government to affect behaviour are: (1) the offender has knowledge about the law and about the likelihood that a violation of the law will be known to the government, and (2) the offender acts rationally and in his own self interest.

Given these two assumptions, government enforcement may become more effective in preventing harm by raising the penalty, by increasing monitoring activities to raise the likelihood that the offender will be caught, or by changing legal rules to increase the likelihood that the offender (if caught) will be punished. Becker's model ultimately leads to an 'efficient' level of crime, whereby the marginal cost of enforcement is equated to the marginal social benefit of crime reduction. Thus, given individual preferences and enforcement technologies, both the crime rate and the level of monitoring and enforcement activities are determined by this model.

In the context of marine oil pollution, there are generally two ways in which spills can occur. First, a firm might intentionally violate the law by not complying with a regulatory standard. For example, a tanker might illegally clean out its bilges or might not maintain

---

[4] A more comprehensive literature review and detailed treatment of the theory of optimal penalties and deterrence is contained in M.A. Cohen, 'Monitoring and Enforcement of Environmental Policy', in T. Tietenberg and H. Folmer (eds), *International Yearbook of Environmental and Resource Economics 1999/2000*, (Edward Elgar Publishers, Cheltenham, 1999), pp. 44-106.

[5] G.S. Becker, 'Crime and Punishment: An Economic Approach', (1968) 76 *Journal of Political Economy* 169-217. Becker's model is explored in more detail in the context of environmental enforcement in M.A. Cohen, 'Environmental Crime and Punishment: Legal/Economic Theory and Empirical Evidence on Enforcement of Federal Environmental Statutes', (1992) 82 *Journal of Criminal Law & Criminology* 1054-1108, and M.A. Cohen, *supra* note 4.

backup or containment systems that catch overflow from oil transfer operations. Second, spills might occur primarily as a result of stochastic events, like a weather-related accident or terrorist attack. Unlike traditional criminal activity where it is assumed that there is criminal 'intent,' oil spills may occur by accident. However, while 'accidental' spills are somewhat outside the control of the oil transporter, only in the rarest of events can one say with complete certainty that the tanker operator bears no responsibility at all. Oftentimes, some degree of negligence is associated with the incident. For example, suppose a tanker operator fails to train its personnel adequately in oil spill containment procedures. In the event of a purely unpredictable event that causes the tanker to leak oil, the *size* of the spill and the resulting damage will likely be larger than it would otherwise be if personnel had been properly trained. Thus, in the case of oil spills, it will often be the case that the degree of negligence will be important in understanding the appropriate government enforcement action.

To complicate the picture further, there are actions that the oil tanker operator can take both to *prevent* a spill from occurring as well as to *reduce the impact* of a spill once it has occurred. While it certainly would seem that prevention would be preferable, once the cost of prevention is taken into account, it is not necessarily the case that from society's standpoint, 'prevention' is always preferable. Thus, there might a trade-off between actions taken to prevent oil spills from occurring and those designed to reduce the impact of a spill that has occurred. To take a simple example, suppose 'small oil spills' cause USD 20,000 in damage, while 'large' spills cause USD 1 million in damage. Suppose further that the firm can only afford USD 100,000 per year on oil pollution expenditures. That money could be used to prevent small spills from occurring – reducing on average five small spills, or it could be used to purchase containment equipment that stops one small spill from becoming a large spill. In the case of preventing small spills, the cost (USD 100,000) is just worth the benefit (5 spills x USD 20,000 each). However, shifting those resources from prevention to containment would have huge expected benefits – USD 1 million.

To formalize these ideas, consider a simple model of firm behaviour. The goal of the firm is to maximize profits, or to minimize the sum of all costs associated with spilling oil (including the cost of prevention). As indicated above, oil spills occur with some randomness; hence, an oil spill is modelled as a random variable with distribution function $F(x,e)$. To simplify, we ignore the firm level of output, which obviously affects the expected level of pollution. It can be thought of as a control variable for most of the analysis.

Vessel operators are required by the government to maintain a certain minimum level of effort, which might include installing and properly maintaining certain pieces of equipment, properly training employees, and so on. With probability $P_I(m_1)$, the vessel will be inspected for compliance, where m_1 is the level of government resources devoted to compliance monitoring. If inspected and found to be in non-compliance (i.e., $e < e^*$), the government will impose a penalty $T_I(e)$. This is an *ex ante* penalty, as it is based on the level of effort devoted by the firm to prevent pollution.

If an oil spill occurs, the government may devote resources to detect and punish the vessel that caused it. Let m_2 be the level of government resources devoted to detecting an oil spill. Then, the probability of detection will be $P_D(x,m_2)$, where the detection probability is increasing in both the level of government detection resources and the size of the spill (larger oil spills are presumably easier to detect). If the pollution is detected and the vessel operator identified, the government imposes a penalty, $T_D(x,e)$. Note, however, that the government does not directly observe the level of effort by the firm. Thus, if the government wants to condition its penalty on the level of effort (a negligence standard), it must expend additional *ex post* monitoring resources, m_3, to determine the culpability of the vessel. Finally, oil spills involve a private loss to the vessel operator who loses the value of the oil, $v(x)$, which would otherwise have been sold.

The vessel operator's expected profit from the oil spill can thus be written as:

$$EU(e) = -P_I T_I(e) - \int_x [v(x) + P_D(x) T_D(x,e)] f(x,e) \, dx - e \qquad [1]$$

The government is assumed to be a social welfare maximizer. It has numerous choices to make, including the level of monitoring expenditures. As such, it wants to minimize the sum of:

$C(rx)$ = cleanup or recovery costs, where r is the fraction of oil cleaned up;
+ $D[(1-r)x]$ = environmental damages;
+ $v(x)$ = private resource loss;
+ e = prevention expenditures.

Combined with government enforcement expenditures, m_1, m_2, and m_3, the welfare maximization problem facing the government is:

$$EW(e, m_1, m_2, m_3, r) = \int_x \{D[(1-r)x] + C(rx) + v(x)\} f(x,e) \, dx - e - m_1 - m_2 - m_3 \qquad [2]$$

Implicit in this formulation is the indifference of the government to the level of the fine paid by the firm, since the fine is a transfer payment. The government has control over the level of monitoring, m_1, m_2, m_3, and can either mandate some level of recovery or cleanup or clean up the damage directly itself. However, the level of effort cannot be observed directly and can only be inferred or imperfectly observed *ex post*. Thus, the government imposes a penalty to induce the firm to take the optimal level of effort. That penalty is:

$$T_D(x) = \frac{D[(1-r)x] + C(rx)}{P_D(x)} \qquad [3]$$

Substituting [3] into [1], the vessel operator's problem becomes one of social welfare maximization, [2] where m_1, m_2, and m_3 are set equal to zero with no need for government monitoring. This penalty is just Becker's optimal penalty equal to the harm divided by the probability of detection, which induces the socially optimal level of effort.

The optimal penalty [3] varies with the probability of detection, a key parameter in the enforcement agency's tool kit. Since increasing the probability of detection requires some expenditure on government monitoring, Becker's policy prescription is to set $P_D(x)$ arbitrarily low, thus raising the penalty. However, there may be limits on how high a penalty is feasible – for political reasons, wealth constraints of polluters, and for purposes of preserving marginal deterrence.[6]

In addition to determining the probability of detection, the optimal penalty [3] requires that the government decide whether to require cleanup of any harm caused by a spill. As Cohen (1987) shows, the optimal recovery/cleanup rule equates marginal damages to marginal cleanup costs, D' [(1-r) x] = C' (rx).[7] The cleanup rule is independent of either the level of care taken by the firm or mandated by the government, and is independent of the optimal penalty. Moreover, if the vessel operator is made responsible for cleanup and held strictly liable for any residual damage, it will determine the optimal level of cleanup by equating marginal cleanup costs to marginal damages.[8]

Note that the optimal penalty [3] does not depend on the level of effort undertaken by the vessel operator. Thus, it is a strict liability standard, whereby the polluter is held liable without regard to his state of mind or to the fact that the spill might have been beyond his control. If penalties are not constrained, such a penalty is best because it economizes on government resources (m_3) that might otherwise be devoted to an *ex post* investigation and potential adjudication or litigation costs associated with determining what level of care the firm actually took (Cohen, 1987). However, one could also specify a technology-based standard and impose a penalty T_D (e), when e < e*. In the parlance of the law and economics literature, instead of a strict liability standard, this would be a negligence-based penalty. A negligence-based penalty is one that is imposed only when it is shown that the polluter did not take an appropriate level of care in preventing the emissions.

If one is concerned about the cost of enforcement, a strict liability standard will generally be less expensive to enforce, as a negligence standard requires additional resources to determine the cause of the spill (and perhaps to litigate over the cause). In addition, a negligence standard results in a lower expected penalty to potential vessel operators – since they are not help liable for pure accidents. Hence, a negligence standard has an advantage over strict liability when regulating oil spills in the presence of risk aversion or a wealth constraint (Cohen, 1987). An alternative approach to dealing with insolvency is to impose non-monetary sanctions on the offender – like imprisonment or restrictions on future business activities.[9]

Thus far, I have limited my analysis to *ex post* penalties and mandates to clean up after a spill has occurred. However, recall that the government might monitor *ex ante*, m_1, to determine the level of effort undertaken by the vessel operator and impose a penalty only if found to be in non-compliance with some optimal standard of care. In theory, either approach can achieve the same level of deterrence. However, we can distinguish instances

---

[6] See A.M. Polinsky and S. Shavell, 'The Optimal Tradeoff Between the Probability and Magnitude of Fines', (1979) 69 *American Economic Review* 880.

[7] See M.A. Cohen, 'Optimal Enforcement Strategy to Prevent Oil Spills: An Application of a Principal-Agent Model with "Moral Hazard"', (1987) 30 *Journal of Law and Economics* 23.

[8] See A.M. Polinsky and S. Shavell, 'A Note on Optimal Cleanup and Liability After Environmentally Harmful Discharges', (1994) 16 *Research in Law and Economics* 17.

[9] For details on when imprisonment might be appropriate instead of a monetary penalty, see S. Shavell, 'Criminal Law and the Optimal Use of Non-Monetary Sanctions as a Deterrent', (1985) 85 *Columbia Law Review* 12. See also M.A. Cohen, *supra* note 5.

when *ex ante* monitoring is preferable to an *ex post* penalty and vice versa.[10] If the solvency of the vessel operator is a problem, *ex ante* monitoring has an advantage, since the optimal penalty for taking an action that increases the probability of harm will be smaller than if that harm has actually occurred. In other words, there is a lower probability of insolvency – thereby preserving the incentive effects of penalties. If the polluter is risk averse, *ex ante* monitoring also has an advantage, since the vessel operator can predict with virtual certainty whether or not he will pass the government's standard as opposed to the uncertainty of being penalized only if some random event occurs.

Under the most simplistic optimal penalty model, decreased monitoring coupled with higher penalties is always beneficial. Indeed, absent externally imposed constraints on penalties, the optimal penalty is arbitrarily high and the optimal expenditure on monitoring approaches zero. As noted above, however, risk aversion and insolvency – not to mention fairness – precludes the use of such draconian policies. Thus, especially in the case of oil spills that might result in very large damages, we are left with a government enforcement policy that requires a significant amount of monitoring expenditures. Three innovations that have been suggested to reduce the need for expensive government monitoring of oil spill operations (and hence to increase deterrence for a fixed budget) are: (1) self-reporting induced by the government – with additional penalties levied against vessel operators that do not voluntarily report their spills, (2) targeted monitoring focusing on vessels thought to have the highest probability of being out of compliance or prone to spilling oil, and (3) differential penalties based on prior compliance history.[11]

Self-reporting is a substitute for government monitoring efforts that may reduce enforcement costs without compromising deterrence. Vessel operators are told they must report any spill. If the government detects a spill that has not been voluntarily reported, the penalty will be higher.

4 EMPIRICAL EVIDENCE ON OIL SPILL PREVENTION AND ENFORCEMENT

While the theoretical model above was based on the actions of a single oil transporter who calculates the probability of detection and penalties and decides what level of compliance or preventive actions to take, not all of the Coast Guard activities described above target a single transporter. For example, the monitoring of port areas looking for oil sheens focuses on all oil transporters in the area. Moreover, if oil transporters are uncertain about the level of Coast Guard activity and they are informed about a recent penalty levied against another ship, they might update their assessment of the likelihood that they too will be detected if they spill oil. Thus, an inspection or penalty targeting one oil transporter might have a deterrent effect on all transporters in the area.

---

[10] See S. Shavell, 'The Optimal Structure of Law Enforcement', (1993) 36 *Journal of Law and Economics* 255.
[11] See for example, W. Harrington, 'Enforcement Leverage When Penalties Are Restricted', (1988) 37 *Journal of Public Economics* 29; and E. Hentschel and A. Randall, 'An Integrated Strategy to Reduce Monitoring and Enforcement Costs', (2000) 15 *Environmental and Resource Economics* 57.

Accordingly, we can distinguish between the *specific* deterrent effect of enforcement and the *general* deterrent effect. *Specific deterrence* refers to the effect that an inspection or enforcement activity targeting a particular firm has on that firm's subsequent performance. *General deterrence* refers to the effect of an enforcement activity on the behaviour of a large number of persons or firms.

It is not always possible to determine whether an estimated effect is due to specific or general deterrence. For example, even though the Coast Guard might target specific firms for an inspection, the researcher may only have aggregate measures of compliance or firm-specific compliance measures without data on which transporters were inspected. In that case, since the researcher observes aggregate spill or compliance data, we do not know if the enforcement policy affects all potential polluters or only those who are targeted. On the other hand, if the policy is not firm specific – like a Coast Guard patrol of a port area looking for spills – any effect we observe must be general deterrence.

The first study to analyze the effectiveness of Coast Guard oil spill policy in the US is Epple and Visscher (1984).[12] As they realized from the outset, while the government might collect data on the number of *detected* spills, they did not have data on the actual number of spills. Moreover, Epple and Visscher noted that while they are interested in estimating the deterrent effect of increased enforcement efforts on oil spill volume, there is another effect that will be difficult to separate out, namely, the fact that increased enforcement will inevitably lead to increased detection. In other words, an increase in government enforcement might have two effects. It might increase the probability of detection (hence leading to more detected spills), and it might decrease the actual number and/or volume of spills (hence leading to fewer actual spills and/or spill volume). Trying to sort out these two effects is a challenge.

Based on the economic theory described in Section 3, Epple and Visscher estimated the volume of oil spilled in US waters as a function of Coast Guard activities, which vary by port and over time. Thus, they had quarterly data on the total number of hours the Coast Guard spent on oil spill prevention activities, and on the total spill volume in each port. Two different measures of effectiveness were examined; frequency of spills and total volume of spills. Both of these measures are calculated on a 'per transfer' basis so that comparability can be maintained across ports. The explanatory variable – Coast Guard enforcement activities – was also measured on a 'per transfer' basis.

The first finding of Epple and Visscher was that an increase in Coast Guard activities led to an *increase* in the observed frequency of spills! Thus, the Coast Guard was now detecting a significant number of oil spills of which they previously were not aware. In fact, a 10% increase in Coast Guard activity led to a 2.1% increase in spill frequency. Put differently, the 'elasticity of spill frequency' with respect to Coast Guard enforcement activity is +0.21. Thus, even though the presence of the Coast Guard might deter and reduce the number of spills, the 'detection' effect of their activities apparently outweighed any 'deterrent' effect. That does not mean that action by the Coast Guard is unjustified – only that we cannot estimate the true impact of their activities on spill frequency.

---

[12] D. Epple and M. Visscher, 'Environmental Pollution: Modeling Occurrence, Detection and Deterrence', (1984) 27 *Journal of Law and Economics* 29.

The second finding of Epple and Visscher was that an increase in Coast Guard activities led to a *decrease* in observed spill volume. In particular, an increase in Coast Guard activities of 10% resulted in a 3.1% decrease in spill volume. Since increased Coast Guard activity causes an increase in the detection rate, this is likely to be an underestimate of the true effectiveness of their activities. Since Epple and Visscher only had aggregate spill volume by port, their study only tests for *general* deterrence.

In Cohen (1987), I extended the Epple-Visscher analysis by comparing the effectiveness of three different types of Coast Guard activities that were specified in the theoretical model in Section 3: (1) actual monitoring of oil transfer operations, (2) random port patrols designed to detect spills, and (3) inspections of vessels to determine whether or not they were in compliance with oil spill prevention regulations. To examine the effectiveness of Coast Guard monitoring in reducing oils spills, I estimated the following regression equation:

OIL SPILLED = F (Price of Oil, Vessel Size, Monitoring Transfer Operations, Patrolling Ports, Compliance Inspections)

The first method – monitoring of oil transfer operations – was found to be an effective deterrent by reducing oil spill volume, with an elasticity of -0.17, indicating that a 10% increase in monitoring of oil transfer operations leads to a 1.7% decrease in spill volume. Since it was not known exactly which transfer operations were monitored and which vessels spilled oil, I do not know if this was a *specific* or *general* deterrent effect. Although I expect there to be a *specific* deterrent effect as the crew might take more care when Coast Guard personnel are present, there might also be a *general* deterrent effect to all vessels that transfer oil. If vessel captains believe there is a higher probability of being monitored in the future, they might take more care in training personnel or keeping equipment properly maintained – regardless of whether or not they are ultimately monitored.

The second method – random port patrols – was also found to be effective, with an elasticity of -0.20. This would likely be a *general* deterrent effect, since it raises the probability of detection for all vessels entering a particular port. Finally, the third method – compliance inspections – was found to be ineffective at the margin in reducing spill volume. Anderson and Talley (1995) found similar results using more recent data.[13] It is important to note that this does *not* imply that compliance inspections are ineffective at all. Instead, the finding is simply that at current levels, a small decrease in compliance inspections would not lead to an increase in oil spill volume. More detailed analysis would be required to determine the 'optimal' level of compliance inspections.

The results in Cohen (1987) illustrate the importance of carefully disaggregating the type of enforcement activity to fully understand the deterrent effect of enforcement. Although Epple and Visscher found an overall deterrent effect, the disaggregation of the same data in Cohen (1987) found that more 'bang for the buck' could be obtained by shifting resources around within the enforcement budget.

---

[13] E.E. Anderson and W.K. Talley, 'The Oil Spill Size of Tanker and Barge Accidents: Determinants and Policy Implications', (1995) 71 *Land Economics* 216.

Of course, finding that one could get more bang for the buck by shifting resources around does not necessarily mean that additional resources should go into any of these Coast Guard activities. To draw such conclusions, a cost-benefit analysis must be conducted in order to compare the social value of the spill reduction to the social cost of increased Coast Guard activities. Such an analysis was conducted in Cohen (1986).[14] On the 'costs' side of the equation were hours spent by Coast Guard personnel, industry prevention activities, and any additional time in port spent by tankers while they are being inspected. On the 'benefits' side were the value of oil not spilled, the cost of cleanup avoided, and the monetary value of any environmental harm averted. It was estimated that at the margin, the cost of preventing a gallon of oil from being spilled was USD 5.50, while the benefit from one less gallon of oil spillage was USD 7.27. Hence, at the margin, the benefit-cost ratio of increased Coast Guard enforcement activity was 1.32 (7.27/5.50 = 1.32).

In Cohen (1987), I not only model and estimate the Coast Guard's *ex ante* monitoring and inspections, but also examine their *ex post* enforcement activities: required cleanup and monetary penalties. A penalty function is estimated based on the actual fines that are meted out for detected oil spills:

PENALTY = F(Vessel Size, Spill Size, Monitoring, Percent Cleaned Up, Location of Spill, Type of Oil, Cause of Spill, Year, Season)

Estimation of the penalty function serves two purposes. First, it addresses the question whether the Coast Guard uses a negligence or strict liability standard in its penalty calculus. Several 'causes' are included as dummy variables, including improper maintenance, personnel error, equipment failure, intentional discharge, and natural causes. All of these variables are found to have significant power in the penalty regression equation. For example, the monetary penalty increases with personnel error, intentional discharges, and improper maintenance, but decreases with natural causes. Thus, the Coast Guard uses a negligence-based standard in its discretionary penalty assessment.

The second reason for estimating the penalty function is to investigate the optimality of the Coast Guard cleanup and penalty policy. Combined with estimates of the probability of detection, the cost of cleaning up a spill, and the environmental damage caused by a spill, I estimate the optimal cleanup costs and penalty as a function of the size of the spill. What I found was that the Coast Guard policy at the time appeared to require excess cleanup for very small spills (which are costly to clean up on a per gallon basis and cause little environmental harm). In addition, statutory maximum penalties are too low for large spills.

Changes in US Coast Guard data collection policies now provide researchers with actual vessel inspection and transfer information. Thus, we can determine how often each vessel is inspected when they are in each port, how much oil is being transferred, and any detected oil spill volume for each transfer. Analyzing these data, Viladrich-Grau and

---

[14] M.A. Cohen, 'The Costs and Benefits of Oil Spill Prevention and Enforcement', (1986) 13 *Journal of Environmental Economics and Management* 167.

Groves (1997) find that Coast Guard monitoring activities have an even larger effect on spill frequency than on spill size.[15]

Viladrich-Grau and Groves also examine the 'expected penalty' facing each vessel to determine whether or not larger penalties deter spills. Thus, they use the average penalty for vessels in each region as an explanatory variable in the regression equations explaining spill frequency and volume. However, they find no deterrent effect of penalties. This may be attributable to the fact that Coast Guard penalties are only a small fraction of the cost of mandatory cleanup. For example, typical fines for most oil spills are in the few hundred dollar range compared to hundreds of thousands of dollars for the cost of cleanup. In fact, the maximum fine possible for spilling oil during the period under consideration was only USD 5,000.

Perhaps the most interesting part of their study, however, examines the Coast Guard's newly implemented monitoring policy of classifying ships into 'low risk' (infrequently monitored) and 'high risk' (always monitored). Beginning in 1983, inspected vessels that were found to be in compliance with safety regulations were deemed to pose the least threat to the environment are were classified as Low Priority (LP) vessels. Those that did not pass the safety inspection were classified as High Priority (HP) vessels. While LP vessels were only sporadically monitored, all transfer operations by HP vessels were monitored.

Viladrich-Grau and Groves find that this two-tiered enforcement policy is effective in reducing the cost of enforcement without having a negative effect on the environment. Thus, identifying the characteristics of firms or vessels that are more likely to spill oil – and targeting them for higher frequency monitoring and inspections – seems like a promising approach.

Gawande and Wheeler (1999) also had vessel-specific details and were able to examine specific deterrence following Coast Guard monitoring and inspections. In addition to looking at the frequency of oil spills, they also examined injuries and deaths following tanker accidents.[16] Unlike Cohen (1987), Gawande and Wheeler actually found routine inspections to be effective in lowering the number of oil spills aboard US flag tankers during the late 1980s. However, Gawande and Bohara (2004) used panel data from the 1990s and confirmed Cohen's earlier conclusion about the ineffectiveness of inspections designed to check compliance.[17]

The first study to find any significant deterrent effect from monetary penalties on oil spills was Weber and Crew (2000).[18] They model the amount of oil spilled in each region as a function of that region's prior year average penalties (per gallon of oil spilled), the probability that a penalty will be assessed, and the number of days between the spill and an assessment of a penalty. Since Weber and Crew are measuring spill volume at each

---

[15] M. Viladrich-Grau and T. Groves, 'The Oil Spill Process: The Effect of Coast Guard Monitoring on Oil Spills', (1997) 10 *Environmental and Resource Economics* 15.

[16] K. Gawande and T. Wheeler, 'Measures of Effectiveness for Government Organizations', (1999) 45 *Management Science* 42.

[17] A.K. Bohara and K. Gawande, 'Inspections, Violations, and Oil Spills: Theory, Evidence, and Policy', *Bush School of Government and Public Affairs Working paper #407*, 2004.

[18] J.M. Weber and R.E. Crew, 'Deterrence Theory and Marine Oil Spills: Do Coast Guard Civil Penalties Deter Pollution?', (2000) 58 *Journal of Environmental Management* 161.

port, their model is one of general deterrence. They find that speed of punishment has a deterrent effect, with a 10-day improvement in speed reducing oil spillage by 0.6%. They also find a deterrent effect from more severe penalties.

Weber and Crew found that fines ranged dramatically from .003/liter to USD 73.35 per litre, with the lowest fines per litre being reserved for the largest spills. They estimate that increasing the fine for large spills from USD 1/gal to USD 2/gal decreases spillage by 50%, but increasing from USD 100 to USD 101 reduces it by 1%. Thus, marginal effectiveness goes down as the penalty increases. They conclude that the current penalty policy that has a relatively high 'per gallon' fine for small spills and very low per gallon fine for large spills undermines deterrence. This is similar to the finding of Cohen (1986) that the statutory maximum penalty of USD 5,000 is too small relative to the optimal penalty required.

5 CONCLUDING REMARKS

Over the past 20 years, empirical research has demonstrated that increased monitoring and enforcement can reduce the frequency and size of oil spills. However, that research has also demonstrated that not all enforcement activities are created equal. Even within the confines of one government agency – the US Coast Guard – shifting resources from one type of enforcement activity to another is likely to have considerable benefits. Given the level of monitoring and compliance, it appears that at the margin, more effort could be devoted to targeted monitoring of transfer operations and random patrols of ports and coastal waterways looking for spills. In return, less attention could be paid to compliance inspections. Having said that, it is important to keep in mind that one cannot necessarily translate these lessons to other countries or to different time frames without knowing more about the legal institutions, compliance history, and regulatory framework within which that country operates. For example, it may be the case that most vessels operating in the US already comply with basic regulations and that little compliance inspection is needed. That might not be the case elsewhere.

The purpose of this chapter was to examine the effectiveness of *government* enforcement activities. However, there are many other legal and extra-legal forms of sanctions and incentives that operate on oil transport vessels outside the Coast Guard. If one is truly interested in developing rational policies and using government enforcement resources wisely, the interdependencies of these different mechanisms should also be considered. Government enforcement does not occur in a vacuum; hence, understanding the motivations and incentives of both polluters and enforcement agencies should be an important component of any study of enforcement.

While few empirical studies of the effectiveness of these alternative institutions exist, it is worthwhile briefly to mention some of these other ways in which oil spills may be deterred. While government regulators (like the Coast Guard or the US Environmental Protection Agency) generally enforce the oil spill laws, there may be a place for private citizens to also bring about enforcement actions. In the US, such 'citizen law suits' are fairly common in both air and water pollution – where private citizens may go to court to ensure that environmental laws are enforced. While I am unaware of any studies that examine the effectiveness of citizen lawsuits, one study found that we are likely to see more private enforcement when environmental groups perceive more of a need for enforcement – i.e. when they believe the government is not doing enough.[19] That study also noted that citizen suits are useful in filling a gap in enforcement against public polluters. Public polluters (e.g., municipal water or sewage facilities) are seldom the target of enforcement actions by public agencies. Private enforcers are apparently less reluctant to file lawsuits against public polluters. To the extent that oil transport vessels or storage facilities in China are government-owned, this approach might be worth investigating.

Another way in which the effectiveness of government enforcement might be enhanced is by the use of the criminal law. While generally reserved for the most serious regulatory violations, in the US, criminal law has been used occasionally in the case of oil spills. While the monetary penalty might not be any larger than one could impose elsewhere in the courts, the criminal law might enhance the negative publicity associated with the action, thereby having a more dramatic effect for a given penalty. It might also be used against individuals who caused a spill – with the threat of incarceration.[20]

Finally, the role of non-regulatory enforcement tools like the impact of information disclosure on firm behaviour is an important emerging topic in the economics of enforcement. Information that a firm has been sanctioned for violating environmental laws may be of interest to employees, shareholders, or lenders for a variety of reasons. Several studies have focused on bad environmental news in the US and Canada, like oil or chemical spills or the announcement of civil or criminal enforcement actions.[21]

Even without the pressure of the stock market, social norms, community pressure and the reputation of the vessel operator might play an important role. Thus, strengthening the vessel operator's ties to the community – as well as ensuring that publicity takes place when an enforcement action is taken – would seem worthwhile as part of an overall social policy of reducing oil spills. This is not only based on US studies and indeed might be particularly useful in countries where government enforcement resources are limited.[22]

---

[19] W. Naysnerski and T. Tietenberg, 'Private Enforcement of Federal Environmental Law', (1992) 68 *Land Economics* 28.

[20] For a thorough treatment of 'environmental crimes' in the US, including legal definitions and citations, see M. Clifford, *Environmental Crime: Enforcement, Policy and Social Responsibility*, (Aspen Publishers, Gaithersburg, 1998) and M.A. Cohen, *supra* note 5.

[21] See for example, R.D. Klassen and C.P. McLaughlin, 'The Impact of Environmental Management on Firm Performance', (1996) 42 *Management Science* 1199 and S. Konar and M. Cohen, 'Information as Regulation: The Effect of Community Right to Know Laws on Toxic Emission', (1997) 32 *Journal of Environmental Economics and Management* 109-124.

[22] For example, a program in Indonesia rated firms by their level of compliance with existing regulations and gave the firms six months advance notice of the rating that would be made public unless they changed their compliance behavior. Afsah, Laplante and Wheeler (1997) report considerable improvements in compliance status both before the initial public announcement (which allowed firms to change their status before the announcement) and following the public announcement. See S. Afsah, B. Laplante and D. Wheeler, 'Regulation in the Information Age: Indonesian Public Information Program for Environmental Management', *World Bank Policy Research Working Paper*, March 1977.

LIST OF REFERENCES

S. Afsah, B. Laplante and D. Wheeler, 'Regulation in the Information Age: Indonesian Public Information Program for Environmental Management', *World Bank Policy Research Working Paper*, March 1977.

E.E. Anderson and W.K. Talley, 'The Oil Spill Size of Tanker and Barge Accidents: Determinants and Policy Implications', (1995) 71 *Land Economics* 216-228.

G.S. Becker, 'Crime and Punishment: An Economic Approach', (1968) 76 *Journal of Political Economy* 169-217.

A.K. Bohara and K. Gawande, 'Inspections, Violations, and Oil Spills: Theory, Evidence, and Policy', *Bush School of Government and Public Affairs Working paper #407*, 2004.

M. Clifford, *Environmental Crime: Enforcement, Policy and Social Responsibility*, (Aspen Publishers, Gaithersburg, 1998).

M.A. Cohen, 'The Costs and Benefits of Oil Spill Prevention and Enforcement', (1986) 13 *Journal of Environmental Economics and Management* 167-188.

M.A. Cohen, 'Optimal Enforcement Strategy to Prevent Oil Spills: An Application of a Principal-Agent Model with "Moral Hazard"', (1987) 30 *Journal of Law and Economics* 23.

M.A. Cohen, 'Environmental Crime and Punishment: Legal/Economic Theory and Empirical Evidence on Enforcement of Federal Environmental Statutes', (1992) 82 *Journal of Criminal Law & Criminology* 1054-1108.

M.A. Cohen, 'Monitoring and Enforcement of Environmental Policy', in T. Tietenberg and H. Folmer (eds), *International Yearbook of Environmental and Resource Economics 1999/2000*, (Edward Elgar Publishers, Cheltenham, 1999), pp. 44-106.

D. Epple and M. Visscher, 'Environmental Pollution: Modeling Occurrence, Detection and Deterrence', (1984) 27 *Journal of Law and Economics* 29.

K. Gawande and T. Wheeler, 'Measures of Effectiveness for Government Organizations', (1999) 45 *Management Science* 42-58.

W. Harrington, 'Enforcement Leverage When Penalties Are Restricted', (1988) 37 *Journal of Public Economics* 29-53.

E. Hentschel and A. Randall, 'An Integrated Strategy to Reduce Monitoring and Enforcement Costs', (2000) 15 *Environmental and Resource Economics* 57-74.

R.D. Klassen and C.P. McLaughlin, 'The Impact of Environmental Management on Firm Performance', (1996) 42 *Management Science* 1996.

S. Konar and M. Cohen, 'Information as Regulation: The Effect of Community Right to Know Laws on Toxic Emission', (1997) 32 *Journal of Environmental Economics and Management* 109-124.

W. Naysnerski and T. Tietenberg, 'Private Enforcement of Federal Environmental Law', (1992) 68 *Land Economics* 28-48.

A.M. Polinsky and S. Shavell, 'The Optimal Tradeoff Between the Probability and Magnitude of Fines', (1979) 69 *American Economic Review* 880.

A.M. Polinsky and S. Shavell, 'A Note on Optimal Cleanup and Liability After Environmentally Harmful Discharges', (1994) 16 *Research in Law and Economics* 17.

S. Shavell, 'Criminal Law and the Optimal Use of Non-Monetary Sanctions as a Deterrent', (1985) 85 *Columbia Law Review* 12.

S. Shavell, 'The Optimal Structure of Law Enforcement', (1993) 36 *Journal of Law and Economics* 255.

M. Viladrich-Grau and T. Groves, 'The Oil Spill Process: The Effect of Coast Guard Monitoring on Oil Spills', (1997) 10 *Environmental and Resource Economics* 315-339.

J.M. Weber and R.E. Crew, 'Deterrence Theory and Marine Oil Spills: Do Coast Guard Civil Penalties Deter Pollution?', (2000) 58 *Journal of Environmental Management* 161-168.

MARINE (OIL) POLLUTION: PREVENTION AND PROTECTION BY CRIMINAL LAW – INTERNATIONAL PERSPECTIVES, CORPORATE AND/OR INDIVIDUAL CRIMINAL LIABILITY

Günter Heine

1 INTRODUCTION

Ensuring the safety of maritime transport and protecting the seas from ship-source pollution is undisputedly one of the hot topics on the agendas of international and national law. On the one hand, it must be stressed that in the last twenty years quite a lot of safety enhancements have been effected through a multitude of international conventions. Good examples of this are the International Convention on Civil Liability for Oil Pollution (CLC), 1969, the International Convention setting up the Oil Pollution Compensation Fund, the Marpol Convention 1973/78 and the Protocol of 1978 related thereto and the UN-Convention on the Law of the Sea of 1982, which entered into force in 1994 (UNCLOS).[1] On the other hand, the general public has been alarmed both by high-profile accidents and by deliberate discharges that have caused massive pollution. In Europe, the most notable of the recent accidents include the sinking of the *Prestige* 2002 and of the *Erica* in 1999 as well as a flurry of further pollution incidents, like the capsizing of the freighter Rockness near Bergen, Norway (2004).[2] However, high-profile accidents are not the only problem. A larger proportion of world-wide pollution by oil and other detrimental substances is the result of deliberate discharges. The unacceptable practice of so-called operational, meaning intentional, discharges from ships, including tank-cleaning operations and waste oil as well as even the release of the whole cargo is still widely practiced for economic reasons, nearly as a rule. A word as to the discharge of the whole cargo: I remember well a case in the Nineties where tons of plastic cups, which had been collected in Germany by citizens as part of their legal duty to recycle, were washed ashore at the South Chinese sea-beaches. Another example speaks to the dimension of deliberate

---

[1] For an overview, see T. Treves, 'The Law of the Sea Tribunal', (1995) 55 *Zeitschrift für ausländisches öffentliches Recht und Völkerrecht* 421; H. Hohmann, 'Weltweiter Schutz der Meeresumwelt', in K.P. Dolde (ed.), *Umweltrecht im Wandel* (Schmidt, Berlin, 2001), pp. 97-127; U. Beyerlin, 'New Delopments in the Protection of the Marine Environment', (1995) 55 *Zeitschrift für ausländisches öffentliches Recht und Völkerrecht* 544-579; COM(2002) 802 final, Commission Communication on a second set of Community Measures, 6 December 2000, 7-10. Concerning the international Conventions see in detail M. Faure and H. Wang, in this volume.

[2] Draft Report of the Temporary Committee on improving safety at sea, 23 February 2004, 2003/2235 (INI), 6 et seq. Cp. also *idem*, 18 March 2004, (PE 339.038/1-208), 5 et seq. COM(2002) 681 final, Commission of the European communities, Communication of 3 December 2002, 4 et seq. See also German Bundesamt für Seeschifffahrt, Marine Pollution (Marpol), Results of Criminal Prosecution and 'Ordnungswidrigkeiten' (kind of administrative penal law), 2001, 1 et seq.

discharges: a study for the Mediterranean Sea reports about 1640 illicit discharges (1999),[3] and in 2001 aerial surveillance detected 390 oil slicks in the Baltic Sea and 596 in the North-Sea.[4] One of the legal remedies against marine pollution is provided by the criminal law. It is this role of the criminal law that I will focus on in this contribution.

First of all, however, what are the reasons for focusing on criminal law (or specific kinds of administrative penal law)? Secondly, is there any need for corporate criminal liability, given that many European legal systems still stick to the 'principle of *societas delinquere non potest*', meaning there is no body to be hit and no soul to be influenced. Does this not mean corporate liability is not a task for criminal law, which is expressly based on sanctioning personal, rather than corporate, guilt?

In a first step, we will analyze problems and shortcomings in international marine pollution laws (section 2). In section 3, criminal law will be come under consideration, and in so doing the possibilities and limitations of the international framework will be discussed before turning to the shortcomings of *individual* responsibility. Next, we will deal with *corporate* criminal liability and corporate sanctions. The concluding remarks will summarize the most important results of our analyses (section 4).

2 PROBLEMS AND SHORTCOMINGS IN INTERNATIONAL LAWS OF MARINE POLLUTION

2.1 GENERAL QUESTIONS

2.1.1 Jurisdiction

One crucial point in preventing and combating marine pollution has to do with a lack of clarity concerning jurisdiction. The internal waters and the territorial seas of a state are surely beyond debate. However, to whose jurisdiction belong the scope of the continental shelf (200-350 nautical miles), the exclusive economic zone (EEZ) (max. 200 nautical miles), straits used for international navigation, and last but not least, the high seas? The Marpol-Convention of 1973 at first explicitly stated that discharges in the high seas are criminal acts and it obligated the parties of the flag State to prosecute masters (Art. 4 Marpol). However, even this competence (and obligation) of the flag States is seldom implemented.[5] UNCLOS goes one step further goes: it is (firstly) an emphatic step stressing the unacceptability of discharges in sea areas beyond the jurisdiction of any state. It also constitutes a major step (in respect of principles of international sovereignty) insofar as it gives the competence of jurisdiction not only to the flag States (as did

---

[3] COM(2003) 92 final, Proposal for a Directive of the European Parliament on pollution (relative à la pollution causée par les navires et à l'introduction de sanctions, notamment pénales, en cas d'infractions de pollution), 5 March 2003, at 2.

[4] Proposal for a Directive (at note 3), 2.

[5] M. Faure and G. Heine, 'The insurance of fines', (1991) 58 *The Geneva Papers on Risk and Insurance* 53-54 with a detailed catalogue. Cp. also D. Ochlcr, 'Internationales Umweltstrafrecht', in O. Kimminich (ed.), *Handwörterbuch des Umweltrechts* (2nd edn, C.H. Beck, Munich, 1994), p. 1136. See Proposal for an EU-Directive, *supra* note 2, 8.

Marpol) but also to the port States, even when the discharge has taken place outside the national boundaries of the states concerned (Art. 218 UNCLOS). However, only a few states have currently implemented this option[6] (it is not obligatory). Therefore, the European Union has obligated its Member States to adopt and implement this jurisdiction in all the above-mentioned geographical scopes, including the high seas (Art. 3 Proposal for a Directive). We will come back to some of the intrinsic problems of this issue, caused by Art. 230 UNCLOS, when discussing the sanctions (see *infra* 3.1).

2.1.2 Technical Safety, Control Management, Compensation

A great many improvements in technical maritime safety have been achieved through international conventions and incentives. For example, Marpol provides for port reception facilities, and this is followed up by EU-Directive 2000/59/EC, which requires ports to provide sufficient port facilities for ship-generated waste and requires vessels subsequently to make use of these facilities. Moreover, the rules of the International Maritime Organization (IMO) can be characterized by their requirements for double hulls, improved certification (CAS), and improved personnel training.[7]

However, even the EU, which is especially eager to optimize technical maritime security within its territories, had to concede terminations and exceptions in Regulation (EC) 1726/2003 of 22 July 2003 concerning the accelerated phasing in of double-hull oil tankers. Specifically, the regulation provides age limits for single hull tankers,[8] and final dates from 2003 up to 2010 resp. 2015 (Art. 4 paragraph 2) are foreseen. Furthermore, even in the EU the transposing of Directives is not adequately guaranteed.[9] Further instruments and rules are, at least, in discussion. In 2001, the IMO has promulgated a monitoring and information system, and especially a voyage date recording system compulsory for certain vessels.[10] The EC Directive 2002/59 has adopted these ideas and tries to establish a Community-wide vessel traffic monitoring and information system, including a 'black box' which records all relevant data,[11] like the speed of the vessel concerned, communication between ship, shore and others. There is no question that these records could play a prominent role in attributing liability in case of illegal discharges; we will come back to this point (see IV. below). As far as compensation of marine pollution is concerned, pollution by tankers is presently regulated at the international level by the regime set up by the CLC and the International Convention, through which the Oil Pollution Compensation Fund (Fund Convention) has been established.[12] The two

---

[6] Proposal for an EU-Directive, *supra* note 2, 8.
[7] On the IMO, cp. H. Hohmann, *Meeresumweltschutz als globale und regionale Aufgabe* (1989), p. 53 et seq. Cp. also *supra* note 1, 110-112. With respect to the role of the IMO see Du, in this volume.
[8] In reaction to the fact that the wrecked ships *Prestige* and *Erika* had been 'in work' for 20 years. See Regulation EC 1726 2003 of 22 July 2003, *OJ* L249/1.
[9] For example, eight Member States (Italy, Portugal, Austria, Luxembourg, Greece, Finland, Belgium and the Netherlands) have still not completely transposed two crucial EU Directives on port state control and classification societies. See Temporary Committee on improving safety at sea, Draft report, *supra* note 2, 7.
[10] *IMO News 2001*, Heft 1, 9.
[11] Cp. also Committee on Regional Policy Transport and Tourism, Draft report of 14 July 2003, 2003/0037 (COD), 11; H. Hohmann, *supra* note 1, 111 et seq.
[12] For details see M. Faure and H. Wang in this volume, section 2.2.-2.4.

conventions establish a two-tier liability system, that builds upon a strict liability for the registered shipowner. They also establish a Fund, financed collectively by oil receivers, that provides supplementary compensation to victims of oil pollution who cannot obtain full compensation for the damage from the shipowner.

In sum, we can state, firstly, that international conventions and recommendations have more and more tried to establish a dense network of safety and management rules. However, *incomplete implementation* by some states or *half-hearted enforcement* by agencies is still a problem. Secondly, the focus of the international regime on marine pollution is mainly on the *compensation of victims*. This orientation leads to the question whether liability of (natural and legal) persons responsible for causing accidental or deliberate marine pollution is adequately provided for.

2.1.3 Compensation Shortcomings

This is not the place to discuss how an optimal combination of liability rules, insurance and fund solutions could work in the future.[13] Vice versa – it is often complained about false effects and misleading issues caused by the existing international regime of compensation.[14] The compensation of the victims is of course a crucial issue. This concern can be addressed straightforwardly in cases where a loss of cargo is in question or when the discharge of detrimental substances is visible in the water and the damage can easily be calculated. Difficulties arise, however, in cases of incidents that do not have direct victims, like deliberate discharges for the sole purpose of saving money on costs for legal disposal (tank-cleaning operations, waste oil disposal, disposal of waste cargo), especially on the high seas where detrimental effects are difficult to calculate. Furthermore, the scope of pollution damage is too narrowly defined at the international level.[15]

Another crucial point refers to the actual international regime, especially in cases of accidents, *as such*. At this level, the personal liability of the persons who caused the environmentally harmful accident is either diluted or even excluded! Firstly, the liability of the direct polluter is diluted: by an apparently unbreakable right of the ship-owner (and of course the master of the ship) to limit his liability (except where the owner has been guilty of actual fault, meaning acting or failing to act recklessly and in full knowledge that pollution damage would probably result), by compulsory insurance, and by collective compensation by cargo receivers through the Funds – irrespective of their actual role in

---

[13] On this topic see M. Faure and H. Wang, in this volume, 5.
[14] Cp. Commission of the European Communities, Proposal for a Directive, *supra* note 3, 2.2. See in detail M. Faure and H. Wang, in this volume.
[15] Cp. The Sea Empress 'Court of Appeal', 17 January 2003, EWCA Civ 6, LLP, 2003, vol. I, 327-340. See M. Faure and H. Wang, 'Liability for oil pollution in Europe: recent developments', (2004) *Environmental Liability* 55 et seq. With respect to Belgium, which incorporated a broader definition ('disruption of the environment') see M. Huybrechts and K. Van Damme, in this volume, section 4.2. On the US's creation of liability for damages to public natural resources through the Oil Pollution Act 1990, which aims at making the environment and the public whole again following a pollution event, see J. Boyd, in this volume, section 2.1.

the accident in question. Secondly, compensation claims for pollution damage against the charterer, manager and operator of a ship are prohibited.[16]

Four major points have to be stressed:

1. The focus on the compensation of the victims.
2. The exclusive channelling of liability to the ship-owner.
3. The limitation of this liability, excessively low amounts of limitation and an all too narrow definition of the damage.
4. Compulsory insurance.

It is often said that the channelling of the liability, combined with the limitation of this liability, compulsory insurance and Fund compensation are the backbones of the international regimes.[17] Indeed, an international compensation regime must of necessity be based on optimal social distribution. However, it has been realized that there is a relationship between the exposure to liability of an operator and the incentives to take adequate preventive measures. As long as the financial limits are set at a lower amount than the real magnitude of the damage, tankers owners may lack sufficient incentives to take necessary steps to reduce technical risks. Furthermore, the channelling of liability hinders effective incentives of other actors involved in incidents. In sum, the actual international regime may be too unbalanced and hinders effective incentives based on the principle of personal self-responsibility, meaning being responsible for disastrous effects which are attributable to incorrect conduct and wrong management decisions management.

2.2 THE NEED FOR REFORM

Therefore, this frame of international liability regime should be open for reforms. For example, the EU Commission highlighted the need for amendments, like the unlimited liability of the ship-owner if it is proved that the pollution damage resulted from gross negligence, and the deletion of the prohibition of compensation claims for pollution damage against the charterer, manager and operator of a ship.[18]

---

[16] CLC 1969 and Protocol of 1992, entered into force 30 May 1996. The development of the international conventions shows that they are always adapted after a new incident has demonstrated that the previously agreed limits were insufficient to meet greater oil spills. Cp. in detail M. Faure and H. Wang, in this volume, section 2.

[17] See, for example, K. Le Couviour, 'Responsabilité pour pollutions majeures', *La Semaine Juridique* (2002) 2271. The argument holds thatinsurance coverage is made possible by the limitation. Moreover, the combination of limited compensation covered by compulsory insurance and the Fund lead to a sharing of the costs of oil pollution damage between the vessels owners and the oil industry. On this, see D. Abecassis, 'IMO and liability for oil pollution from ships: a rectrospective', (1983) *Lloyds Maritime and Commercial Law Quaterly* 47; E. Gold, *Handbook on marine pollution* (Gard, Arendal, 1985), p. 115.

[18] Erika II Communication COM(2000) 802 final. For a model of full liability of tankerowners combined with a limitation of the duty to insure and fund backing, see M. Faure and H. Wang, in this volume, section 4.3. Concerning the US's drastic imposition of liability on the private sector, based on liability for natural resource damage and mandatory asset, insurance requirements and financial responsibility rules, see J. Boyd, in this volume, 5. Concerning the broad definition of damage in Belgium and Italy, see M. Huybrechts and K. Van Damme, in this volume, 4.2.1.

Indeed, a well functioning legal regime of compensation, fund solutions and liability rules must be promulgated. Rules that emphasize the liability of the actors concerned and thereby indirectly support politics preventing marine pollution should play an important role. On these grounds, criminal law as well should come into play.

3 COMBATING AND PREVENTING MARINE POLLUTION THROUGH CRIMINAL LAW

3.1 INTERNATIONAL FRAMEWORKS: OPPORTUNITIES AND LIMITATIONS

Before implementing the United Nations Convention on the law of the sea, UNCLOS, (1994), the rules of the States concerning criminal jurisdictions on the seas made up something of a patchwork. Some states limited their jurisdiction to their territorial seas, while others looked to the continental shelf or to their exclusive economic zone. Still other states, like Greece and Germany, extended the national criminal jurisdiction to the high seas and made it a crime to pollute the high seas, insofar this would be in violation of their respective national (!) prohibitions.[19] Indeed, for these last states, this jurisdiction was sometimes implemented without any restriction whatsoever, meaning that it was not limited to the jurisdiction of the states whose flags were flown by the vessels concerned.

The enactment of UNCLOS has established some important landmarks on the international landscape of preventing and combating marine pollution through criminal law.

On the one hand, UNCLOS strengthens jurisdiction and competence, especially on the part of the port States: The State is considered a port State in the sense of UNCLOS (when a vessel is voluntarily within a port or at an off-shore terminal, according to Art. 218 No. 1 UNCLOS). This port State has jurisdiction (meaning it is able to undertake investigations and commence proceedings) over any discharge from a vessel, even outside the internal waters, territorial sea or exclusive economic zone. Therefore, discharges are covered in the high seas as well as in foreign waters, and even irrespective of any damage (whereas the coastal State's jurisdiction is restricted to the territorial sea and the exclusive economic zone, Art. 220 UNCLOS). What is required, however, is that applicable international rules and standards, like Marpol, have been violated.[20] Corresponding to this extensive jurisdiction, the port State has to comply with requests from any state afflicted by discharge (Art. 218 No. 3 UNCLOS) or competent by law (for example by flag States).

---

[19] See Art. 13 and Art. 3 Act. 743/1977 on the Protection of the Seas and regulation of related issues (Greece). Cp. E. Fitrakis and C.P. Kareklas, in G. Heine (ed.), *Umweltstrafrecht in mittel- und südeuropäischen Ländern* (Edition Iuscrim, Freiburg, 1998), p. 139 et seq. On this Act see M. Faure and G. Heine, *Environmental Criminal Law in the European Union* (Edition Iuscrim, Freiburg, 2000), p. 177. To Germany see P. Cramer and G. Heine, in A. Schönke and H. Schröder (eds), *Strafgesetzbuch Kommentar*, (26th edn, C.H. Beck, Munich, 2001), § 324, note 6. German Environmental Criminal Law (§ 324-330 d German Criminal Code) covered the High Seas, limited, however, by the rules of the law on jurisdiction (§ 3-9 German Criminal Code).

[20] Therefore, some countries had to *restrict* their criminal marine laws by adopting a breach of 'international rules and standards' (instead of national rules), cp., for example, to Germany BT-Drs. 13/193, 23 et seq., BT-Drs. 12/7829, 270 et seq. and BT-Drs. 13/696, 21 et seq.

On the other hand, the jurisdiction over discharges is restricted. *Only monetary penalties* can be imposed in respect of violations by ships flying a foreign flag (Art. 230 No. 1, 2 UNCLOS), except in the case of a wilful and serious act of pollution in the territorial sea.

Some states complain about this restriction imposing merely monetary penalties[21] because there are specific doubts about the dissuasive nature of fines (we will come back, *infra* 3.2). Indeed, in current maritime insurance practices the insurance cover provided to ships through the policies of mutual Protection and Indemnity Clubs (which provide cover for some 90% of the world's storage) include monetary penalties, including sanctions of penal nature related to pollution offences.[22] A solution could be to provide that fines shall not be insurable, as is set out in the EU Commission's Proposal (Art. 6 No. 6).[23] However, as long as insurers are based abroad, for example in the Bahamas or Bermudas, or offshore, regional prohibitions will not function very well[24] – and for the time being there is less hope for a global solution.

Nevertheless, this restriction (that only monetary penalties can be imposed) merely concerns the jurisdiction over vessels flying a *foreign* flag. To the contrary, the flag State is not bound by this restriction,[25] resulting out of limited sovereignty-rights of the port or coastal State. Indeed, under UNCLOS (Art. 217 No. 8), it is the duty of the flag States to ensure that penalties provided for by the laws of States for vessels flying their flag 'shall be adequate in severity to discourage violations wherever they occur.' In other words, international public law allows the deprivation of liberty alongside other serious sanctions, like the confiscation of the proceeds, a ban on engaging in commercial activities, or placement under judicial supervision or judicial winding-up of a legal entity – irrespective of their nature being criminal law.[26]

According to this comprehensive power in criminal affairs, UNCLOS (Art. 228 No. 3) provides the right of the flag State 'to take any measures ... to impose penalties, irrespective of prior proceedings of another State.' Furthermore, to support this *mighty role of the flag State*, all other States involved in the pollution case promptly have to notify the flag State concerned of any measures taken against foreign vessels (Art. 231 UNCLOS). On the other hand, however, it still falls to the criminal jurisdiction of the port or coastal State to prosecute such cases if the flag State does not initiate proceedings within six months, or if the proceedings relate to a case of major damage of the coastal State, or if the flag State has repeatedly disregarded its international obligations (Art. 228 UNCLOS).

In sum, it must be conceded that international public law tries to balance a bulk of different interests (sovereignty-rights, effectiveness, enforcement, mutual assistance and

---

[21] See to Germany BT-Drs. 13/193, 23 et seq.; cp. also M. Faure and G. Heine, *supra* note 5, 39 et seq.
[22] See European Commission, Proposal for a Directive, *supra* note 3, 4.6. To the problems see M. Faure and G. Heine, *supra* note 5, 53 et seq.
[23] European Commission, Proposal for a Directive, *supra* note 3. M. Faure and G. Heine, *supra* note 5, 53.
[24] See M. Faure and G. Heine, *supra* note 5, 55 et seq.
[25] To this opinion see BT-Drs. 13/193, 24 (Germany); K. Ambos, in B. von Heintschel-Heinegg, *Münchner Kommentar zum Strafgesetzbuch* (C.H. Beck, Munich, 2003), § 5, note 31.
[26] Cp. BT-Drs. 13/193, 24 (Germany), EU-Commission, Proposal for a Directive, *supra* note 3, 4.6. and art. 6; Council Framework Decision to strengthen the criminal-law framework for the enforcement against ship-source pollution of 2 May 2003 (COM(2003) 227 final), art. 3 comment.

information). However, in seeking to taking effective, proportionate and dissuasive sanctions as a common standard for a convincing strategy of combating and preventing marine pollution, it must be noted that the major weakness of the system lies in the contents of the criminal law of the flag State as well as the (political willingness) to enforce.

This willingness is not always obvious – taking into account the actual situation. The case of the disastrous accident of the tanker Prestige off the coast of Galicia in November 2002 highlights the problems.[27] The master and operators of the tanker were Greek, while some owners were from various other European countries. Part of the owners' firms were located in the principality of Liechtenstein, and the crew was drawn from different Third-world-countries, yet the ship was registered to a Liberian company. Finally, even though no one from the crew came from the Bahamas and the vessel had never been in that country, the Prestige nevertheless flew the Bahamian flag.[28]

The first question was whether there was a sufficient 'genuine link between the state (of Bahamas) and the ship' (Art. 91 No. 1 UNCLOS). The legal answer would affirm this. However, it was a flag of convenience. In that there are complaints about these flags from 'open registry' countries, the traditional point of criticism is the lax state of control (because substandard ships might escape the port State control, and ships flying a flag of convenience may be crucial for maritime safety).[29] In criminal law, the practical answer will be guided by a principle inspired by John Wayne: The first (state) who moves has lost! In other words, the flag State that should have to play a primary role in combating and preventing marine pollution will fail. Moreover, one must again take into consideration that in this situation, where no flag State commences proceedings, the subsidiary jurisdiction of port or coastal States seems to be restricted.[30] The only sanctions available to them (excepting those mentioned above) are monetary sanctions, and they are insurable!

It is obvious that the valid international regime providing criminal powers cannot function well – the port and coastal States are encouraged to warp international law by widening the exceptions (wilful and serious pollution in the territorial sea, Art. 230 No. 2 UNCLOS).

Indeed, in the case of the shipwreck of the tanker Prestige, criminal proceedings were initiated by the coastal State, and were focused only on the master of the ship, in the suspicion that there was wilful pollution and that he was the main actor of the high-profile incident. In the process, he was detained for about twelve weeks and had to pay a bond of USD 3.2 million (paid via the London Club P and I).[31] On the other hand, deeper analysis has revealed that a variety of faulty decisions, on the part of a number of many actors involved, all 'caused' the sinking of the 26 year-old Prestige. These were, among other things, an insufficient level of technical control and supervision, pre-existing damages incurred during a longer period of port lay-days and caused by faulty mooring, insufficient

---

[27] *Supra* note 2.
[28] Cp. Temporary Committee on improving safety at sea, Draft report 18 March 2004, PE 339.038/1-208, at 9.
[29] Cp. V. Power, *EC Shipping Law* (2nd edn., LLP, London, UK, 1998), p. 25 et seq.
[30] See, however, for a preferable interpretation *infra* at 4.
[31] See Temporary Committee, *supra* note 28, at 5 et seq.

investment decisions in improving the ship's maritime safety, problematic orders by the coastal authorities in the actual case of shipwrecking, an overexerted master, and so on.[32]

3.2 CRIMINAL LIABILITY: SHORTCOMINGS OF INDIVIDUAL RESPONSIBILITY

Empirical data shows that the typical offender convicted is the master of the ship, and that rather light financial penalties are imposed.[33] Therefore, it seems that in actuality criminal law runs the risk being ineffective and unfair.

It runs the risk being *ineffective* and of lacking any deterring consequences because the principle of the strict personal nature of the criminal sanction imposed on human beings demands that fines are measured to the capacity of the perpetrator, the master, and to his financial situation. Furthermore, these petty fines are usually reimbursed out of the large corporate coffers or, as we have seen, paid by the P and I Clubs. Increasing the fines in order to cure these inefficiencies is not a realistic approach.[34] Moreover, the risk being *unfair* results from the fact that the actual situation[35] reinforces the impression that criminal law enforcement uses someone as a *scapegoat*, in that a member of the crew or the master of the ship all follow instructions from 'behind' – or have been influenced by a corporate culture.[36] Indeed, most of the time theses persons are the last link in a long chain, and thus may be responsible only for a single faulty aspect of the development of the whole incident. If we do not wish to accept the proverb, 'the last will be bitten by the dogs,' as a leading principle of criminal policies, we have to be aware of a multitude of actors that bear full or partial responsibility, namely, (only to mention the most important ones) the operators, the chartering parties, the owner of the cargo, the ship-owners, the classification societies (which failed to maintain good safety and pollution prevention performance records).[37] Sometimes even the port authorities may be seen to bear some responsibility, in that they, for example, failed to provide the waste reception facilities in ports that are prescribed by law or refused to support a distressed vessel by hindering the ship from calling at the port and thereby promoted pollution on the high seas.[38]

Therefore, new tasks have to be shouldered by criminal law and its accessory systems (administrative penal law, etc). It will be a challenge to keep the liability of these actors

---

[32] See *Neue Zürcher Zeitung am Sonntag*, 30 March 2003, no. 13, at 71, giving informations about the analysis of the American Bureau of Shipping (ABS), the competent classification society. See also *Der Bund*, 22 April 2003, at 2.

[33] See to Germany, German Bundesamt für Seeschifffahrt, *supra* note 2; EC-Commission, Proposal for a Directive, *supra* note 3, 2.1.

[34] See M. Faure and G. Heine, *supra* note 5, 51, 55 on an economic and legal basis.

[35] See specifically the proposals in the European Parliament for a Draft report on improving safety at sea concerning the Prestige incident, 18 March 2004, *supra* note 28, at 5 et seq., 9 et seq. Cp. also Int. Salvage Union, 'Associate Member's Day', *Maritime Risk International*, 18 April 2004, 3.

[36] On this problem, see G. Heine, 'Criminal Liability of Enterprises and New Risks', (1995) 2 *Maastricht Journal of European and Comparative Law* 110 et seq.; G. Heine, 'New Developments in Corporate Criminal Liability in Europe', (1998) *St. Louis-Warsaw Transatlantic Law Journal* 176 et seq.

[37] See, for example, EC-Commission, COM(2002) 681 final (7), 3 December 2002 and *supra* note 32. It seems to be a special need to control the activities of classification societies more than before by law, because 'the classification society sometimes seems to judging in its own case' (EC-Commission, *Ibid.*).

[38] On this problem, see Temporary Committee on improving safety at sea; *supra* note 28, 5 et seq.

from being diluted by gaps in the compensation systems, as was mentioned above,[39] whereby actors have typically been able to turn illegal discharges to their own profit without any direct action referring to the pollution.

However, a criminal law approach based only on *personal* responsibility of human beings, must, of necessity, be limited.

There are five reasons for this.

First, *criminal sanctions* imposed on individuals has proven to be rather *limited* in this context. Fines must be individually tailored with respect to the degree of the crime and the defendant's financial situation. Moreover, they are often reimbursed by corporate funds, or covered by insurances without any real chance of adequate increasing.[40]

Second, individual responsibility is tailored to a *single faulty act*. The whole harmful process that ends in a deliberate or accidental discharge typically results from *incorrect developments* carried out by *many different actors* over a period of time. Thus, this process demands that the law finds an approach different from the one it has for cases an individual's single faulty act. That answer has to be founded on a system of *collective responsibility*.

Third, collective responsibility does not replace individual responsibility. Collective responsibility guarantees, however, that effective, proportionate and dissuasive sanctions can be imposed in cases where the capacities of a human being are restricted, compared to the *more comprehensive liability of corporations*. The latter represent an incomparably greater level of social might, they act continuously over time and space, and they are disposing powerful legal and technical departments.

Fourth, in light of the new tasks on the agenda of criminal law, namely, in the aim of influencing and steering global players to create good corporate citizen culture[41] and to implement organizational structures that would be adequate for risk management,[42] criminal law requires something else than just the *sanctions* traditionally imposed on individuals.

Fifth, economic and globalized power needs equivalent counterparts in the criminal law. As a last resort, one of these should be *corporate criminal liability*. Although this might seem as if we are instigating a long debate about the necessity of a dense cluster of international safety regulations, at least one crucial point should be clear: neither a governmental agency nor an international institution has enough detailed information about modern innovation risks, like marine pollution by vessels. To clarify, this does not

---

[39] See above 2.1.
[40] See above 3.1 and G. Heine, *Die strafrechtliche Verantwortlichkeit von Unternehmen* (Nomos Verlag, Baden-Baden, 1995), pp. 75-78; G. Heine, *supra* note 36, 176; K. Seelmann, 'Unternehmensstrafbarkeit' in J.B. Ackermann, A. Donatsch and J. Rehberg, Wirtschaft und Strafrecht: Festschrift für Niklaus Schmid, (Schulthess, Zurich, 2001), p. 173; G. Dannecker, 'Zur Notwendigkeit der Einführung kriminalrechtlicher Sanktionen gegen Verbände', (2001) *Goltdammer's Archiv für Strafrecht* 102 et seq.; M. Löschnig-Gspandl, *Die strafrechtliche Verantwortlichkeit von Unternehmen und Verbänden mit Rechtspersönlichkeit in Österreich* (Karl Franzens-Universität Graz, Institut für Strafrecht, 2004), p. 76 et seq., p. 147.
[41] See G. Dannecker, 'Das Unternehmen als 'Good Corporate Citizen'', in H. Alwart (ed.), *Verantwortung und Steuerung von Unternehmen in der Marktwirtschaft* (Hampp Verlag, Munich/Mering, 1998), 5 et seq.
[42] See G. Heine, *supra* note 40, p. 248 et seq. To an international overview, G. Heine, 'Kollektive Verantwortlichkeit als neue Aufgabe', in D. Dölling (ed.), *Jus humanum: Grundlagen des Rechts und Strafrecht: Festschrift für Ernst-Joachim Lampe* (Duncker & Humblot, Berlin, 2003), p. 588 et seq.

mean that international regulations concerning double hulls, for example, are useless. Rather the opposite. Prevention of incidents can, of course, be achieved through regulation. Nevertheless, *detailed* knowledge about specific risks is mostly created over time through entrepreneurial experience, which means that the more uncertainty there is concerning specific operations, the less the state or international agencies are able to achieve this through regulations.[43] As a consequence, entrepreneurial freedom and responsibility for operational decisions are necessarily linked. Otherwise, international institutions will simply have to forbid certain activities, meaning transporting dangerous substances by vessels older than twenty or more years, providing security for the next years by double hulls, and for the future by triple hulls and so on. Obviously, constraining a bureaucracy and allowing market-control mechanisms to play a greater role inevitably leads to strengthening of entrepreneurial collective self-responsibility.[44] Therefore, in order to avoid an international bureaucracy inducing more and more regulations and direct state surveillance, it is in the interest of the private actors involved to strengthen their self-responsibility. This self-responsibility may compensate for a reduced State's bureaucracy. In order that the notion of self-responsibility does not remain just a notion, but backs safety, an effective Corporate Criminal Liability has to be established.

The crucial question concerns how entrepreneurial responsibility, which can greatly influence the character and direction of environmental protection of the seas, can be meaningfully built into the legal systems.

3.3 CORPORATE CRIMINAL LIABILITY

The international crusade against marine pollution essentially demands that legal persons can be held liable and that appropriate and effective sanctions can be imposed.[45] Indeed, whereas some years ago it seemed to be an unchangeable principle of criminal law in many countries that a corporation cannot be held criminally responsible, the trend has totally changed in the meantime. Nowadays, most countries have incorporated distinct forms of corporate liability in their respective national legal systems and most international conventions emphasize the need for (criminal) sanctions to be imposed on legal persons.[46] However, the various international approaches diverge quite a lot (*infra* a). Modern principles refer to the faulty risk management of the enterprise as such to be found in a defective corporate culture. In a second step of analysis, there will be an examination of the current relevenace of the classic British maxim, 'Since a company cannot be put behind bars, the only possible sanction is a fine.' If this were true, it would be absolutely counterproductive in respect of the actual situation in marine pollution, and indeed it can be shown it has been overruled by international developments (*infra* b).

---

[43] See in detail G. Heine, *supra* note 36, 179 et seq. Cp. as an example the detailed circumstances in the case of 'Prestige', and *supra* note 28.
[44] G. Heine, *supra* note 36, 178 et seq. with further references.
[45] See art. 6 no. 5 Proposal EU-Directive of 5 March 2003, *supra* note 3, Proposal Council Framework Decision of 2 May 2003, *supra* note 26, art. 3 no. 2; cp. furthermore M. Löschnig-Gspandl, *supra* note 40, at pp. 12-60 to European Incentives concerning the (Criminal) liability of legal persons.
[46] For a comparison of recent international developments see G. Heine, *supra* note 42, 577-596.

3.3.1 Approaches by an International View

The classical identification theory as well as the traditional *respondeat superior* rule (and approaches influenced thereby) are no longer a suitable starting point for questions as to how to ensure that there is broader liability for entities (rather than for individuals) and to influence and steer global players in order to improve their corporate culture and risk management.[47] Such laws, created around a hundred years ago, are focused on small enterprises dominated by a *bonus pater familias*. They do not recognize, firstly, the important difference between individual responsibility, which may belong to the *pater familias* itself, and collective responsibility, because they merely impute responsibility for wrongdoing to the enterprise involved. Nevertheless, this approach might be helpful if one would impute the offence of a deliberate discharge by the master of a ship to the company owners, insofar as the master acted on their behalf. However, as many courts have emphasized,[48] it is obvious that this sort of corporate liability is too strict. Accordingly, because the possibility to avoid the offence is replaced by a mere imputation of an illegal action, economic theory and principles of Criminal Law (which takes the wrongdoing of the corporate actor as a pre-condition) refrains from this sort of automatic or strict liability.[49] Second, in cases of accidental discharges, where there can be a longer chain of faulty actions, this approach would promote (and justify) a judicial witch hunt focused on the master and the crew (who, it should be remembered, probably only made small mistakes) for the sole reason of being in a position to seek retribution from the company. This is because in order to impute liability to the enterprise, one needs to prove that there was an offence by a corporate agent!.

Therefore, special criteria differentiating between the responsibilities of natural persons, on the one hand, and corporate liability of the entity, on the other, have to be established.

These criteria must stem from the organizational structures of the enterprise, its corporate culture, its business policy and strategy avoiding risks of marine oil pollution.

Modern laws make corporate liability into a matter of a deficient corporation, meaning one where improper risk management is in place as a result of faults in the organization or deficiencies in the corporate culture. This is true, for example, in Australia or Italy. With these forms of corporate liability, the misconduct of subordinates is no longer a solid base for collective responsibility. This solid base should rather be grounded in a type of organizational blame placed on the entity as such for neglecting its collective duty to concern itself with adequately balancing the dangerous risks that arise from the transport of hazardous substances on waters and seas, or the classification and controlling of

---

[47] On this, see in detail G. Heine, *supra* note 36 and 42, and G. Heine, 'Corporate Criminal Liability', in G. Weick (ed.), *Competition or Convergence* (Peter Lang, Frankfurt, 1999), pp. 97-110; G. Heine, 'Rethinking Criminal Liability of Enterprises', in M. Faure and C.A. Schwarz (eds), *De strafrechtelijke en civielrechtelijke aansprakelijkheid van de rechtspersoon* (Intersentia, Antwerp, 1998), pp. 163-178.

[48] With respect to France, see 'Tribunal corr. Lyon 9 October 1997', *Journal Pénal*, 1997, décembre, 11; see, however, 'Cour de Cassation, 2 Decembre 1997, *J.C.P.* 1998-II, 10.023, annotation M.F. Desportes and M. Löschnig-Gspandl', 'Bestrafung von Unternehmen' in W. Hochreiter (ed.), *Bestrafung von Unternehmen, Informationen zur Umweltpolitik, No. 157* (Bundeskammer für Arbeiter u. Angestellte, Vienna, 2003), p. 61.

[49] W.M. Landes and R.A. Posner, 'Tort Law as a regulatory regime for catastrophic personal injuries', (1984) 13 *Journal of Legal Studies* 421. Cp. M. Faure and G. Heine, *supra* note 5, 46 et seq.

tankers, and so on. It is no longer a question of the personification of responsibility for single faulty decisions for deliberate or accidental discharges. To the contrary, as a result of this shift in perspective on corporate liability, these incidents would be shown for what they really are: social disturbances, namely, through marine oil pollution, which under certain circumstances fall the liability of the entity. The Australian legislature addresses just this point: it is decisive whether the *(company) as such* expressly, tacitly or implied authorized or permitted the incident, irrespective of personal responsibility for this wrongdoing. That means proving that there was a corporate culture within the corporation that encouraged or tolerated the incident or that the entity failed to create and maintain the corporate culture.[50]

Of course, rules concerning the legally postulated risk management are still sometimes lacking. However, the IMO, its branches and other international organizations are filling up this gap in the rules more and more.[51]

3.3.2 Sanctions

One can find a broad scope of sanctions nowadays imposed on enterprises can be found.[52] These sanctions are meant to incite the company to provide adequate risk management and thus to prevent environmental disasters.

The conventional corporate sanction is the *fine*. To ensure that fines are dissuasive and in proportion with the economic capacity of the enterprise, EU regulations make reference to a certain percentage of turnover or assets.[53] The assets approach is included as an appendix to the turnover approach. In the domain of ship-source pollution, two levels are envisaged.[54] The first, from 1 % to 10 % of turnover from the previous year, concerns is for less serious cases, whereas the second, 10 % to 20 %, concerns the most serious cases. Furthermore, these regulations envisage that fines shall not be insurable.[55] However, this stipulation cannot be enforced in cases where the P and I Clubs are located abroad.

Therefore, sanctions other than fines are often available. Thanks to such sanctions, a new prominent role of corporate criminal law has been achieved. These include compensation by emphasizing the status of victims (if there are such victims), as in the

---

[50] See Part. 2.5 Div 12 para. 12.2, 12.3 (2) (c) (c/) Criminal Code Act 1995 Australia.
[51] Cp. The International Safety Management Code, IMO Resolution A. 741 (18) 1993, explanatory notes, marine orders, part 58, issue 2, order no 10 of 2002. Cp. also N. Bosch, *Organisationsverschulden in Unternehmen* (Nomos Verlag, Baden-Baden, 2002), p. 411 et seq.
[52] To the following see G. Heine, 'Sanctions in the Field of Corporate Criminal Liability', in A. Eser, G. Heine and B. Huber (eds), *Criminal Responsibility of Legal and Collective Entities* (Edition Iuscrim, Freiburg, 1999), pp. 237-254.
[53] See Council Framework Decision of 2 May 2003, *supra* note 26, at 6; see also COM(2004), 334 final, Greenbook of the EC-Commission of 30 April 2004, 16. Cp (crit.) M. Löschnig-Gspandl, *supra* note 40, 445. See also § 4 Draft Corporate Liability Law Austria (Verbandsverantwortlichkeitsgesetz), IMZ 318.017/0001-II 2/2004. Cp., however, the Baltic Marine Environment Protection Commission (HELCOM) which has envisaged a minimum level of penalty for each offence provided for in the annexes to the Marpol-Convention, HELCOM-Recommendation 19/14, based on so-called Special Drawing Rights, the unit of account defined by the International Monetary Fund.
[54] Art. 3 no. 2 Council Framework Decision, *supra* note 26.
[55] Art. 6 no. 6 Proposal EC-Directive, *supra* note 3.

case of Zeebrugge Ferries in the UK,[56] or the establishment of a whole package of compensations, as in the famous case of the Exxon Valdez,[57] whose wreckage in Prince William Sound, Alaska (1989), caused an environmental catastrophe. The latter settlement had three distinct parts. In a criminal plea agreement, Exxon was fined USD 150 million. As criminal restitution caused to the wildlife, natural environment, and so on, the enterprise paid USD 100 million. Finally, in a civil settlement, Exxon agreed to pay USD 900 million for injuries, and this even contained even a 'reopener window'. Furthermore, the Netherlands have shown that restitution is not restricted to mere financial compensation but can also allow for the enforcement of the 'good corporate citizen'. Therefore, these developments have opened the door for the imposition of *operational duties*, like the improvement of entrepreneurial risk management, of the organization or of the technical classification system.

Indeed, *restrictions of entrepreneurial liberty* aiming at influencing the entities behaviour in future are quite common, sometimes by imposing judicial directives regulating adequate organization or product safety (as is provided, again, in the Netherlands), sometimes by restricting certain activities, like using tankers more than 15 years old or using straits that are known to be especially dangerous, or banning access to public assistance or subsidies (see Art. 6 No 5 lit. f Proposal EC-Directive of 5 March 2003) or, at least, placing them under judicial supervision.

Therefore, criminal law (or administrative penal law, etc.) has armed itself more and more, and even sharper weapons are available to it. These include, for example, a total confiscation of property, as is used in the US, which may have the same consequences as the closure of the enterprise or a judicial termination of its charter (see Art. 6 No. 5 lit. e Proposal EC-Directive of 5 March 2003). These kinds of 'corporate death penalties' are mostly provided for 'criminal purpose organizations,' that is, for those corporations that should not only be put out of business but also out of action.

4 SUMMARY AND CONCLUDING REMARKS

We have seen that international law (UNCLOS) grants jurisdiction not only to the flag States but also to the port States, even when the discharge has taken place outside national boundaries. The international network of technical safety, control management and compensation has been greatly improved. However, the international regime on marine pollution aims mainly and rather solely at compensating victims. This orientation has certain shortcomings. The liability of the polluters runs the risk being diluted, by a right of the ship-owner to limit his liability, by compulsory insurance and by collective compensation by cargo receivers through Funds. Moreover, compensation claims for pollution damage against the chartering parties, manager and operator of a ship are prohibited. Therefore, this international liability regime should be opened up to reforms, as is for example proposed by the EU Commission. Indeed, a well-functioning legal

---

[56] See G. Heine, *supra* note 47, 98.
[57] See settlement of 9 October 1991 among the State of Alaska, the US government and Exxon Corporation, Exxon Shipping Company Civil Action no. A 91-082 CIV, A 90-015 CR.

regime of compensation, fund solutions and liability rules must be promulgated. In mentioning legal rules that focus on the liability of the actors involved and thereby promoting the prevention of marine pollution there can equally be a need to discuss the role of the criminal law.

However, the international framework is ambivalent. On the one hand side, UNCLOS strengthens the jurisdiction of the port States. They have the right to start up criminal investigations and to institute proceedings against any discharge from a vessel, even outside their internal waters, territorial sea or exclusive economic zone. Moreover, this criminal jurisdiction falls to the port State (or the coastal State) if the flag State does not initiate proceeding. However, only monetary penalties can be imposed with respect to violations by ships flying a foreign flag (except in cases of wilful and serious act of pollution in the territorial sea). Furthermore, current maritime insurance practices cover such monetary penalties, including sanctions of a penal nature related to pollution offences. At the same time, the flag State has a broad jurisdictional mandate, and is not limited imposing to financial penalties. Two aspects of this situation have to be emphasized.

First, the flag State has in some cases been given a bum deal, and often there is no political willingness to prosecute. In this respect one only needs to think about the problem of flags of convenience. Second, the port and coastal States are encouraged to warp international law by widening the exceptions – for the sole reason of being effective at punishing the persons 'behind', that is, responsible for, the pollution.

Therefore, for the time being, typically the captain of a ship runs the greatest risk being caught and criminally sanctioned. As a scapegoat, they are unfortunately the last link of a long chain of possible actors who may share responsibility for the pollution, like the operators, the chartering partiers, the owner of the cargo, the ship-owner, the classification society, and sometimes even the port authorities.

However, imposing criminal sanctions on *individuals* has proven to be rather *limited* in this context. Individual responsibility is of necessity tailored to a single faulty act and the same has to be done with respect to the amount of sanctions, which have to be set in proportion to the financial capacity of an individual. The new task on the agenda of criminal law is for it to steer global players towards good corporate citizen culture and adequate risk management. Whereas collective responsibility does not remove individual responsibility, these new aims cannot be achieved solely through sanctions imposed on individuals. Furthermore, it is in the interest of such corporate entities to plea for adequate forms of corporate criminal liability, in order to avoid overwhelming bureaucratic measures. However, on the other side of the coin, entrepreneurial freedom entails collective liability.

Modern principles of law refer to *faulty risk management* of the enterprise as such based on a *defective corporate culture*. On this legal basis, it is no longer a question of the personification of responsibility for single faulty decisions for deliberate or accidental discharge. Rather, these incidents have to be shown for what they are: social disturbances, namely through marine pollution, inside or outside territorial waters, which under certain legal circumstances fall under the liability of those entities that have encouraged the incident and have typically benefited from the incident or the circumstances concerned.

To ensure that *sanctions* are effective for the new tasks, the international view offers a broad spectrum of sanctions aimed at stressing the entities' risk management and its corporate culture, for example by providing and applying package sanctions covering settlements, compensation, restitution and fines, as was the case with the Exxon Valdez, settlement in 1991 or by judicial directives restricting the entrepreneurial freedom and aiming to improve tanker safety.

Furthermore, even sharper weapons are available to criminal law, like the total confiscation of property, the closure of the enterprise or judicial termination.

However, the actors concerned might try to escape from corporate criminal liability. A prominent example comes from the so-called *one ship corporations*[58] that organize their own insolvency or corporate dissolution. Legal remedies might be found in combinations of corporate law, financial responsibility rules, and codes of good behaviour for corporations and corporate criminal liability. For financial responsibility rules, US law may have an answer. There are four types of 'allowable mechanism' that can be used by firms to demonstrate the existence of coverage: insurance, surety bond, self-insurance and financial guaranty.[59] Corporate law may allow victims to seize the assets of a mother company in cases of damage caused by a subsidiary.[60] Moreover, even corporate criminal liability may be helpful for piercing the corporate veil. The above-mentioned approach, based on the idea that enterprises as such are supervisory guarantors for operational risks, creates a solution for full or partial liability for the entities 'behind' an incident, and creates an opportunity for sanctions to be imposed on those entities that even indirectly, tacitly or implicitly permitted the social disturbance.[61]

In sum, criminal law stands ready for action!

As to whether it should be brought into action, the answer depends on the possibilities for reforming civil liability laws, insurance law and economic incentives. Criminal law and its ancillary partners, like administrative penal laws and so on, should occupy the role of a last resort.

In any case, we should think about finer tweaks to international law that, in cases of subsidiary jurisdiction (when the flag State does not initiate proceedings), would transfer the competence of the flag State to the port and coastal State, in accordance with general principles of representation. In other words, the port and coastal States' jurisdiction should not be limited to monetary penalties. Furthermore, I cannot recognize anything that would reasonably contravene the establishment of a general vessel traffic monitoring and information system, including the duty to install a 'black box' recording all relevant data – think of the efforts western countries have been made concerning trucks on roads – if we really want to change the actual deficits!

---

[58] See K. Le Couviour, *supra* note 17, 2273.
[59] See J. Boyd, in this volume, section 5.
[60] See M. Faure and H. Wang, in this volume, section 5.
[61] See G. Heine, 'Modelle originärer (straf-)rechtlicher Verantwortlichkeit von Unternehmen', in M. Hettinger (ed.), *Reform des Sanktionenrechts, III, Verbandsstrafe* (Nomos Verlag, Baden-Baden, 2002) 145 et seq.

LIST OF REFERENCES

D. Abecassis, 'IMO and liability for oil pollution from ships: a rectrospective', (1983) *Lloyds Maritime and Commercial Law Quaterly* 47.

K. Ambos, 'Commentary of §§ 3 ff.', in B. von Heintschel-Heinegg, *Münchner Kommentar zum Strafgesetzbuch* (C.H. Beck, Munich, 2003).

U. Beyerlin, 'New Delopments in the Protection of the Marine Environment', (1995) 55 *Zeitschrift für ausländisches öffentliches Recht und Völkerrecht* 544-579.

N. Bosch, *Organisationsverschulden in Unternehmen* (Nomos Verlag, Baden-Baden, 2002).

P. Cramer and G. Heine, 'Commentary of § 324', in A. Schönke and H. Schröder (eds), *Strafgesetzbuch Kommentar*, (26th edn, C.H. Beck, Munich, 2001), § 324.

G. Dannecker, 'Das Unternehmen als 'Good Corporate Citizen'', in H. Alwart (ed.), *Verantwortung und Steuerung von Unternehmen in der Marktwirtschaft* (Hampp Verlag, Munich/Mering, 1998), 5 et seq.

G. Dannecker, 'Zur Notwendigkeit der Einführung kriminalrechtlicher Sanktionen gegen Verbände', (2001) *Goldammer's Archiv für Strafrecht* 102 et seq.

M. Faure and G. Heine, 'The insurance of fines', (1991) 58 *The Geneva Papers on Risk and Insurance* 53-54.

M. Faure and G. Heine, *Environmental Criminal Law in the European Union* (Edition Iuscrim, Freiburg, 2000).

M. Faure and H. Wang, 'Liability for oil pollution in Europe: recent developments', (2004) *Environmental Liability* 55 et seq.

E. Fitrakis and C.P. Kareklas, 'Landesbericht Griechenland', in G. Heine (ed.), *Umweltstrafrecht in mittel- und südeuropäischen Ländern* (Edition Iuscrim, Freiburg, 1998), p. 139 et seq.

E. Gold, *Handbook on marine pollution* (Gard, Arendal, 1985).

G. Heine, 'Criminal Liability of Enterprises and New Risks', (1995) 2 *Maastricht Journal of European and Comparative Law* 110 et seq.

G. Heine, *Die strafrechtliche Verantwortlichkeit von Unternehmen* (Nomos Verlag, Baden-Baden, 1995).

G. Heine, 'New Developments in Corporate Criminal Liability in Europe', (1998) *St. Louis-Warsaw Transatlantic Law Journal* 176 et seq.

G. Heine, 'Rethinking Criminal Liability of Enterprises', in M. Faure and C.A. Schwarz (eds), *De strafrechtelijke en civielrechtelijke aansprakelijkheid van de rechtspersoon* (Intersentia, Antwerp, 1998), pp. 163-178.

G. Heine, 'Corporate Criminal Liability', in G. Weick (ed.), *Competition or Convergence* (Peter Lang, Frankfurt, 1999), pp. 97-110.

G. Heine, 'Sanctions in the Field of Corporate Criminal Liability', in A. Eser, G. Heine and B. Huber (eds), *Criminal Responsibility of Legal and Collective Entities* (Edition Iuscrim, Freiburg, 1999), pp. 237-254.

G. Heine, 'Modelle originärer (straf-)rechtlicher Verantwortlichkeit von Unternehmen', in M. Hettinger (ed.), *Reform des Sanktionenrechts, III, Verbandsstrafe* (Nomos Verlag, Baden-Baden, 2002) 145 et seq.

G. Heine, 'Kollektive Verantwortlichkeit als neue Aufgabe', in D. Dölling (ed.), *Jus humanum: Grundlagen des Rechts und Strafrecht: Festschrift für Ernst-Joachim Lampe* (Duncker & Humblot, Berlin, 2003), p. 588 et seq.

H. Hohmann, *Meeresumweltschutz als globale und regionale Aufgabe,* (1989).

H. Hohmann, 'Weltweiter Schutz der Meeresumwelt', in K.P. Dolde (ed.), *Umweltrecht im Wandel* (Schmidt, Berlin, 2001), pp. 97-127.

W.M. Landes and R.A. Posner, 'Tort Law as a regulatory regime for catastrophic personal injuries', (1984) 13 Journal of Legal Studies 421.

K. Le Couviour, 'Responsabilité pour pollutions majeures', (2002) *La Semaine Juridique* 2271.

M. Löschnig-Gspandl, 'Bestrafung von Unternehmen', in W. Hochreiter (ed.), *Bestrafung von Unternehmen, Informationen zur Umweltpolitik, No. 157* (Bundeskammer für Arbeiter u. Angestellte, Vienna, 2003), p. 61.

M. Löschnig-Gspandl, *Die strafrechtliche Verantwortlichkeit von Unternehmen und Verbänden mit Rechtspersönlichkeit in Österreich* (Karl Franzens-Universität, Institut für Strafrecht, 2004).

D. Oehler, 'Internationales Umweltstrafrecht', in O. Kimminich (ed.), *Handwörterbuch des Umweltrechts* (2nd edn, C.H. Beck, Munich, 1994), p. 1136.

K. Seelmann, 'Unternehmensstrafbarkeit', in J.B. Ackermann, A. Donatsch and J. Rehberg, *Wirtschaft und Strafrecht: Festschrift für Niklaus Schmid*, (Schulthess, Zurich, 2001), p. 173.

T. Treves, 'The Law of the Sea Tribunal', (1995) 55 *Zeitschrift für ausländisches öffentliches Recht und Völkerrecht* 421.

INTERDEPENDENCIES BETWEEN CRIMINAL LAW AND OIL POLLUTION REGULATION IN CHINA

Thomas Richter

1 INITITAL POSITION IN CHINA

1.1 MARINE OIL POLLUTION IN THE CHINESE REALITY

China is the third biggest country on the planet, lying at the east of the Eurasian continent. It has a coast line of more than 18.000 km – the coast line of the islands not included[1] – starting from the (North) Korean border and ending at the Vietnamese border. The coast line meets the Bohai Sea, the Yellow Sea, the East China Sea and the South China Sea. This coast line is where China's economic rise started in the late Seventies and is where we can find a relatively advanced economy. As well, the coastal provinces belong to the areas of China with the large populations. The sea is not only important for the population as a traffic route to other coastal areas in China and foreign countries, but also for fishing and more and more for domestic and foreign tourism.

1.2 ECONOMIC DEVELOPMENT AND ENERGY IN CHINA

The population in China has not only more than doubled in the last 50 years, but the demographic and economic development has had extreme consequences for energy production and energy consumption.[2]

---

[1] That would be another 3.200 km of coast line.
[2] All data in this and the following graphs were collected as official statistical data.

Michael G. Faure and James Hu (eds), Prevention and Compensation of Marine Pollution Damage. Recent Developments in Europe, China and the US, 61-81.
© *2006 Kluwer Law International. Printed in the Netherlands.*

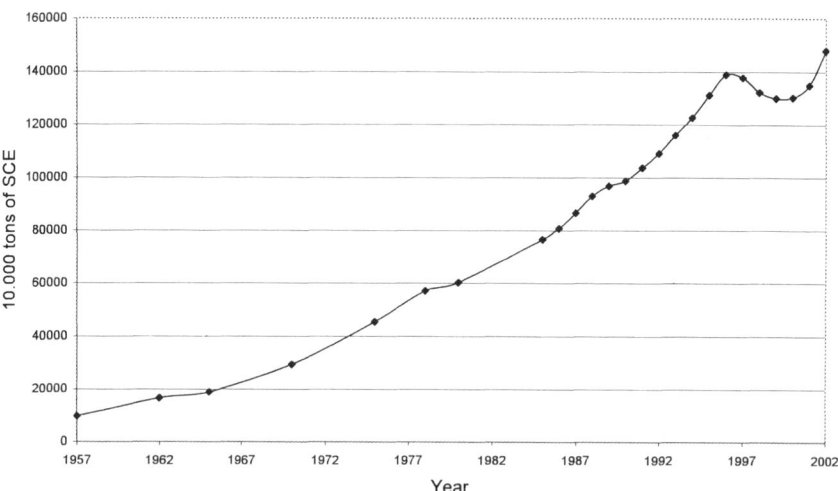

Total Consumption of Energy in the PRC
(China Statistical Yearbook 2000, P. 239; 2003, P. 265)

Nowadays, China is called the 'world's factory.' Graph 1 shows that the consumption of energy has steadily risen and has been in higher proportion to the population during the last half century. In the second half of the Nineties, the statistics show a decline of energy consumption, but since 2000 the appetite for energy has been increasing once again and faster than ever. In the forty years from 1962 to 2002, the consumption of energy increased almost with the factor of 9 and still rose to more than the double consumption within the period between 1980 and 2000.

The statistical data about energy production is rather similar. China has basically used its coal and oil to satisfy its own demands. At the moment, China is the fifth largest producer of crude oil in the world. The government plans to increase the volume of oil production, but the reserves of crude oil are rather limited.[3] It is important to know that since the beginning of the Nineties, the production of energy could no longer satisfy the demand. China has lost its self-reliance in this field. Since then, there has been a certain deficiency of energy production that had its recent climax around the millennium, but which has been getting smaller since then. This development can be reviewed in the following graph.

---

[3] Only 2 % of the global reserves on crude oil lie in China, *Neue Zürcher Zeitung*, 11 May 2004, 17.

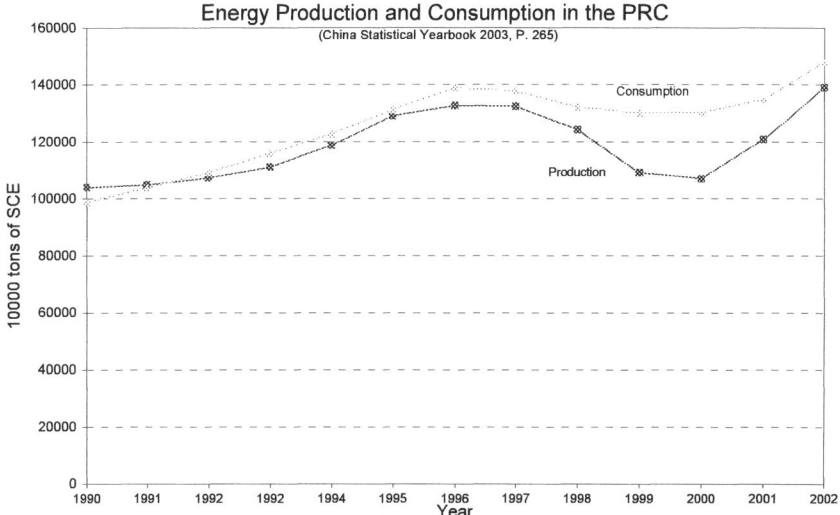

The deficiency of energy production accelerated the need of importing energy from abroad, in order to satisfy the demand. This increased the importance of all forms of energy, but especially the importance of oil for China. At the time, oil is the second largest energy source, with involement in around 25 % of all energy sources. The primary source of energy is still coal, but this has seen a steady decline. It is doubtful whether the Chinese government has not underestimated, concerning the next few decades, the amount of oil that has to be imported while overestimating the possibility of using alternative energy sources.[4]

---

[4] See S. Hieber, 'Energiesicherheit in China: Instrumente zur Versorgungssicherung', (2004) *China Aktuell* 399.

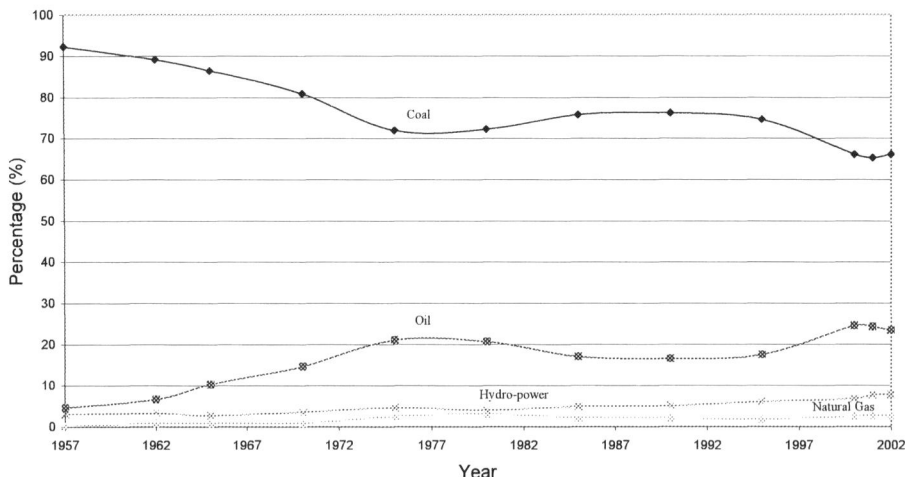

Composition of the Total Consumption of Energy
(China Statistical Yearbook 2003, P. 265)

According to estimations of the International Energy Agency, China consumed 5.46 million barrels[5] of crude oil per day in 2003, which – for the first time in history – was more than the Japanese consumption.[6]

With the new economic possibilities since the late Seventies, the exchange of goods within the country and with other countries has increased dramatically. As China is a coastal country, transportation of goods on water routes is common. However, the volume of freight transported on water routes increased faster in the Eighties and – even more – in the Nineties. This can be derived from the data of major coastal ports in China.

---

[5] 1 barrel corresponds to 159 litres.
[6] *Neue Zürcher Zeitung*, 11 May 2004, 17.

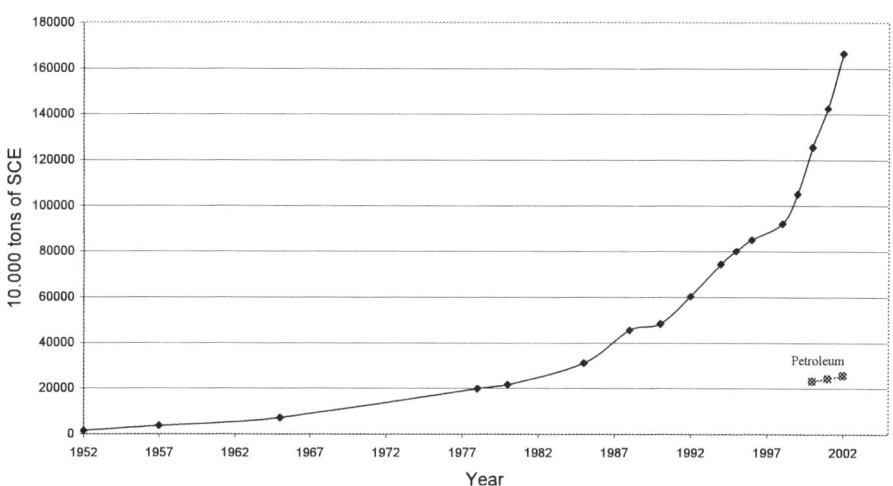

Volume of Freight Handled in Major Coastal Ports
(China Statistical Yearbook 2000, P. 536; 2003, P. 597 etc.)

The volume of freight handled in those ports in 2000 was about five times the volume compared to 1980. Statistical data about the volume of oil handled in those ports is only available since 2000. However, the volume of oil handled in 2000 is more than the volume of all freights handled in 1980. The current Chinese demand to increase the import of oil is regarded as one reason for the high price of crude oil on the world market.[7] In the future, one part of the imported oil shall come from Kazakhstan and also maybe Siberia via pipeline,[8] but this is not likely to decrease the number of tankers carrying oil from the Middle East.

1.3 MARINE OIL POLLUTION AND FOCUS OF THE REPORT

Thinking of marine oil pollution, the man on the street is likely to remember serious tanker accidents, like the accidents with tankers named Torrey Canyon, Exxon Valdez, Erika or Prestige. These accidents present us with dramatic footage of broken or sinking tankers, dead animals, polluted beaches and politicians who shovel black mud. Such footage can be regarded as breaking news that is interesting enough to be shown for a couple of days, but which is then forgotten about, even when the disaster continues to have serious effects. However, assessments tell us that major and minor tanker accidents are responsible for less then 10 % of marine oil pollution.

On the other side, around 70 % of marine (oil) pollution is caused by land-based activities (LBA). In China, more than 4 billion tons of harmful substances per year are

---

[7] F. Vorholz, 'Lunte an der Weltwirtschaft', (2004) 24 *Die Zeit* 21.
[8] *Neue Zürcher Zeitung*, 21 May 2004, 15.

released into the sea by land-based industries.[9] Operational shipping traffic and dumping cause another 10 % respectively.

For this report, I have selected two of the above-mentioned phenomena for analysis. The first is the phenomenon of land-based activities on account of their great importance for polluting the sea. The second concerns tanker accidents: although they are much less responsible for marine oil pollution and no major tanker accident along the Chinese coastline has been reported so far, they play a major role in the mind of the public at both the national and the international level. Besides that, they can have disastrous effects for a certain coastal areas. It is not only that persons on the ship might be injured or killed, or that the tanker or the transported oil will be lost. On top of all this, plants and animals living in or near the sea will be harmed by the crude oil in the water and on the beaches, to say nothing of the fact that the coastal population might be injured as well. Another economic effect of a tanker accident can be the bad influence it may have on tourism in the affected coastal areas. Even though it may be accidental, the reproach remains that some tankers do not take enough precautions and foresee enough money for safety, which is to the detriment of a large number of people and the environment.

2 THE LEGAL MILIEU

2.1 INTERNATIONAL LAW

Since the end of the Cultural Revolution, China has more and more abandoned its policy of self-isolation and has taken on more and more responsibilities in the international scene. The PRC participated in the 1992 Earth summit, and has signed the UN Law of Sea Convention[10] as well as a number of other treaties in the field of environmental protection.

As far as marine pollution is concerned, the first thing to be mentioned is the International Convention for the Prevention of Pollution from Ships (MARPOL 73/78, 国际防止船舶造成污染公约) with 6 annexes and a series of amendments since 1984. The PRC adopted the convention and its Annexes I, II, III and V. In this context, the obligatory Annex I is of particular interest, because it regulates the prevention of pollution by oil.

2.2 NATIONAL LAW: OIL POLLUTION REGULATIONS

2.2.1 The Relevant Legislation

The PRC adopted three major laws on water resources and their protection. The first of these laws was enacted in 1982 and is at the same time the main national law for preventing and fighting marine oil pollution. It is, namely, the Marine Environment Protection Law (海洋环境保护法), which had been revised in 1999. Neither the Water

---

[9] L. Fu, *Huanjing xingfa xue* (Zhongguo Fangzheng Chubanshe, Beijing, 2001), p. 307.
[10] The Convention dates to 10 December 1982. The PRC adopted it on 28 July 1994 and ratified it on 15 May 1996.

Law of 1988 (水法, revised in 2002), nor the Law for Prevention and Control of Water Pollution (水污染防治法 enacted in 1984, revised in 1996) are directly applicable to marine pollution. The Law for Prevention and Control of Water Pollution aims at protecting surface and ground water, but not sea water (Art. 2 para. 2). Only indirect marine pollution through the pollution of rivers can be treated by this law.

The Law on Pollution through Solid Wastes (固体废物污染环境防治法, 1995) is applicable neither to crude oil nor to used oil. In case of tanker accidents, it is evident that crude oil will not considered as waste. In case of land-based activities dumping used oil, this law is not applicable for two main reasons. Firstly, the law mainly focuses on solid wastes and excludes explicitly ordinary sewage (Art. 75). Secondly, the law aims at protecting water, air and soil, but does not include protection of the sea. Therefore, this law is not concerned with marine pollution by solid wastes (Art. 2).

Concerning the Marine Environment Protection Law, this law includes several general chapters, but also divides the various sources of pollution in five separate chapters:

- Pollution caused by land-based pollutants (Chapter IV),
- Pollution caused by coastal construction projects (Chapter V),
- Pollution caused by marine construction projects (Chapter VI),
- Pollution caused by dumping of wastes (Chapter VII),
- Pollution caused by vessels and their related operations (Chapter IIX).

Chapters IV and V can be considered as regulations on pollution through land-based activities, whereas Chapter IIX includes regulations on pollution through tanker accidents.

2.2.2. Administrative Sanctions and Other Forms of Sanctions

The administrative measures taken against offenders of regulations of the Marine Environment Protection Law are focussed in Chapter 9 of the Law ('Legal Liabilities'). Depending on the violation, the administration might:

- give a warning (art. 74, 75, 85, 86, 87, 88, 89).
- impose a fine (art. 73, 74, 75, 76, 77, 78, 79, 80, 81, 83, 84, 85, 86, 87, 88, 91).
- confiscate illegal gains (art. 76).
- order a correction (art. 73, 76, 89) or order cessation of a law-breaking act (art. 80, 81, 83, 84).
- order the adoption of remedial measures (within a prescribed time limit) (art. 76, 80).
- order the closure of any illegally polluting project (art. 77, 82) or order the dismantling of any illegal installation (art. 80).
- order the reinstallation and use of any environment protection installation (art. 78).
- detain or revoke permits temporarily or definitely (art. 86).
- order a vessel to withdraw from the sea areas under the jurisdiction of the PRC (art. 79).
- impose administrative sanctions (行政处分) on civil servants (art. 91, 94).

Some of the measures can be taken instead of others, and some might be combined. Almost in all cases, a fine will be imposed on the offender. The amount to be paid has to be determined within the frame given in the relevant provision.

Besides administrative measures taken by a range of different authorities, civil liability is not excluded, but the law itself makes a general statement: Persons polluting the marine environment must repair the damage and compensate for losses incurred (Art. 90). It is interesting to note that even the administration is encouraged to claim civil compensation for certain damages to the State (Art. 90 para. 2)

Art. 91 para. 3 of the Marine Environment Protection Law is the junction between this law and criminal law. In cases of 'accidents that cause major marine environment pollution and that thus result in grave consequences [...], criminal responsibility shall be investigated according to law.' As in the former tradition of environmental laws,[11] the questions of criminal responsibility is not regulated in this law itself, but is handed over to penal law. Whether a corresponding regulation can be found in the penal law, is entirely independent from the Marine Environment Protection Law. However, it should be rather clear that the legislation wants serious accidents with grave consequences to be punished through criminal sanctions.

As far as civil servants are concerned, it is remarkable that the Marine Environment Protection Law is already mindful of criminal responsibility in serious cases according to the code of penal law (Art. 94 MEPL). This might be helpful for the later interpretation of penal law.

3 APPROACH OF THE CRIMINAL LAW

3.1 PRECONDITIONS

Criminal law is the part of the legal system where a prohibition of certain behaviour is linked with a so-called criminal sanction. The prohibition might be embedded in the criminal law itself or may be situated somewhere else in the legal system. The criminal sanction is generally the most serious State sanctions that can be applied against an individual. Within the different frameworks for criminal sanctions, imprisonment and death penalty are the most serious in China. For the moment, the core crimes against the environment have fixed-term imprisonment as the greatest possible punishment. As an instantiation of the general principle of proportionality, criminal sanctions should only be used if there are no other, less vigorous adequate measures for fighting certain behaviour. Hence, criminal punishment is found in the field of subsidiary application.

3.2 DEVELOPMENT OF CRIMES AGAINST THE ENVIRONMENT

The first Penal Law of the PRC was enacted in July 1979 as one of the first acts of legislation after the third rehabilitation of Deng Xiaoping. Shortly afterwards, a series of

---

[11] A similar provision already existed in Art. 44 of the Marine Environment Protection Law of 23 August 1982.

criminal laws followed until 1997, when a large-scale revision took place. This lead to a major modification of the criminal code, especially with respect to its specific offences: the number of criminal acts in the code more than tripled. Chapter 6 of the Specific Part of the criminal code includes a section with crimes against the environment. This chapter, called 'Crimes of Obstructing the Administration of Public Order,' already existed in the Penal Law of 1979, was but expanded from twenty-one to ninety articles in nine sections. Section 6, comprised of nine articles, is called 'Crimes of Impairing the Protection of the Environment and Resources'.

Chinese criminal law does not approach the field of environmental protection from an ecological perspective. The protected interest or object (客体) is seen in the 'Administration of Public Order.' The laws on environmental protection are considered as a part of the administration of public order. In broader terms, a violation of these laws would be understood as a violation of '(socialist) social relations'.[12] As an instantiation of these principles, Section 6 dealing with environmental crimes in Art. 338 to Art. 345 PL should safeguard the state management on environmental protection.[13] Most of the articles deal with crimes with respect to the protection of nature and wildlife, and are the so-called 'Crimes of Impairing Resources' (破坏资源犯罪). Of these, only very few are crimes penalizing pollution (污染环境犯罪). For all such offences, it is possible to have economic and other entities (单位) as subject. If an entity commits such an offence, a fine might be imposed as the sole possible sanction (art. 346, art. 31 PL).

3.3 CRIMES HAVING TO DO WITH OIL POLLUTION

The revision of the Penal Law in 1997 did not provide an offence of polluting the sea. Although a number of authors recommended introducing such an offence,[14] further reforms of the environmental offences since 1997 have not included the pollution of the sea. However, there are some offences with relation to marine pollution.

3.3.1 Art. 338 PL (Considerable Accident with Environmental Pollution)

Art. 338 holds that whoever, in violation of the State regulations, discharges, dumps or treats radioactive wastes, wastes containing pathogens of infectious diseases, toxic substances or other hazardous wastes on the land or in the bodies of water or the atmosphere, thus causing a major environmental pollution accident that leads to the serious consequences of heavy losses of public or private property or injuries or deaths of persons, shall be sentenced to fixed-term imprisonment of not more than three years or criminal detention and shall also, or shall only, be fined; if the consequences are

---

[12] Cp. Y. Han, 'Fanzui keti', in G. Mingxuan (ed.), *Xin bian Zhongguo xingfa xue volume I* (Zhongguo Renmin Daxue Chubanshe, Beijing, 1993), p. 472.

[13] Cp. T. Richter, *Umweltstrafrecht in der Volksrepublik China* (Edition Iuscrim, Freiburg, 2002), p. 147 with more details.

[14] C, Yang, Z. Xiang and S. Liu, *Weihai huanjing zui de lilun yu shiwu* (Gaodeng Jiaoyu Chubanshe, Beijing, 1999), pp. 222-228; L. Tong, 'Shi xi pohuai huanjing ziyuan baohu zui' (1998) 7 *Xingshi Faxue* 42-43; W. Jiang, 'Lun huanjing fanzui de xingfa kongzhi', (2000) 2 *Jiangsu Gong'an Zhuanke Xuexiao Xuebao* 25; S. Chen, 'Dui huanjing fanzui kongzhi duice de jingji fenxi' (2001) 2 *Xingshi Faxue* 72.

especially serious, he shall be sentenced to fixed-term imprisonment of not less than three years but not more than seven years and shall also be fined.

Art. 338 PL is first mentioned in Section 6, called 'Crimes of Impairing the Protection of Environment and Resources' of Chapter VI. Together with Art. 339 PL,[15] it is one of two 'crimes fighting pollution'. All other crimes in this Section are 'crimes of impairing resources.' Art. 338 PL solely has to do with the discharge, dumping and treatment of hazardous wastes with great effect on the environment. Apart from water pollution, Art. 338 PL covers air and soil pollution. The polluter can be an individual or an economic or other entity, following Art. 346 PL.

As far as *mens rea* is concerned, there is no consensus amongst the jurists. Although a majority of authors regard Art. 338 PL as a crime for which negligence and (lower) recklessness (过失) is sufficient, many of them declare that in cases when there is intention (故意), Art. 338 PL can be applied as well.[16] The uncertainty about *mens rea* is a general problem in Chinese criminal law and has to do with the understanding of Art. 15 PL.[17] This interpretation leads to an extreme expansion of the crime of Considerable Accident with Environmental Pollution. Regarding *mens rea*, it would include negligence, recklessness and intention within one simple sentencing framework. However, the interpretation of Art. 338 PL is also an extreme example and should not be over evaluated.

The sanctions for Considerable Accident with Environmental Pollution are divided into two ranges of punishment. In cases where there are serious consequences, an imprisonment of up to 3 years can be imposed. The imprisonment can be combined with, but also replaced, by a fine.

As far as tanker accidents are concerned, Art. 338 PL is not applicable because the transported crude oil cannot be considered as waste, but is one of the most important raw materials for industry, transport and other public and private uses in daily life. However, in cases of LBA, used oil might be considered as waste. This sort of waste is dangerous to human beings or the environment. It does not matter whether the used oil enters the ocean directly or via rivers. The question is whether this dangerous waste causes serious consequences for public or private property or causes human casualties. Vessels or oil platforms in the sea do not pose such a great risk to cause suffering, even when they would spill large amounts of used oil. Although Art. 9 of the Constitution of the PRC says that waters are public property, but the mere pollution of the water and its ecological balance is likely not to be regarded as a 'serious consequence'.

3.3.2 Art. 408 PL (Misconduct in Office while Supervising and Controlling the Environment)

Art. 408 holds that any functionary of a State organ who is responsible for environmental protection, supervision and control, through his gross neglect of duty, causes a serious environmental pollution accident, which leads to the serious consequences of heavy losses

---

[15] Art. 339 PL includes the Illegal Treatment of Imported Solid Waste (art. 339 §1 PL) and Unauthorized Import of Solid Waste (art. 339 §2 PL).
[16] Cp. T. Richter, *supra* note 13, 83-84.
[17] Cp. T. Richter, *supra* note 13, 188-190.

of public or private property or injuries or deaths of persons, shall be sentenced to fixed-term imprisonment of not more than three years or criminal detention.

Art. 408 PL is one of several Crimes of Dereliction of Duty (Chapter IX) that has to do with environmental protection. It corresponds more or less to Art. 94 of the Marine Environment Protection Law.[18]

The subject has to be a perpetrator who is responsible for environmental protection. This includes State Marine Administration, Harbour Survey and other authorities. Gross neglect of duty by this perpetrator has to lead to serious environmental pollution accident with serious consequences for property or persons. Concerning *mens rea*, negligence or (lower) recklessness is needed.[19] Therefore, the offence of Misconduct in Office while Supervising and Controlling the Environment is very similar with Art. 338 PL. The most important difference can be found in the subject. Both tanker accidents and land-based activities might fall under this article.

3.3.3 Art. 136 PL (Trouble with Dangerous Objects)

Art. 136 holds that whoever violates the regulations on the control of explosive, inflammable, radioactive, poisonous or corrosive materials and thereby causes a serious accident during the production, storage, transportation or use of those materials, if there are serious consequences, shall be sentenced to fixed-term imprisonment of not more than three years or criminal detention; if the consequences are especially serious, he shall be sentenced to fixed-term imprisonment of not less than three years but not more than seven years.

The offence of Trouble with Dangerous Objects lies outside of the core of environmental crimes and belongs to the Crimes of Endangering Public Security (Chapter II). Unlike in Art. 338 PL, the object here is not waste, but is harmful material in one form or another. This material has to be produced, stored, transported or used in violation of certain regulations and with serious consequences. If the consequences of the illegal treatment of harmful materials are not serious enough, no criminal punishment will be given. Only individuals, but no economic or other entities can be the subject of the crime. Negligence or (lower) recklessness is needed. If there are serious consequences, the punishment will be up to three years of imprisonment, and with especially serious consequences, the range of punishment goes up to three to seven years of imprisonment.

In case of a tanker accident, Art. 136 PL might be applied, at least in theory. Crude oil can be seen as a poisonous material that is transported by the tanker and that leads to serious consequences. However, the offence is only applicable if the perpetrator violates the regulations on the control of explosive, inflammable, radioactive, poisonous or corrosive materials.

---

[18] See above section 2.2.2.
[19] D. Zhou, C. Shan and S. Zhang (eds), *Xingfa de xiugai yu shiyong* (Renmin Fayuan Chubanshe, Beijing, 1997), p. 815.

4 INTERDEPENDENCIES BETWEEN ENVIRONMENTAL CRIMINAL LAW AND OIL POLLUTION REGULATION

4.1 GENERAL REMARKS

Especially since the Twentieth century, oil has become a material necessary for daily life, for private and public use, in transportation, heating and production. Oil has become a sort of symbol of the power of industrialized countries. The lack of oil in the early Seventies shocked the Western industrialized countries and still today, the world economy is highly dependent on oil for its own stable proliferation.

Loosely speaking, the burning of oil by cars, buses, and heating systems can be said to cause pollution. This is pollution that we sometimes recognize immediately and sometimes discover only months, years or decades later. A great deal of this pollution, maybe most of it, is tolerated by society and is not even regulated at all. Some explain this acceptance with the capability of nature to get clean by its own forces.[20] Others think that people have gotten used to taking the car or bus and that a reduction of traffic would result in major economic losses. However, in light of more negative effects on human health, the environment or property, more and more countries have been trying to reduce the pollution by (administrative) regulation. Standards have been set out in order to avoid certain forms of pollution by penalizing to those who still pollute. In certain major cases, pollution has been prohibited through criminal sanctions. This situation hence demonstrates the interdependence of administrative and criminal law.

In one way, it seems quite normal that this kind of administrative law should have created criminal laws. This was exactly how the Chinese environmental law developed in the Eighties and Nineties. The enactment of criminal laws while depending on administrative law is not a new phenomenon,[21] but has been widely broadened with the revision of the Chinese Penal Law in 1997, in general and more precisely in the field of environmental crimes.

Environmental law is a complex matter of public and private affairs, and of administrative and criminal law. This interdisciplinary character of environmental law is, however, not reflected in the way environmental law has been codified in China. China chose the Penal Code Model, where most of criminal offences are focused on penal law, which includes offences committed with respect to the environment. There are different codifications for environmental law, which make it necessary to have links between administrative and criminal (environmental) law.

---

[20] Cp. L. Fu, *supra* note 9, 213.
[21] Cp. Art. 128 to 130 Penal Law of 1979.

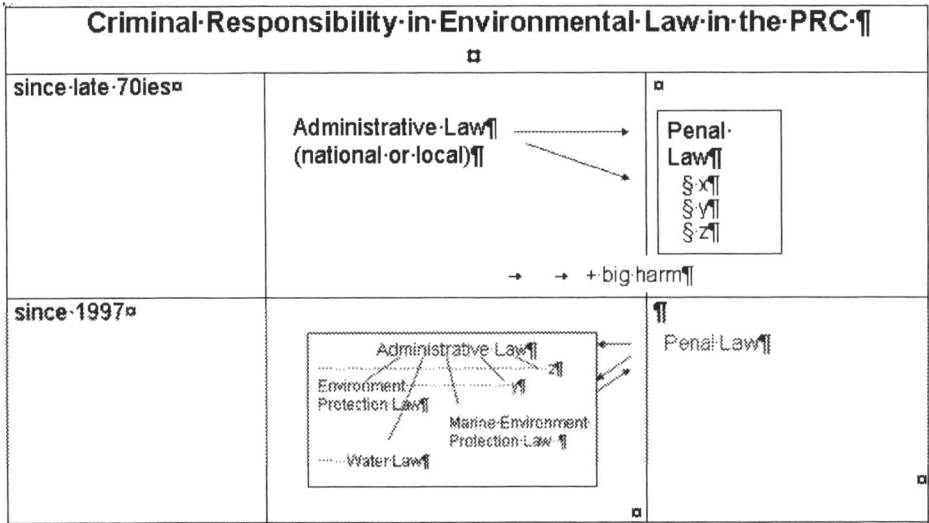

4.2 DEPENDENCE OF CRIMINAL LAW ON ADMINISTRATIVE REGULATION

4.2.1 Function

Making a form of criminal law link with and depend on administrative regulations is a sort of legal method or legislative technique. The advantages of the dependence are that firstly, this method allows there to be a rather effective form of criminal liability for someone who does not follow administrative laws. Secondly, there is no (flagrant) contradiction within the unity of the legal system (法秩序一致性, 'Einheit der Rechtsordnung')[22] and especially in the field of administrative and criminal law. Some say that this unity is a command of the rule of law. Thirdly, the accessory status of criminal law allows for a flexible development of substantial environmental law without being obliged to revise the whole Criminal Code at the same time.

Problems with or disadvantages attached to the accessory status of criminal law lie first of all in the lack of transparency with this kind of regulation. Criminal law itself does not say anything. Rather, the core of the conditions for punishment must be found in other laws, and these are often dispersed in various laws and regulations. Therefore, the Criminal Code no longer seems to be the *Magna Charta* of the offender. Rather, a possible perpetrator has to look into other laws, which are often administrative laws with difficult legal contexts and which sometimes only legal experts are able to understand.

---

[22] This is sometimes referred to in different terms, like 'freedom from contradiction' ('Widerspruchsfreiheit'). Compare M. Baldus, *Die Einheit der Rechtsordnung. Bedeutungen einer juristischen Formel in Rechtstheorie, Zivil- und Staatsrechtswissenschaft des 19. und 20. Jahrhunderts*, (Duncker & Humblot, Berlin, 1995). The decision as to whether certain behaviour is regarded legal or not is an important example in German law of the unity of the legal system. For instance, justifications in criminal law might be drawn civil law or administrative law.

The criminal code no longer establishes the offence, but just seems to be a sort of covering. Another inconvenience of the dependence on administrative regulations lies in the difficulty of steering criminal responsibility. It is often a balancing act for criminal policy to focus only on the serious cases, while still being sure to phrase the criminal acts in such a way that all these serious cases are effectively covered. The legal system is likely to run into difficulties if broad portions of the population are criminalized. On the other hand, if even cases with serious harm for the environmental cannot be prosecuted, although such offences exist, the legal basis has to be studied as well. Finally, the accessory status of criminal law may raise some specific problems concerning *mens rea* and incorrect decisions by State authorities.

4.2.2 Form

In the Chinese Penal Law, different forms of the accessory status of criminal law are presented:

1 It is referred to in a very general way: 'in violation of the State regulations' (违反国家规定). The legal definition of 'violation of State regulations' is given in Art. 96 PL, which specifies it as being 'the violation of the laws enacted or decisions made by the National People's Congress or its Standing Committee and the administrative rules and regulations formulated, the administrative measures adopted and the decisions or orders promulgated by the State Council.' This is a very broad definition. It is a challenge for every lawyer to check whether any regulation has been violated. Even if there are a couple laws that tend to be violated more often in this context, almost every national law might be violated. The notion 'illegally' (违法性) that is also used in the delineation of other environmental crimes[23] is even broader, since it even includes regulations at the local level. Art. 338 PL (Considerable Accident with Environmental Pollution) requires the violation of State regulations.
2 A second way of referring to administrative law is to restrict the general reference to a certain field of regulations, as can be detected in the offence of Trouble with Dangerous Objects, Art. 136 PL. The criminal law here refers to the regulations on the control of explosive, inflammable, radioactive, poisonous or corrosive materials. This link already makes it rather clear which regulations have to be violated.
3 The third manner is the reference to an administrative decision, like a permission of licence given by any state authority. Examples in environmental crimes can be found, but do not concern marine pollution.[24]

---

[23] Art. 341 § 1 PL, art. 345 § 3 PL.
[24] Art. 339 § 2 PL, art. 343 § 1 PL.

Techniques in Criminal Law		Examples (Penal Law 刑法)
General Reference	illegally (非法)	Art. 341 para. 1
	In violation of the regulations of the State (违反国家规定，...)	Art. 338, Art. 339 para. 1
Specified General Reference	In violation of the law or regulations on land administration etc. (违反土地管理法规等) also: in violation of the provisions of the Forestry Law (违反森林法的规定)	Art. 340 Art. 341 para. 2 Art. 342 Art. 344, 345 para. 2
Reference to Administrative Act	without permission of the competent administration department under the State Council (未经国务院有关主管部门可、...)	Art. 339 para. 2 Art. 343 para. 1

4.2.3 Integration in the Structure of Criminal Law

It is not evident how the dependence on administrative regulations can be integrated into the structure of Chinese criminal law. Criminal theory in China still is greatly influenced by the so-called socialist criminal law theory, which recognizes four main positive elements for the constitution of an offence:

- First, an (abstract) object that for the core of environmental crimes in Section 6 would be the State management on environmental protection.[25]
- Secondly, objective requirements ('*actus reus*') which can be found in the Penal Law and other laws.
- Thirdly, a subject which is treated apart from objective requirements and
- Fourthly, *mens rea*. *Mens rea* is divided into 故意 (guyi) and 过失 (guoshi). Guyi includes intention, while guoshi includes negligence. Recklessness falls between the two.

As a fifth requirement, one could add the requirement of a certain seriousness of the act, according to Art. 13 PL; if the act's seriousness is not given, the act would not be regarded as a criminal act, but might be punished in other forms. Defences are known in the exclusion of social harmfulness (排除社会危害性).

Without commenting too much, the textbooks on criminal law all treat the references to administrative regulations as objective requirements.[26] This means that the objective non-violation of administrative regulations is not a defence, but that there is no criminal act at all.

---

[25] See above section 3.2.
[26] For instance M. Zhang, *Xingfa Xue* (2nd edn, Falü Chubanshe, Beijing, 2003), p. 858.

4.3 MENS REA

The Chinese criminal law recognizes the necessity of personal guilt, that is, there is no criminal liability without intention, recklessness or at least negligence. According to Art. 16 PL, strict liability does not exist.

A major question in the context of the interdependency between criminal law and (marine) pollution regulations is whether *mens rea* is related to the administrative regulations. Is it necessary that the perpetrator violates any administrative prohibition with intent or is it sufficient that the administrative regulation is objectively violated without regard to the perpetrator's knowledge and will? The next question is even more complicated: What about those offences where *mens rea* demands negligence or (lower) recklessness (过失)? Is an objective violation of administrative regulations sufficient there? For instance, in the case of a perpetrator of a Considerable Accident with Environmental Pollution (Art. 338 PL), is he to be held criminally responsible if he does not know that his behaviour violates certain regulations on marine oil pollution?

The answer cannot easily be found. It is necessary to take a general approach to criminal law theory. Traditional criminal law theory in China requires that there be (1) harmfulness to society, (2) criminal illegality, that is, the crime has to violate the criminal law (or in other words the behaviour must be defined as an offence in the Penal, Laws) and (3) that this offence is to be criminally punished.[27] Socialist law theory focuses on the concept of social harm (社会危害性) as the crucial point for crimes. The notion of illegality (违法性) is, first of all, strange to the Chinese criminal law theory. However, it has been widely accepted in civil and administrative law. Hence, with the interdependency of criminal and administrative law, the issue of illegality has entered into the criminal law discussion. The discussion concerning the difference between social harm and illegality is far from being concluded.[28] The question whether intention is related to the illegality (or the harm to the society) of the behaviour is significantly discussed in China. In the Penal Law, there is no specific rule about mistake of law (法律错误). In general, the requirements for legal evaluation of behaviour are not given much importance.[29] The leading opinion says that, in principle, knowledge of illegality is not necessary for intent. However, in cases where the illegality of certain behaviour is only addressed in specific laws, knowledge of such prohibitions is necessary to punish an intentional crime. As in the field of environmental crimes, the prohibitions are laid down in specific laws, as the Marine Environment Protection Law. Knowledge of such prohibitions is necessary, at least as far as intentional crimes are concerned.

When a mistake of law is recognized, it is still not clear which legal conclusions might be drawn. Some authors speak about mitigated punishment. Others argue that if the perpetrator does not know about the prohibition, he will be regarded as not having committed an intentional offence. If the criminal law offers a similar offence where negligence or (lower) recklessness (过失) is required, the perpetrator might be punished

---

[27] M. Gao, 'Fanzui gainian', in M. Gao (ed.), *Xin bian Zhongguo xingfa xue* (Zhongguo Renmin Daxue Chubanshe, Beijing, 1998, 3rd print 2000), p. 69.

[28] Cp. M. Liu, *Xingfa zhong cuowu lun* (2nd edn, Zhongguo Jiancha Chubanshe, Beijing, 1999), pp. 201-202.

[29] Cp. X. Li, *Xingzheng xingfa xue daolun* (Falü Chubanshe, Beijing, 2003), p. 357.

for the latter offence. If there is not a similar offence, he will not be punished at all.[30] This result might be considered as coherent with the principle of personal guilt.

The leading opinion is remarkable because it might be of great importance to a certain group of possible offenders who have no knowledge of administrative regulations. For instance, it might apply to international traders, foreigners[31] and even Chinese living abroad, who travel to China or even through it, and in so doing produce pollution that affects the Chinese territory.[32] Whereas all Chinese citizens are generally obliged to learn the laws, foreigners will often have a certain language barrier that keeps them from being well-informed about Chinese legislation.

As far as an offence of negligence or (lower) recklessness (过失) is concerned, there might be the case of a perpetrator who acted with inadvertent negligence. That is, he could not even think about a possible prohibition. It might be possible that the perpetrator would not have known the prohibition, if he was aware of the situation. Should this mental situation influence judgements about the commitment of the offence or, at least, about its punishment? If we follow the above-mentioned opinion, which rejects intentional offences in cases of mistake of law and replaces it with negligent offences,[33] a mistake of law would never play any role in negligent offences.

4.4 FALSE ALLOWANCES

Criminal and administrative law each follow their own rules and at least in some areas, certain contradictions seem to be unavoidable. A number of problems arise when authorities do not apply environmental (administrative) law according to law. In China, false allowances of authorities can be withdrawn. If a permission is withdrawn, the effect will be *ex tunc*, that is, the allowance seems never to have existed, never to have been given. In other words, even after having received an administrative allowance, this person might be punished according to criminal law, if the administrative authority later withdraws the allowance. Therefore, nobody can rely on an administrative permission. The only way to avoid a hardly convincing result is the acceptance of a mistake of law (法律错误), which is not included in the Chinese Penal Law, but which is accepted by academics and courts.

Another issue of possible contradiction between criminal and administrative law would be the case of a formally illegal, but substantially legal act. For instance, one can think of an enterprise putting sewage into a river without a legally required licence. According to administrative law, the licence should have been given to the enterprise. Here, Chinese authors tend to give criminal liability to the offender, because the required administrative allowance has not been given. An exception is to be made for offenders who already applied for such an administrative allowance, but the authority has not yet

---

[30] M. Liu, *supra* note 28, 197.
[31] Cp. G. Sun, *Guoshi fanzui daolun* (Nanjing Daxue Chubanshe, Nanjing, 1991), p. 106.
[32] Art. 6 § 3 PL, art. 2 para. 3 Marine Environment Protection Law, art. 22 UN Convention on Sea Law.
[33] See *supra* note 30.

made a decision. The reason is that, in reality, there is no social harm in such behaviour and it is only due to the authority that an allowance has not yet been given.[34]

5 CONCLUSIONS

5.1 CONCLUSION ON THE GENERAL INTERDEPENDENCIES BETWEEN CRIMINAL LAW AND ENVIRONMENTAL LAW

The issue of environmental protection has to be regulated on a primary, and not necessarily criminal, stage first. Administrative regulations will play a major role. Criminal law should be enacted only as a supplementary means, where there are no better methods. Therefore, the dependence of criminal law on administrative regulation generally seems to be an adequate method. The link between criminal law and oil pollution regulation, for instance in Art. 338 PL, does not seems to be a major obstacle for the efficiency of criminal law. There are other problems, like the high requirements for the result of the criminal behaviour.

A balancing act between the effectiveness of the criminal regulations and justice has to do with *mens rea* in the context of administrative prohibitions. Even if the Chinese population might be forced to learn all environmental regulations, especially foreigners are likely to give rise to some problems with such an obligation. Even more specifically, the question arises whether a mistake of law is possible in the context of offences with (lower) recklessness and especially negligence.[35] A final answer cannot be given in this report. At the moment, there seems to be few authors in China who have analyzed this problem. Many might even say that this is a very theoretical problem with little or no practical relevance. Even if its practical relevance is in short supply, I think studies on this question might be of interest especially for the criminal law theory in China and its further development.

To improve the prevention of environmental pollution through criminal law, one possibility could be to transpose the requirement of a disastrous result into a requirement for merely jeopardizing of the environment (危险犯). In my opinion, Chinese criminal law tends to overstress the result of an action. However, criminal law should focus on the action itself and try to influence it. The result of the action has a lot to do with coincidence.

Another and much broader issue is the approach of environmental crimes. Due to the general criminal law theory in China, the Penal Law wants to safeguard the 'state management on environmental protection.'[36] It is doubtful whether this theoretical approach is convincing for the protection of the environment.

---

[34] C. Yang, Z. Xiang and S. Liu, *supra* note 14, 121.
[35] See above section 4.3.
[36] See above section 3.2.

5.2 CONCLUSION ON THE INTERDEPENDENCIES BETWEEN CRIMINAL LAW AND OIL POLLUTION REGULATION

Especially in the case of tanker accidents, since they usually happen without intent, technical solutions like the introduction of double-hulled tankers shall be made obligatory by administrative regulations. It is evident that international solutions in this field are more promising than national solutions, although national regulations are better than no regulations at all. Besides, the adoption of an international treaty does not help much if the treaty is not implemented and enforced. The PRC has already undertaken a large number of efforts, but the process must be continued. It seems, for instance, that MARPOL 73/78 with Annex I has more detailed regulations than the Marine Environment Protection Law.

In the field of marine oil pollution, criminal law and oil pollution regulation do not really seem be linked together. Whereas the offence of Considerable Accident with Environmental Pollution (Art. 338 PL) is the only one with reference to oil pollution regulation, neither tanker accidents nor land-based activities fall under this article. On the other hand, Misconduct in Office while Supervising and Controlling the Environment might be applicable to civil servants in oil pollution cases, but this is independent from oil pollution regulation. The offence of Trouble with Dangerous Objects refers to administrative regulations in the context of public security and might be able to take oil pollution regulation up into this approach. However, oil pollution was not to the impetus for creating this offence. If Art. 136 PL is applied to cases of tanker accidents or land-based activities, it is worth rethinking about a more specific offence and, systematically, in an environmental context of the Penal Law. Chinese writers have already made such propositions to create an offence of polluting the sea (污染海洋罪).[37] At the moment, there seems to be a dearth of coordination in the two legal fields of legislation. To sum up, Marine Environment Protection Law and Penal Law correspond better as far as civil servants are concerned (art. 94 MEPL, art. 408 PL) than compared to other persons (Art. 91 para. 3 MEPL, art. 338, 136 PL). At least in the field of oil pollution, there is no risk of legally incorrect administrative decisions,[38] as neither Art. 338 nor Art. 136 PL refer to an administrative decision.

Besides measures taken by the legislative organs, some legal areas might be open to modification even without revision of the law. Art. 9 para. 1 of the constitution counts waters and beaches as public property. If 'waters (水流)' include seawater, this is open to interpretation. In my opinion, seawater within the territory of the PRC would fit into the legal system. Why should its pollution not be considered as a (serious) attack on public property? Such an interpretation could be a pragmatic solution in the sense of a better form of protection of natural resources, even if it is part of an approach not aimed at the environment as such, but at the environment as part of public property.

---

[37] L. Fu, *supra* note 9, 307.
[38] See above section 4.4.

List of References

M. Baldus, *Die Einheit der Rechtsordnung. Bedeutungen einer juristischen Formel in Rechtstheorie, Zivil- und Staatsrechtswissenschaft des 19. und 20. Jahrhunderts*, (Duncker & Humblot, Berlin, 1995).

Q. Chen and Z. Zhang, *Xianfa yu xingzheng fa de shengtaihua* (Falü Chubanshe, Beijing, 2001).

S. Chen, 'Dui huanjing fanzui kongzhi duice de jingji fenxi', (2001) 2 *Xingshi Faxue* 69-72.

W. Frisch, *Verwaltungsakzessorietät und Tatbestandsverständnis im Umweltstrafrecht. Zum Verhältnis von Umweltverwaltungsrecht und Strafrecht und zur strafrechtlichen Relevanz behördlicher Genehmigungen* (Müller, Heidelberg, 1993).

L. Fu, *Huanjing xingfa xue* (Zhongguo Fangzheng Chubanshe, Beijing, 2001).

M. Gao, 'Fanzui gainian', in M. Gao (ed.), *Xin bian Zhongguo xingfa xue* (Zhongguo Renmin Daxue Chubanshe, Beijing 1998, 3rd print 2000), pp. 59-71.

Y. Han, 'Fanzui keti', in M. Gao (ed.), *Xingfa xue yuanli* (Zhongguo Renmin Daxue Chubanshe, Beijing 1993, vol. 1, 2nd print 1994), pp. 471-505.

S. Hieber, 'Energiesicherheit in China: Instrumente zur Versorgungssicherung', in *China aktuell* (Institut für Asienkunde, Hamburg 2004) pp. 398-405.

H. Hohmann (ed.), *Basic Documents of International Environmental Law* (Graham & Trotman, London/Dordrecht/Boston, 1992).

H. Hohmann, 'Weltweiter Schutz der Meeresumwelt unter besonderer Berücksichtigung des Schutzes von Nord- und Ostsee (einschließlich des Wattenmeeres)' in Gesellschaft für Umweltrecht (ed.), *Umweltrecht im Wandel*, (Erich Schmidt, Berlin 2001) pp. 99-128.

W. Jiang, 'Lun huanjing fanzui de xingfa kongzhi. Jiangsu Gong'an Zhuanke', (2000) 2 *Xuexiao Xuebao* 22-27.

S.P. Johnson, *The Earth Summit: The United Nations Conference on Environment and Development (UNCED)* (Kluwer Academic, Dordrecht, 1993).

M. Kloepfer, 'Aspekte eines Umweltstaates Deutschland. Eine umweltverfassungsrechtliche Zwischenbilanz' in Gesellschaft für Umweltrecht (ed.), *Umweltrecht im Wandel* (Erich Schmidt, Berlin, 2001), pp. 745-765.

X. Li, *Xingzheng xingfa xue daolun* (Falü Chubanshe, Beijing, 2003).

Y. Li, *Guoji huanjing fa yuanqi* (Guangzhou, 2002).

M. Liu, *Xingfa zhong cuowu lun* (2nd edition, Zhongguo Jiancha Chubanshe, Beijing, 1999).

C. Pi and Z. Ren, *Xingzheng fa jiaocheng* (Wenhua Yishu Chubanshe, Beijing, 1989).

S. Qiao, *Huanjing sunhai yu falü zeren* (Zhongguo Jingji Chubanshe, Beijing, 1999).

T. Richter, *Umweltstrafrecht in der Volksrepublik China* (Edition Iuscrim, Freiburg im Breisgau, 2002).

K. Rogall, 'Umweltschutz durch Strafrecht – eine Bilanz' in Gesellschaft für Umweltrecht (ed.), *Umweltrecht im Wandel* (Erich Schmidt, Berlin, 2001), pp. 795-835.

G. Sun, *Guoshi fanzui daolun* (Nanjing daxue Chubanshe, Nanjing, 1991).

L. Tong, 'Shi xi pohuai huanjing ziyuan baohu zui' (1998) 7 *Xingshi Faxue* 38-43.

F. Vorholz, 'Lunte an der Weltwirtschaft' (2004) 24 *Die Zeit* 21-22.

C. Wang, 'Lun Woguo xian you chengzhi huanjing fanzui lifa de queshi ji qi wanshan', (2002) 1 *Guoji Huanjing Fa Yu Bijiao Huanjing Fa Pinglun* 392-409.

Z. Wang, *Zhongda zeren shigu diaocha yu dingzui liangxing* (Qunzhong Chubanshe, Beijing, 2002).

C. Yang, J.W. Head and S. Liu, 'China's Treatment of Crimes Against the Environment: Using Criminal Sanctions to Fight Environmental Degradation in the PRC', (1994) 2 *Journal of Chinese Law* 145-184.

C. Yang, Z. Xiang and S. Liu, *Weihai huanjing zui de lilun yu shiwu*, (Gaodeng Jiaoyu Chubanshe, Beijing, 1999).

M. Zhang, *Xingfa xue* (2nd edition, Falü Chubanshe, Beijing, 2003).

D. Zhou, C. Shan and S. Zhang (eds), *Xingfa de xiugai yu shiyong* (Renmin Fayuan Chubanshe, Beijing, 1997).

JURISDICTION OF THE PEOPLE'S REPUBLIC OF CHINA OVER POLLUTION FROM VESSELS IN ITS EXCLUSIVE ECONOMIC ZONE

Chen Xuebin

1 INTRODUCTION

Economic globalization and liberalization leads to closer cooperation of world economies. The development of international trade could not operate without transportation, and transportation by vessel from one country to another could not operate without the sea. Conversely, though, marine pollution from vessels will destroy the resources of the sea, and the livelihoods connected to it. More attention should therefore be drawn to the prevention, reduction and control of pollution damage resulting from vessels discharge into the marine environment.

It is obvious that many States are concerned with whether or how they exercise their jurisdiction over the pollution of the marine environment. This is especially so considering the jurisdiction of coastal States on pollution in their exclusive economic zone (henceforth referred to as the EEZ). In this regard, many States have made extensive regulatory efforts to control such situations, through negotiation of international conventions and by the enactment of national laws and regulations. Such enactments are abundant. There are some provisions in the 1982 'United Nations Convention on the Law of the Sea' (henceforth referred to as the UNCLOS or the Convention), which China entered into in 1996, and provisions in the 'Law of the People's Republic of China on Protection of Marine Environment' (henceforth referred to as the Marine Protection Law), the 'Law of the People's Republic of China on the Exclusive Economic Zone and the Continental Shelf' (henceforth referred to as the EEZ Law) and the *'Law of the People's Republic of China on Maritime Special Proceedings'*.

In this paper, the topic of the jurisdiction of the People's Republic of China as to pollution discharged by vessels in the EEZ shall be considered. The paper is divided into the following five parts:

After this introduction, first the legal resources for China to exercise its jurisdiction on discharge pollution from vessels in the EEZ will be examined (2). Some international conventions are introduced, like the UNCLOS, and Chinese laws and regulations, like the Marine Protection Law, the EEZ Law and other laws and regulations.

Second and thirdly, substantive and procedural requirements for China in the exercise of its jurisdiction over discharge pollution from vessels in the EEZ will be examined (3 and 4). Under UNCLOS, there are four substantive requirements and three procedural requirements. The paper will discuss them respectively.

Fourth, the jurisdiction exercised by China in civil proceedings against the polluting vessel responsible for discharge in the EEZ will be examined (5). According to the

Michael G. Faure and James Hu (eds), Prevention and Compensation of Marine Pollution Damage.
Recent Developments in Europe, China and the US, 83-96.
© *2006 Kluwer Law International. Printed in the Netherlands.*

UNCLOS and Chinese laws, China may especially exercise such rights and jurisdiction vis-à-vis civil proceedings on discharge pollution by vessels in its EEZ.

Fifth, the revision of Chinese laws on marine environment protection to comply with the international convention will be discussed (6). The paper provides some comments and suggestions on the issue (7).

2 LEGAL RESOURCES FOR CHINA TO EXERCISE ITS JURISDICTION

This century is prime for the world to advance the development of the marine economy. However, such development requires people to respect the marine environment and to take care of marine protection. However, discharge from vessels introduces substances or heat energy into the marine environment, including estuaries, either intentionally or unintentionally, directly or indirectly. Such an act of pollution would, or may possibly, result in such deleterious effects as harm to living resources and marine life and to the marine environment generally. Pollution like this often happens and thus becomes a serious problem. Due to this, the theme of the 2004 World Environment Day thus became 'Wanted! Sea and Oceans – Dead or Alive'.

Environmental protection is one of China's fundamental policies. As part of this policy, China has enacted over thirty laws and regulations concerning marine environmental protection, including, for example, the Law of the People's Republic of China on the Protection of Marine Environment, promulgated in 1983, and revised in 1999. As a coastal State, China exercises its rights of jurisdiction over vessels in its marine area in accordance with international conventions and its own laws and regulations. Thereby, it orders vessels in Chinese waters to take responsibility for the protection of the marine environment and imposes liability for any occurrent marine pollution.

If vessel pollution occurs in the territorial sea of a coastal State, there is no problem in the State enforcing its jurisdiction on the vessels. Although there is no separate act concerning compensation for vessels' pollution damage, there are particular provisions under Chinese laws and regulations governing principles of protection of the marine environment. Meanwhile, there are some provisions in relevant international conventions, for example, UNCLOS. However, particularly for vessels' pollution in the EEZ, there is some disagreement in respect of those people and states that have rights to exercise jurisdiction and how to do so.[1]

Differently than the territorial sea, the EEZ has its own specific legal regime, the principles of which are the rights and jurisdiction of a coastal State on marine resources and certain other special matters. Under UNCLOS, in the EEZ, the coastal State has:

a   sovereign rights for the purpose of exploring and exploiting, conserving and managing the natural resources, whether living or non-living, of the waters superjacent to the sea-bed and of the sea-bed and its subsoil, and with regard to other activities for the

---

[1] For example, the USA, with a long coastline, has not entered and refuses to enter into the UNCLOS, and there are different views between flag States and coastal States.

economic exploitation and exploration of the zone, like the production of energy from the water, currents and winds;[2]

b   jurisdiction as provided for in the relevant provisions of the Convention with regard to the protection and preservation of the marine environment.[3]

The breadth of the EEZ under the provision of China's EEZ Law[4] is the same as the provision of UNCLOS. Section 5 of the latter states:

a   The exclusive economic zone is an area beyond and adjacent to the territorial sea,[5] and
b   The exclusive economic zone shall not extend beyond 200 nautical miles from the baselines from which the breadth of the territorial sea is measured.[6]

Meanwhile, in the EEZ, there are some rights and freedoms held by other States under UNCLOS. In exercising its rights and performing its duties in the EEZ, the coastal State must have due regard for the rights and duties of other States in that area and act in a manner compatible with the provisions of the Convention.[7]

Besides UNCLOS, there are several other international conventions concerning marine pollution from vessels, eight of which China is, or has been, a party to. In 1973, the 'International Convention for the Prevention of Pollution from Ships' was promulgated, although no provisions are applicable to jurisdictional areas.[8] In 1969, the 'International Convention on Civil Liability for Oil Pollution Damage' (henceforth referred to as 1969 CLC), with specific application to oil pollution in the territory and territorial sea (as opposed to the EEZ) of a contracting State, gave States the right to take necessary measures to prevent or reduce oil pollution.[9] From another angle, the 1969 'International Convention Relating to Intervention on the High Seas in Cases of Oil Pollution Casualties' affirms the right of a coastal State to take measures on the high seas, rather than the EEZ, as may be necessary to prevent, mitigate or eliminate danger to its coastline or related interests from pollution by oil or the threat thereof following a maritime casualty.[10] It should be mentioned that differently to the 1969 CLC Convention, the 1984 and 1992 CLC Protocols extended their application to oil pollution within the EEZ of the coastal State,[11] in addition to oil pollution in its territory and territorial sea. It might be

---

[2]   The United Nations Convention on the Law of the Sea, 10 December 1982, §1, art. 56.
[3]   Ibid.
[4]   The Law of the People's Republic of China on the Exclusive Economic Zone and the Continental Shelf (1998), art. 2.
[5]   The United Nations Convention on the Law of the Sea (concluded in 1982), art. 55, *supra* note 2.
[6]   The United Nations Convention on the Law of the Sea, art. 57, *supra* note 2.
[7]   The United Nations Convention on the Law of the Sea, §2, art. 56, *supra* note 2.
[8]   (CLC), The International Convention for the Prevention of Pollution from Ships, (Marpol 73/8/8) 1973, as amended by the 1978 Protocol, often referred to as MARPOL 73/78.
[9]   The International Convention on Civil Liability for Oil Pollution Damage, 29 November 1969, art. 2. The Convention is often referred to as CLC Convention, which was modified in 1976, 1984, 1992, 2000, and most recently in 2003.
[10] The International Convention Relating to Intervention on the High Seas in Cases of Oil Pollution Casualties, 29 November 1969, art. 1.
[11] CLC, 1984 Protocol and 1992 Protocol of the International Convention on Civil Liability for Oil Pollution Damage, art. 2.

noted that such an extension was created to meet the provisions of UNCLOS, which provides the *minimum* standards for contracting parties in terms of their duties to conform, and give effect to, generally accepted international rules and standards. In order to implement such measures, the coastal State is empowered to enact its own laws and regulations regarding the EEZ in order to prevent, reduce and control pollution resulting from vessels, and, indeed, other sources.

On 15 May 1996, when it approved the UNCLOS,[12] the National People's Congress of the PRC made four points of declaration, one of which states that China has rights and jurisdiction over the EEZ. This is of principal importance, as, besides the preservation of its rights, China has no particular provisions regarding marine pollution resulting from vessels in its EEZ under the EEZ Law, which was promulgated on 26 June 1998. That law simply states that: 'The People's Republic of China shall exercise the rights and jurisdiction on its exclusive economic zone and on its continental shelf, if not provided in this law, in accordance with international laws and other relevant laws and regulations of the People's Republic of China.'[13] A certain amount of detailed explanation is needed to clarify the sense of such a statement. It is fortunate that the Marine Protection Law, as amended in 1999, makes the law apply to China's domestic waters, territorial sea, contiguous zones, the exclusive economic zone, the continental shelf, and other maritime regions under the jurisdiction of the People's Republic of China.[14] Furthermore, the Chinese Maritime Court has jurisdiction to hear any case in relation to vessels pollution in China's EEZ.[15] Therefore, it may be noted that, in respect of jurisdiction over vessels pollution in its EEZ, China now has its own legal resources.

3 SUBSTANTIVE REQUIREMENTS FOR CHINA TO EXERCISE ITS JURISDICTION.

Regardless of all other acts of vessels at sea, discharge by a vessel nevertheless causes marine pollution. Such a discharge refers to the release into the sea of pollutants, including pumping, spilling, releasing, spouting, and pouring,[16] though must be distinguished from an act of dumping by the vessel. However, an act of dumping does not as such include:

i   the disposal of wastes or other matter incidental to, or derived from, the normal operation of vessels, aircraft, platforms or other man-made structures at sea and their

---

[12] Reservation 1, decision by the 19th Session of the 8th National People's Congress of the People's Republic of China on 15 May 1996: China has rights and jurisdiction over the Exclusive Economic Zone up to 200 miles and Continental Shelf in accordance with the provisions of the United Nations Convention on the Law of the Sea.

[13] The Law of People's Republic of China on the Exclusive Economic Zone and the Continental Shelf, art. 13, 26 June 1998.

[14] The Law of the People's Republic of China on the Protection of Marine Environment (effective as of 1 April 2000, Art. 2.

[15] The Law of the People's Republic of China on Maritime Special Proceedings, art. 7.

[16] The Law of the People's Republic of China on the Protection of Marine Environment, §8, art. 95, 1982.

associated equipment, other than wastes or other matter transported by or to vessels, aircraft, platforms or other man-made structures at sea, operating *for the purpose* of disposal of such matter or derived from the treatment of such wastes or other matter on such vessels, aircraft, platforms or structures;

ii placement of matter for a purpose other than the *mere* disposal thereof, provided that such placement is not contrary to the aims of this Convention.[17]

Therefore, this paper focuses solely on acts of discharge and consequential pollution, and not on pollution by dumping, though there are many provisions governing dumping and, further, preventative measures regarding dumping pollution in the marine environment.

As, then, to discharge from a vessel in China's EEZ, China, as a coastal State, has the right to exercise jurisdiction in response. Additionally, UNCLOS contains provisions covering discharge pollution in the territorial sea of a coastal State. In exercise of such jurisdiction in instances of pollution discharge by vessels in the EEZ, though, UNCLOS provides some substantial requirements. There are four aspects to be considered.

Firstly, one should consider what is a vessel according to UNCLOS. UNCLOS requires that the subject responsible for the discharge should be a 'vessel,' no matter how many tonnes it has, and no matter whether it is a vessel flagged under the coastal State itself, or under another State. There is no differentiation between whether the vessel navigates in the EEZ or territorial sea of the coastal State. However, the discharge must occur within the area of the EEZ of the coastal State. The flag flown by a ship shows the country in which the ship is registered, and shows the ownership and nationality of the ship. Nowadays, the ownership or national origin of a ship is not a simple matter. Sometimes, the nationality of the master or operator of a ship may be different from the nationality of the owner of the ship, or of the ship itself, and thus the ship will use a convenient flag in order to facilitate the freedom of international transportation. As an example, in the case of the Prestige, the master and operators of the tanker were Greek, while some owners were from various other European countries. Part of the owners' firms were located in the principality of Liechtenstein, and the crew was drawn from different Third-world-countries, yet the ship was registered to a Liberian company. Finally, even though no one from the crew came from the Bahamas and the vessel had never been in that country, the Prestige nevertheless flew the Bahamian flag.[18]

Secondly, there has to be an act of discharge from the vessel. UNCLOS requires that there be a discharge from the vessel and that the act occurs in the EEZ of the coastal State. The discharge must cause marine pollution in the sea area of the coastal State, such that there is a resultant violation to a provision under the Convention. In the Prestige incident, the vessel flew a Bahamian flag and was laden with 77,000 tonnes of heavy fuel oil, but the disastrous accident occurred off the coast of Galicia, Spain, in November 2002, and spilled there a substantial quantity of its cargo. It should be noted that the discharge may be due to the collision of vessels, mechanical error or other reasons, but is not, normally, an act of dumping. Under UNCLOS, 'dumping' means (i) any deliberate disposal of

---

[17] The United Nations Convention on the Law of the Sea, art. 1, §5(b), *supra* note 2.
[18] See Draft Report of the Temporary Committee on improving safety at sea, 23 February 2004, 2003/2335 (INI), 6 et seq.

wastes or other matter from vessels, aircraft, platforms or other man-made structures at sea or (ii) any deliberate disposal of vessels, aircraft, platforms or other man-made structures at sea.[19] Here the phrase 'deliberate' is key. Normally, dumping is intentionally done by someone on the ship. The act of discharge is different, yet it is still usually hard to identify whether an act is intentional or otherwise. Another relevant factor is that pollutant spills are not always solely, or even partly, oil spills. If the vessel specifically spills oil in the EEZ, such a discharge may be governed by the 1992 CLC. On the other side of this requirement, the discharge causing pollution of the marine environment must violate the convention provisions on the protection of that environment. That is to say, there must be a violation of the applicable international rules and standards for the prevention, reduction and control of pollution from vessels or alternately of any laws and regulations of the relevant coastal State that are meant to conform to and effect those rules and standards.[20]

Thirdly, UNCLOS requires causal injury damage. While not usually describable in detail, generally speaking, the core damage that results from the vessel discharge is the pollution of the sea area of the coastal State, that is, to pollute the marine environment of the coastal State. The act of discharging is the introduction, directly or indirectly, of substances or heat energy into the marine environment, including estuaries, which would be harmful to the coastline of the State, or to certain related interests, or the resources of the territorial sea or EEZ, of the coastal State. In other words, the causal injury resulting from the vessel's discharge is a form of damage 'to living resources and marine life, hazards to human health, hindrance of marine activities, including fishing and other legitimate uses of the sea, impairment of quality for use of sea water and reduction of amenities'.[21] For instance, when the oil tanker 'Exxon Valdez' ran aground in Prince William Gulf in Alaska, USA, in 1989, it leaked a mass of petroleum, leading to losses of eight billion US dollars. However, the damages that were caused to the marine environment of the coastline and the territorial sea area of the USA took many years to fully rectify. The immanent danger, then, of damage to the marine environment in the EEZ from vessel pollution often impacts the stability of the marine environment of coastal States. Thus, the exploitation and utilization of the ocean by the coastal State is limited and the State's interests are consequentially harmed.

Fourthly, the result should be sufficiently serious. That is to say, such an introduction by discharge from vessels has results or is likely to result in such deleterious effects as harm to living resources and marine life, hazards to human health, hindrance to marine activities[22] in the EEZ or other sea areas of the coastal State. Under the Convention, the phrase 'results or is likely to result in such deleterious effects' shows that it covers cases of damage or possibilities of damage from vessel discharge polluting the marine environment in the EEZ or other sea areas of the coastal State. However, damage merely resulting from a vessel discharge will not be sufficient. UNCLOS also requires that the result should be serious. It uses the word 'major' to describe that such a degree of damage is required. It requires that the damage or threat of damage from the pollution discharged

---

[19] The United Nations Convention on the Law of the Sea, §(5)(a), *supra* note 2.
[20] The United Nations Convention on the Law of the Sea, §3, art. 220, *supra* note 2.
[21] The United Nations Convention on the Law of the Sea, art. 1.§1(4), *supra* note 2.
[22] *Ibid.*

by vessels be major, rather than simply average or minimal. That is, the Convention's threshold is that the pollution's *result* should be major damage or that it should constitute a threat thereof. It may be easy to understand when major damage has occurred, but it is not so easy to identify the threat of major damage. The latter's importance, however, can be shown in the example of the Bo Sea in China. This sea has been seriously polluted over the past twenty years, for many reasons, including the discharge of pollutants from vessels into the sea. Thus, some experts now warn that if there are no decisive and effective measures taken by the government immediately to reduce and control marine pollution, then in about ten years, the Bo Sea will gradually become the first 'dead sea' on earth. This situation reflects the fact that sometimes the serious results of pollution may be latent for a long period of time. Therefore, since the result of pollution from vessels' discharge into the sea may not be immediately apparent, there needs to be a period of time delay and, moreover, a certain period of scientific evaluation and assessment. Under the Convention, then, the requirement is satisfied if there is a threat of, and not simply the occurrence of major damage.

4 PROCEDURAL REQUIREMENTS FOR CHINA TO EXERCISE ITS JURISDICTION

When the coastal State exercises its jurisdiction over the discharge of pollution from a vessel in the EEZ, the State must not only satisfy certain substantive requirements of UNCLOS. It also provides procedural requirements for the coastal State in such exercise.[23] According to the Convention, China, as a coastal State, may exercise its jurisdiction providing satisfaction of the following three requirements;

1 There must be clear objective evidence showing the causal relationship between the act of discharge from the vessel and any damage. Such evidence should be objective, as opposed to transient or insubstantial. Thus, it should be clear and obvious that there has emanated from a vessel, navigating in the EEZ or the territorial sea of a coastal State, a discharge into the EEZ, violating a Convention provision and causing major damage or a threat thereof to the coastline or related interests of the coastal State, or to any resources of its territorial sea or EEZ. If so, then the coastal State may require the vessel to give information regarding its identity and port of registry, its last and next port of call and any other relevant information it requires to establish whether a violation has occurred.[24] For the purpose of evidence collection, the State may, if necessary, undertake physical inspection of the vessel for matters relating to the

---

[23] The United Nations Convention on the Law of the Sea, art. 220, §6, Where there is clear objective evidence that a vessel navigating in the exclusive economic zone or the territorial sea of a State has, in the exclusive economic zone, committed a violation referred to in §3, resulting in a discharge causing major damage or threat of major damage to the coastline or related interests of the coastal State, or to any resources of its territorial sea or exclusive economic zone, that State may, subject to section 7, provided that the evidence so warrants, institute proceedings, including detention of the vessel, in accordance with its laws. See *supra* note 2.

[24] The United Nations Convention on the Law of the Sea, §3, art. 220, *supra* note 2.

violation, should the vessel refuse to hand over information or if the information supplied by the vessel is manifestly at variance with the evident factual situation, and if the circumstances of the case justify such inspection.[25]

2   The coastal State may exercise its jurisdiction over pollution from a vessel only when the State has sufficient evidence to warrant such action. Here, UNCLOS does not mention the extent to which the requirement for evidence is set so as to 'warrant' intervention. Therefore, in determining the relevant standard for 'warranting' action, it may be necessary to refer to Chinese laws. In China, there is a strict liability regime applicable to compensation for damage to the marine environment,[26] and there is a reversed burden of proof system. Henceforth, two elements are required in this instance. First, the fact of damage to the marine environment, in violation of Chinese laws, and second, a causal relationship between the act of discharge from the vessel and pollution damage to the marine environment. Under Chinese law, responsibility for determining the environmental pollution is borne by the plaintiff, thus the injured party provides certain evidence as to the damage resulting from the marine pollution, and its requisite extent. Correspondingly, if the injurer wishes to escape liability for his pollution, he must provide evidence to prove that there is no causal link between the act of discharge from the vessel and damage to the marine environment.[27] The State must present evidence of the damage and there is a reversed burden to disprove causality upon the purported injurer. It should be noted that the Convention requires that the evidence of discharge pollution on which the coastal State founds its jurisdiction not only rest on clear and objective bases, but also that that evidence duly warrant exercise of that jurisdiction. In the author's opinion, the Chinese legal system is in accordance with such a requirement.

3   If all the above conditions are satisfied, then the coastal State may exercise its jurisdiction over the pollution discharged by vessels in its EEZ, and institute certain proceedings, including detention of the vessel, in accordance with its laws. Such 'proceedings' cover civil as well as administrative proceedings, but exclude the institution of criminal proceedings. Before ratification of UNCLOS in 1994, the rules of the States concerning criminal jurisdiction over marine pollution were rather piecemeal. Some States limited their jurisdiction to their territorial sea, while others limited it to the continental shelf or to the EEZ. Nevertheless, UNCLOS empowers the coastal State to exercise its jurisdiction over pollution in its EEZ, subject to certain restrictions. Initially, it should be noted that the Convention gives the flag State much broader rights and jurisdiction. Under the UNCLOS there are at least two kinds of restriction: first, the coastal State has no rights or jurisdiction for criminal cases due to discharge of pollution in its EEZ. If there is serious marine pollution caused by vessel discharge in the EEZ, the Chinese Procurator therefore has no right or jurisdiction in China to institute criminal proceedings against the owner or captain of the vessel or to claim criminal liability against him. Second, only monetary penalties may be imposed

---

[25] The United Nations Convention on the Law of the Sea, §5, art. 220, *supra* note 2.
[26] The General Rules of Civil Law of the People's Republic of China, art. 124.
[27] The Provisions of Supreme Court of the People's Republic of China on Evidence in Civil Proceedings, §1.3, art. 4.

with respect to violations by vessels flying a foreign flag.[28] In the case of the Prestige, the question of whether there was a sufficient link between the State (of the Bahamas) and the ship[29] would, for legal purposes, be answered in the negative, because it was a flag of convenience. In this regard, there is no flag State to undertake criminal proceedings, since the Bahamas was just used for the convenience of the ship. Which manner of proceeding will then be taken? Under the Convention, the only available sanctions are monetary sanctions, but such sanctions are often insufficient, and may fail to punish the illegal injurers appropriately. Furthermore, such monetary sanctions are insurable, raising a moral hazard issue. One exception to these provisions is when discharge causes pollution in the territorial sea, when the discharge is wilful and when the pollution is serious.[30] This goes some way towards addressing the balance between the different interests under the Convention. Moreover, in the conduct of proceedings in respect of violations committed by a foreign vessel which may result in the imposition of penalties, there are certain recognized rights of the accused that are to be observed.[31]

5 JURISDICTION EXERCISED BY CHINA IN CIVIL PROCEEDINGS

The mode in which China exercises its jurisdiction for civil proceedings against vessels causing injury in its EEZ is very important. It should be noted that institution of civil proceedings for losses and damage are fundamental. Under section 7 of the UNCLOS, some safeguards are applicable, and whatever they are, nothing in the Convention would affect the institution of civil proceedings in respect of a claim for any loss or damage resulting from the pollution of the marine environment.[32] Such civil proceedings in China include judicial authority to hear such cases, the bringing of suits, jurisdiction on specific cases, and the trial of cases, as well as the necessary enforcements.

In cases relating to marine pollution by discharge from vessels in the EEZ, injured legal or natural persons have the right to be the named plaintiff in a suit against such vessels. However, whether the State or Chinese government also has rights to bring a suit is sometimes at issue. Normally, the State is an entity, and the state government is competent to represent the State. The problem arises insofar as the State government has certain public rights as well as being responsible for the interests of the State in dealing with political and foreign affairs, and to conclude international deals and conventions. However, it is not a 'person' who has the civil right to bring a civil suit. Under the continental legal theory of the 'Direct[ly] Interested Party', the qualified litigant parties in civil proceedings should have direct rights, interests or obligations with respect to the specific case, and only a person who has substantial rights or interests in this respect is entitled to bring a civil action. Under Chinese law, the plaintiff in civil actions is either a

---

[28] The United Nations Convention on the Law of the Sea, §1, art. 230, *supra* note 2.
[29] The United Nations Convention on the Law of the Sea, §1, art. 91, *supra* note 2.
[30] The United Nations Convention on the Law of the Sea, §2, art. 230, *supra* note 2.
[31] The United Nations Convention on the Law of the Sea, §3, art. 230, *supra* note 2.
[32] The United Nations Convention on the Law of the Sea, art. 229, *supra* note 2.

citizen, a legal person or another organization who has direct interests in the case.[33] Based on such a traditional theory, the State itself never suffers any direct losses, and so in fact there have to be different procedures in practice. In some cases, the State government can be a plaintiff that is competent to raise an action, though sometimes the court will not accept the case. Notwithstanding those considered above, this issue is worth further discussion. In the author's opinion, the 'Theory of Action for Public Interests' should be adopted. Under this theory, the scope of parties who qualify as litigants has been expanded. They not only include persons who have direct interests in the civil disputes, but also cover persons without direct *personal* stakes, who may bring an action on behalf of other persons' civil rights and interests. The latter could, then, be public organizations, or indeed government offices and organizations. In other words, the Chinese State government would have the right to sue injurers in the courts, since in China, the Chinese sea areas belong to the People's Republic of China, and thus the State Council may exercise ownership and incumbent rights over the sea areas on behalf of the State.[34] Such sea areas cover the territorial sea, the continental shelf, the EEZ, and other sea areas. Should this be adopted, then in the case of marine pollution, the 'person' with the right to bring the case to court may be the competent administrative department in charge of environmental protection, or marine affairs, or fisheries, and so on. One relevant legal resource is the Civil Proceedings Law and General Principles of Civil Law, under which the scope of a legal person does in fact include a 'Legal Person of Authorities', i.e. government authorities, or departments. Moreover, under the Law of the People's Republic of China on Maritime Special Proceedings, torts arising from pollution to marine environment fall under the jurisdiction of the maritime court.[35]

In China, the main civil liability penalties are the following: an order to cease infringements, removal of obstacles, elimination of dangers, return of property, restoration of the original condition, repair, reworking or replacement, compensation for losses, breach of contract damages, elimination of ill effects and rehabilitation of reputation and extension of an apology.[36] The above forms of penalties for civil liability may be applied either exclusively or concurrently.[37] Compensation for damage to the marine environment often possesses its own characteristics, however, and it is vital for the coastal State to take measures to avoid any enlargement of the pollution, including enabling cleanup of the pollution. Essentially, those who cause pollution damage to the marine environment *should* pay the clean-up charges and costs, compensate the State for losses, and compensate injured parties or individuals for their costs and losses.

In civil proceedings, on the other hand, possible actions are property preservation measures, evidence preservation, and of course, very importantly, detention of the injuring vessel. That said, whenever appropriate procedures are established, either by the competent international organization or as otherwise agreed, whereby compliance with

---

[33] The Law of the People's Republic of China on Civil Proceedings, art. 108.
[34] The Administrative Provisions on Measures to Use Chinese Sea Areas, art. 3.
[35] The Law of the People's Republic of China on Maritime Special Proceedings, art 7, *supra* note 15.
[36] The General Principles of the Civil Law of the People's Republic of China, art. 134.
[37] *Ibid.*

requirements for bonding or other appropriate financial security is assured, the coastal State, if bound by such procedures, shall allow the vessel to proceed.[38]

As for oil pollution by vessels in the EEZ, the 1992 CLC, as effective in China from 5 January 2000, has set up a unified compensation system. The 1992 CLC has definite provisions regarding determination of the injurer's liability and also on proper compensation(s) for the injured. It was unfortunate that China did not join the 1971 International Convention on the Establishment of an International Fund for Compensation for Oil Pollution Damage (referred to as Fund Convention), which has entered into force in Hong Kong only. A consequence of this is that if there is oil pollution damage caused by discharge from a vessel in the EEZ, the injured party, in China, can only get compensation under limited liability levels in accordance with the 1992 CLC. Any other or further losses are not compensated under the Fund Convention, which may provide a solution to such limitations.

6 REVISION OF MARINE ENVIRONMENT PROTECTION LAWS

It is not surprising that people have called the UNCLOS 'the Constitution of the Sea,' since it has detailed stipulations on almost every aspect of sea and marine activities, and provides for the rights and liabilities of contracting parties on the sea, as well as those of international organizations. The Convention defines the territorial sea and EEZ of contracting coastal States, while protecting freedom of vessel navigation and aviation. In preventing or resolving some disputes arising from pollution of the seas, the Convention takes a major role in the protection of world safety and in the promotion of economic development. That said, it should be made clear that there is still much remaining work to do, and that contracting States should make great national efforts to protect fishery resources and the marine environment.[39]

Indeed, the UNCLOS represents the first time in history that States have been *obligated* to protect and preserve the marine environment.[40] Further, in Articles 194 and 195, it requires States to take all necessary measures to prevent, reduce and control marine pollution resulting from any source. In conforming to Articles 198 and 199 of the Convention, China's Marine Protection Law provides its contingency plans for responding to pollution incidents in the marine environment.[41]

Since contracting countries bear their duties under an international convention, in general they are responsible for making national laws that comply with the provisions of that international convention. Therefore, it will be necessary for China to revise its Marine Protection Law and EEZ Law in order to make them consistent with the requirements of the UNCLOS, even though the parts on conflicts of law in the Convention are, in fact, in reference to flag States. However, relevant clauses of the Convention may be used as

---

[38] The United Nations Convention on the Law of the Sea, §7, art. 220, *supra* note 2.
[39] Speech by Kofi Annan, General Secretary of United Nations, at the 20th Anniversary Meeting for the United Nations Convention on the Law of the Sea, <http://www.un.org>.
[40] The United Nations Convention on the Law of the Sea, art. 192.
[41] The Law of the People's Republic of China on the Protection of Marine Environment, art. 18, *supra* note 16.

models for such revision. As a coastal State, China has rights as well as obligations for the prevention, reduction and effecting of measures to control marine pollution from vessels. They include, but are not limited to, the following:

1. China should take, individually, or jointly with other States as may be appropriate after notifying other States it deems likely to be affected by any damage (as well as any competent international organizations), all measures that are necessary to prevent, reduce and control pollution of the marine environment from vessels. It must use, for this purpose, the best and most practicable means at its disposal and in accordance with its capabilities. Finally, China shall endeavour to facilitate harmonization of its policies regarding such duties.
2. While taking all measures necessary to eliminate the effects of pollution and prevent or minimize the damage from such pollution, China should ensure that pollution arising from incidents or activities under its jurisdiction does not spread beyond the areas where China exercises sovereign rights in accordance with the Convention and Chinese laws.
3. The measures taken pursuant to the Convention and Chinese laws should deal with all sources of pollution of the marine environment. These measures include, *inter alia*, those designed to minimize to the fullest possible extent pollution from vessels, in particular measures for preventing accidents and dealing with emergencies, ensuring the safety of operations at sea, preventing intentional and unintentional discharges, and regulating the design, construction, equipment, operation and manning of vessels.
4. In taking measures to prevent, reduce or control pollution of the marine environment, China must refrain from unjustifiable interference with the activities of other States in the exercise of their rights and in pursuance of their duties in conformity with the Convention.
5. Inclusion, also, of measures taken in accordance with the Convention and Chinese laws necessary for the protection and preservation of rare or fragile ecosystems as well as the habitat of depleted, threatened or endangered species and other forms of marine life.[42]

In taking measures to prevent, reduce and control pollution of the marine environment, China must act so as not to transfer, either directly or indirectly, damage or hazards from one area to another or transform one type of pollution into another.[43]

China is a coastal State, but provisions of its EEZ Law and its Marine Protection Law are too abstract and broad to be applied in practice. When there is no specific relevant provision under Chinese laws, like when, for example, there is no national definition of terms like 'threat of major damage' or 'be likely to result in such deleterious effects'

---

[42] The United Nations Convention on the Law of Sea, art. 194 could be a reference here, referring, as it does, to 'Measures to prevent, reduce and control pollution of the marine environment', *supra* note 2.

[43] The United Nations Convention on the Law of the Sea, art. 195: duty not to transfer damage or hazards or transform one type of pollution into another, *supra* note 2.

under Chinese laws,[44] then the interpretation of such terms falls back on the international convention. If, however, the convention has no such provision either, there is a question of how to deal with the problem. It will result in an unguided application of law, something especially prevalent in non-contracting States, like the US. This problem arises despite the fact that UNCLOS requires coastal States to promulgate their own detailed laws and regulations for compliance with the provisions of the Convention. Thus, China should reconsider the protection of the marine environment, and do further research on the prevention and control of marine pollution in particular. Furthermore, China should, in respect of their exclusive economic zones, adopt and revise the laws and regulations for the prevention, reduction and control of pollution from vessels so that they conform to and effect the generally accepted international rules and standards established through both competent international organizations or general diplomatic conferences.[45]

7 CONCLUSIONS

It is obvious that clearly defined rights and jurisdiction over pollution of the marine environment is vital to many States, especially in the territorial sea area or the exclusive economic zone of the coastal State. Many States have made distinct efforts to ensure such stability through the conclusion of international conventions and the enactment of their national laws and regulations. The United Nations Convention on the Law of the Sea, concluded in 1982, is a model for this process, though it should be noted that it is unfortunate that there are still some shortcomings in the Convention. UNCLOS provides both substantive and procedural requirements for the coastal State in the exercise of its rights and jurisdiction over pollution from discharges of vessels in their EEZ, while also stipulating certain limitation provisions on the jurisdiction of the coastal State. Thus, a balancing of voices and interests of different States is effected.

There can be no doubt that China has rights and jurisdiction in its EEZ in accordance with the Convention and its national laws. Although there are some provisions that impose limitations on the coastal State, China still can exercise its jurisdiction on the discharge of pollution from vessels in its EEZ, for instance, by instituting civil proceedings against the injuring vessel, including, potentially, retention of the vessel. However, China should revise its Law on the Exclusive Economic Zone and the Continental Shelf and its Law of the People's Republic of China on Protection of Marine Environment, as well as other relevant laws and regulations, to ensure that they comply with UNCLOS and other international conventions so that the legal system is improved and perfected, especially in relation to laws for the prevention or reduce of and control over marine pollution discharged by vessels in the EEZ.

---

[44] In the Marine Environment Protection Law of the People's Republic of China, pollution of the marine environment does not cover the instance of activities 'likely' result in damage to marine environment or to a 'threat of major damage' to the marine environment, *supra* note 16.

[45] The United Nations Convention on the Law of the Sea, §5, art. 211, *supra* note 2.

LIST OF REFERENCES

K. Annan, General Secretary of United Nations, at the 20th Anniversary Meeting for the United Nations Convention on the Law of the Sea, <http://www.un.org>.

PART II: COMPENSATION OF MARINE POLLUTION DAMAGE: CASE STUDIES

INDONESIA'S LAWS AND REGULATIONS CONCERNING POLLUTION OF THE SEA BY OIL: CASE STUDIES ON COMPENSATION FOR OIL POLLUTION DAMAGES

Etty R. Agoes

1 INTRODUCTION

Indonesia is the largest archipelagic State in the world, sprawling over 17,508 islands occupying a land area of 1,826,440 square kilometres, and with a coastline of over 81,290 kilometres in length. It has a total national territory of about 7.8 million sq. km, out of which more than five million sq. km. are sea area (including the exclusive economic zone). The total sea area can be divided into 0.3 million sq. km. of territorial sea, 2,905,743 sq. km. archipelagic waters, and 2,707,092 sq. km. of the exclusive economic zone.[1] With a population of over two hundred million scattered across some six thousand inhabited islands, Indonesia is divided into about thirty-two provinces, three special regions (Aceh, Yogyakarta and Papua) and one special capital city district (Jakarta). The provinces are divided into some three hundred municipalities (districts and cities), most of which are located along the coastal areas. The main islands are Java, Sumatra, Sulawesi, Kalimantan (the Indonesian portion of Borneo island) and West Irian (now known as the Indonesian Papua). The economy is dominated largely by agriculture, industry and mining, with the main exports being petroleum and natural gas, textiles, minerals, manufactured goods and timber. The largest cities are the national capital, Jakarta, the East Java capital city of Surabaya, the West Java capital of Bandung and the northern Sumatran capital city of Medan.

For some years, Indonesia has been experiencing severe environmental problems resulting from uncontrolled deforestation and burning of forests, degradation of its land and marine resources, air and water pollution from industrial wastes and contamination from the mining industry, and pollution of the marine environment. The importance of natural resources to Indonesian society is substantial. Approximately forty million people depend directly on the forest, marine, coastal and agricultural ecosystems for their livelihood. In their daily lives, Indonesian communities use more than six thousand species of plants and animals. Timber taken from naturally regenerated forests comprises more than 10% of Indonesia's annual non-oil export earnings.

Indonesian biodiversity is among the richest in the world. There are at least forty-two distinct natural terrestrial ecosystems and five marine ecosystems. These various ecosystems harbour a wide diversity of different species. While Indonesia occupies only 1.3% of the global land area, it is home to a full 17% of total world species. Due primarily

---

[1] Based on information given by the Naval Hydro-Oceanographic Office (DISHIDROS).

to globally significant coral reef ecosystems, aquatic biodiversity is particularly striking. Indonesian fish represent 37% of the world species total. Given the isolated and highly specialized nature of many Indonesian species, endemism is also high.[2]

As in most developing countries, rapid industrialization and the opening up of the economy have resulted in unsustainable developmental practices in many instances. Furthermore, the devastating effects of the economic turmoil and social unrest in Indonesia have led to worries that environmental protection would be relegated to a concern of secondary importance. In addition, the Indonesian territory consists of large marine areas that for quite a long time have been used as international routes for navigation, including routes for oil transportation.

The Straits of Malacca and Singapore, for example, have been the busiest sea-lane in the world for quite a long time. More than fifty thousand vessels per year transit the 621-mile long Straits. Linking the Indian and Pacific Oceans, the Straits of Malacca and Singapore are the shortest sea route between three of the world's most populous countries – India, China, and Indonesia – and is therefore considered to be the key bottleneck in Asia. Narrow channels, shallow reefs, thousands of tiny get-away islands, and slow traffic, with some 900 commercial vessels passing through each day, make the waters around Singapore, Malaysia and neighbouring Indonesia prone to maritime disasters.

The narrowest point of this shipping lane is the Phillips Channel in the Singapore Strait, which, at its narrowest point, is only 1.5 miles wide. This creates a natural bottleneck, with the potential for a collision, grounding, or oil spill. Some four hundred shipping lines and seven hundred ports worldwide rely on the Malacca and Singapore straits to get to the port of Singapore. For example, 80% of Japan's oil comes from the Middle East via the Malacca Straits. To skirt the straits would force a ship to travel an extra 994 miles from the Gulf. In such an instance, all excess capacity of the world fleet might be absorbed, with the effect being strongest for crude oil shipments and dry bulk like coal. The closure of the Strait of Malacca would immediately raise freight rates worldwide. With Chinese oil imports from the Middle East increasing steadily, the Straits of Malacca and Singapore are likely to grow in strategic importance in the coming years.

2 INSTITUTIONAL ARRANGEMENTS

The institution responsible for environmental management and coordination is the Office of the State Minister for the Environment (sometimes known as the State Ministry for the Environment). This office, being non-departmental (i.e. not possessing any office in the provinces), has substantially less mandates than a full-fledged Ministry. The Office plays more of a coordinating role from Jakarta, and has no actual enforcement competence in the provinces. The areas of responsibility of The Office of the State Minister of Environment include, among other things: formulating the Government of Indonesia's

---

[2] Indonesia, Country Profile-Implementation of Agenda 21: Review of Progress Made Since the United Nations Conference on Environment and Development, 1992, Information Provided by the Government of Indonesia to the United Nations Commission on Sustainable Development Fifth Session, 7-25 April 1997, New York.

environmental policies; planning national implementation programmes; coordinating all environmental activities carried out by government institutions, and enhancing the people's participation in environmental programmes and activities.

In 1990, the Indonesian government established the Environmental Impact Management Agency (*Badan Pengendalian Dampak Lingkungan/BAPEDAL*) through Presidential Decree No. 23/1990. This agency reports directly to the President, and is headed by the State Minister for the Environment himself. Its main task is to exercise functions and activities in environmental impact control. Its strategies include developing environmental compliance, strengthening of institutional capacity and strengthening relationships within the community. To fully understand the task of environmental management in Indonesia, the jurisdiction and activities of both the Office and *BAPEDAL* must be appreciated.

3 NON-GOVERNMENTAL ORGANIZATIONS

The environmental NGO movement in Indonesia is significant. In 1980, the then State Minister for the Environment, Emil Salim, promoted the creation of WALHI, a forum for environmental NGOs in Indonesia. WALHI was created out of the first national meeting of seventy nine environmental organizations. These organizations subsequently became participants in the WALHI network. There are now over three hundred and thirty environmental groups affiliated to WALHI. In total, there are over six hundred NGOs working on environmental issues throughout Indonesia.

In 1988, WALHI became the first NGO to obtain legal standing to sue upon an environmental issue, in the case of *WALHI v. Indorayon*. Following this, WALHI obtained standing as a community group in two other suits: *WALHI et al v. Mojokerto District Prosecutor* and *WALHI et al v. President of Indonesia*. These three cases were the genesis of the development of public interest environmental litigation in Indonesia.

4 ENVIRONMENTAL LEGISLATION IN INDONESIA

There are several types of national legislation in Indonesia which are ranked in order as follows:

1. The 1945 Constitution (which has recently been amended);
2. Decisions passed by the People's Consultative Assembly (recently there has been a draft law which will eliminate regulations issued by this organ, with certain exceptions, from this order);
3. 'Law of Parliament' or 'Law' passed by the legislature;
4. 'Government Regulation' issued by the government;
5. 'Presidential Decisions' (formerly in the form of decrees or instructions) issued by the President through the State's Secretariat;
6. Implementing regulations like Ministerial Regulations and Decisions.

At the local and regional level, there are regulations issued at various levels of local/municipal government which are usually classified as Local/Municipal Regulations (*Peraturan Daerah*) issued either by the governor, regents or mayors. Usually, these local regulations serve as enabling instruments to the national laws and regulations, guidelines for which implementation are usually given at the ministerial level.

For some time, the framework for environmental legislation in Indonesia was the Environmental Management Law No. 4 of 1982. The enactment of this Law and the establishment of the Office of the State Minister for the Environment followed the emergence of environmental interest and activity generated by the 1972 UN Stockholm Conference on the Environment. On 19 September 1997, a new Law, No. 23 of 1997, concerning the Management of the Living Environment was passed to replace Law No. 4 of 1982.[3] Noting that this Law requires the establishment of other implementing instruments, this Law must today be read in conjunction with the numerous implementing regulations that had been passed under the former law.

4.1 THE 1982 UNCLOS (AS RATIFIED BY LAW NO. 17 OF 1985) AND THE MARINE ENVIRONMENT

In the history of marine law, regulations for the protection of the marine environment date back to the 1958 Geneva Convention on the High Seas, wherein Article 24 of the said Convention states that:

> 'Every State shall draw up regulations to prevent pollution of the seas by the discharge of oil from ships or pipelines or resulting from the exploitation and exploration of the seabed and its subsoil, taking account of existing treaty provisions on the subject'.

Through this article, the obligation to prevent pollution of the seas was put on to coastal States as well as the flag States.

Some of the primary components of the 1982 Convention dealing with the protection and preservation of the marine environment lie within the provisions of Part XII of the 1982 Convention, in Articles 192 to 237. Other rules and regulations dealing with the subject matter can also be found throughout the Convention, like in the provisions concerning territorial sea and the contiguous zone,[4] straits used for international navigation,[5] archipelagic States,[6] the exclusive economic zone,[7] high seas,[8] enclosed and semi-enclosed seas,[9] the Area,[10] and settlement of disputes.[11]

Part XII also consists of a section on safeguards with regard to additional enforcement regulations against foreign vessels, specifically: measures to facilitate proceedings; the

---

[3] Hereinafter referred to as the '1997 EMA'.
[4] Unclos, United Nations Convention of the Law of the Sea, 10 December 1982, arts. 21, 22 and 23.
[5] Arts. 42 and 43, *supra* note 4.
[6] Arts. 42 and 43 *mutatis mutandis, supra* note 4.
[7] Arts. 61-67, *supra* note 4.
[8] Arts. 117-120, *supra* note 4.
[9] Art. 123, *supra* note 4.
[10] Arts. 145 and 150, *supra* note 4.
[11] Arts. 290 and 297, *supra* note 4.

exercise of enforcement powers and the duty to avoid adverse effects of such powers; procedures for investigation, bonding and other financial security, release of vessels;[12] establishment of monetary penalties; notification to flag States and other states concerned; and recourse through national courts for damage or loss for unlawful enforcement measures exceeding those reasonably required, attributable to States.[13]

The 1982 Convention has introduced a new global legal regime that provides a solid legislative basis for governing international relationships in ocean management, and particularly of the oceans and coastal areas. In coastal areas and the exclusive economic zone, States have certain rights and responsibilities within national jurisdiction.

Under Article 237, specific obligations assumed by States under special conventions and agreements concluded previously with respect to the protection and preservation of the marine environment, are to be carried out in a manner consistent with the general principles and objectives of this Convention. In December 1985 Indonesia ratified the 1982 UNCLOS through the enactment of Law No. 17 of 1985. Under the Indonesia system the 1982 UNCLOS is, then, binding. However, in some cases, the entry into force of such international instruments needs further implementing regulations to be established.

4.2 IMO CONVENTIONS, AS RATIFIED BY PRESIDENTIAL DECREES OF 1978, 1986 AND 1999

It should be noted here that with regard to regulations concerning pollution of the sea by oil, the International Maritime Organization (IMO) has played an important role by establishing various agreements in the area. The IMO is thereby a leading institution with regard to the establishment of various regulations dealing with marine pollution, and especially pollution of the sea by oil. Some of these conventions are:

1. International Convention for the Prevention of Pollution of the Sea by Oil (OILPOL), 1954, amended and revised in 1962, 1969 and 1971;
2. International Convention Relating to Intervention on the High Seas in Cases of Oil Pollution Casualties (INTERVENTION), 1969;
3. Convention on the Prevention of Marine Pollution by Dumping of Wastes and Other Matter (LDC), 1972 and its Protocol of 1996;
4. International Convention for the Prevention of Pollution from Ships, 1973, as modified by the Protocols of 1978 and 1997 (MARPOL 73/78); and
5. International Convention on Oil Pollution Preparedness, Response and Co-operation (OPRC), 1990.

As a direct result of the Torrey Canyon disaster, the IMO established additional conventions regarding liability, and within the shipping and oil industries there emerged arrangements for compensation schemes, such as:

---

[12] Arts. 223-226, *supra* note 4.
[13] Arts. 230-232, *supra* note 4.

1 Convention on the Civil Liability for Oil Pollution Damage (CLC) 1969, as amended by Protocols of 1976 and 1992;
2 Convention on the Establishment of an International Fund for Compensation for Oil Pollution Damage (FUND), 1971 as amended by the Protocols of 1976, 1992 and 2003;
3 Convention on Limitation of Liability for Maritime Claims (LLMC), 1976;

In addition, tanker owners also took on initiatives to give additional compensation coverage through:

1 Tanker Owners' Voluntary Agreement concerning Liability for Oil Pollution (TOVALOP); and
2 Contract regarding an Interim Supplement for Tanker Liability for Oil Pollution (CRISTAL), 1971.

At present Indonesia is a party to MARPOL 73/78 (Annex I), CLC Convention 1969, CLC Protocol 1992, and Fund Convention 1971. Indonesia's ratification is carried out through the establishment of Presidential Decrees. Similar to other ratifications to international instruments, the provisions of these IMO Conventions became binding. However, as was mentioned, their implementation usually also needs implementing regulations.[14]

To understand further Indonesia's environmental regulations relating to oil pollution, and in particular pollution of the sea by oil, the above regulations must also be read in conjunction with other national laws and regulations established through various sectors and based on the international instruments, like those issued by the Departments of Communication (particularly the Directorate General for Sea Communication), and the Department of Energy and Mineral Resources (formerly the Department of Mining). The recently established Department of Marine Affairs and Fisheries has not yet shown any significant contribution to the issuance of laws or regulations relating to this matter.

4.3 RELEVANT NATIONAL LAWS AND REGULATIONS

4.3.1 Law No. 6 of 1996 on Indonesian Territorial Waters

As part of the country's effort towards the implementation of the 1982 UN Convention on the Law of the Sea (UNCLOS), the government has enacted Law No. 6 of 1996, on the Indonesian Territorial Waters.[15] The Law revised Law No. 4 of 1960, using principles embodied in the 1982 Law of the Sea Convention. Basically, the new Law upholds some old principles like that on the 12 nautical mile breadth of the territorial sea, while incorporating new ones derived from UNCLOS.

---

[14] Presidential Decrees no. 18 and 19 of 1978, Presidential Decree no. 46 of 1986, and Presidential Decree no. 52 of 1999.
[15] *State Gazette* no. 73 of 1996, *Additional State Gazette* no. 3647 of 1996.

With regard, then, to the protection and preservation of the environment, Article 23 of this Law stipulates that it shall be conducted in conformity with the national laws and regulations in force, as well as international law. When circumstances require, the Government may, through a Presidential Decree, establish a coordinating institution or - council, to enhance protection and preservation of the environment in the territorial waters of Indonesia

Several Governmental Regulations have followed, like for example Government Regulations No. 36 and 37 of 2002 that regulate the rights and obligations of foreign vessels exercising the rights of innocent passage and archipelagic sea-lanes passage, and Government Regulation No. 38 of 2002 depicting the base points of the Indonesian baselines.

4.3.2 Law No. 23 of 1997 concerning the Management of the Living Environment

The basis, objective and target of the 1997 EMA is laid out in Article 3. It is namely that of 'environmental management consistent with national responsibility and sustainable development', and 'exploitation within the framework of the holistic development of the Indonesian individual and community in its entirety'. Some significant features concerning the rights and obligations within the 1997 EMA are as follows:

- everybody is to have the right to enjoy a clean and healthy environment;[16]
- everybody is also to have the right to participate in the environmental decision-making process;[17]
- everybody is subject to the obligation to prepare an environmental impact assessment (EIA) for activities that have a significant and important impact on the environment;[18]
- licensing institutions or agencies are to be put under an obligation to incorporate in the relevant licenses environmental conditions for activities which have a significant and important impact.[19]

1997 EMA also provides for provisions that recognize:

- the voluntary choice to use environmental mediation and arbitration;
- the application of the principle of strict liability (liability without fault) for certain activities;
- that NGOs have legal standing in environmental matters;
- procedures for class actions;
- the appointment of Special Environmental Investigators;
- corporate criminal liability that imposes criminal sanctions against corporations (as legal entity), corporate officer(s) and instruction giver(s).

---

[16] Art. 5 paragraph (1).
[17] Art. 5 paragraph (3).
[18] Art. 15 paragraph (1).
[19] Arts. 18 and 19.

Other significant features of the 1997 EMA appear in Chapter VII: first, strict liability is prescribed for violations involving hazardous and toxic materials which cause significant environment impacts. Second, following recent decisions in the courts, community and environmental organizations are explicitly given standing to bring class actions in court or to report environmental violations.

By contrast, with regard to the issue of compensation, the 1997 EMA does not adequately address the issue in either the public or private spheres. It is worth noting, though, that some provisions can be regarded as relevant to compensation. Some of these are:

- Articles 30-33 regarding settlement of disputes out of court;
- Article 34 on compensation obtained through court's settlement;
- Article 35 with regard to compensation based on strict liability;
- Articles 37-38 on the procedure for class action; and
- Article 38 paragraph (1) on the legal standing of NGO's

Other interesting features of the 1997 EMA are as follows:

1 Chapter VIII – on the potentiality for investigations to be carried out by National Police Investigators and relevant Civil Investigation Officials, in conformity with provisions under the Criminal Procedures Law.
2 Penalties for environmental violations are regulated in Chapter IX:
   a. Intentional offences – maximum imprisonment of 10 years and a fine of 500 million rupiah; for death or serious injury, maximum imprisonment of 15 years and fine of 750 million rupiah.
   b. Negligent offences – maximum imprisonment of 3 years and a fine of 100 million rupiah; for death or serious injury, maximum imprisonment of 5 years and fine of 150 million rupiah.
   c. Intentional release of toxic or hazardous materials into the environment – maximum imprisonment of 6 years and a fine of 300 million rupiah; for death or serious injury, maximum imprisonment of 9 years and fine of 450 million rupiah.
   d. Negligent release of toxic or hazardous materials – maximum imprisonment term of 3 years and a fine of 100 million rupiah; for death or serious injury, maximum imprisonment of 5 years and fine of 150 million rupiah.
3 Fines are increased by a third if the offender is a company or a corporate body. More importantly, penalties are also visited upon the individuals who gave the order to commit the violation or who acted as leaders in the commission of the violation. Thus, individuals will not be able to hide behind the facade of a company committing an environmental offence.
4 Other penalties include 'seizure of profits arising from criminal acts'; 'closure of all/part of business'; 'reparation of consequences of action'; 'carrying out of what was wrongfully neglected'; 'destroying what was wrongfully neglected' and/or 'placing the business under administration for a maximum of three years'.

4.3.3 Government Regulation No. 19 of 1999 on Marine Pollution Control

With regard to control of marine pollution, on 27 February 1999 the government issued Government Regulation No. 19 of 1999 on Marine Pollution Control. Under Article 9, any pollution of the marine environment is strictly prohibited. Meanwhile, those responsible in the execution of any activities (including business activities) have an obligation to prevent pollution of the marine environment.

Article 24 of Chapter XI stipulates an obligation to provide compensation for pollution or destruction of the marine environment, including rehabilitation costs. Calculation, claim, and payment of compensation will further be regulated by the Minister, indicating that such a procedure requires the further issuance of a decree from the Office of State Minister for the Environment.

4.3.4 Law No. 21 of 1992 on Shipping

This law contains several articles dealing with pollution arising from a vessel's activities. Article 66 sets out obligations for every vessel to be equipped with instruments to prevent pollution and to maintain the vessels' seaworthiness. Furthermore, the vessel's captain and crew have an obligation to prevent the vessel from polluting the environment. Article 67 lays down an obligation for the captain also to tackle any pollution caused by his vessel, and to report any pollution of the marine environment to the nearest authority, no matter whether it is caused by his own vessel or someone else's. Under Article 68, the vessel owner or operator becomes liable for pollution caused by its vessel(s), and both are under the obligation to effect the necessary insurance.

4.3.5 Decree of the Minister for Communication No. 86 of 1990 on the Prevention of Pollution of Oil from Vessels

To prevent pollution of the sea by oil from ships, the Minister for Communication has enacted a Ministerial Decree, No. 86 of 1990, whereby oil tankers are subjected to certain requirements to prevent pollution of the sea by oil.

According to this Decree, all vessels are prohibited from discharging oil or oily waste within the Indonesian Territorial Waters and the Exclusive Economic Zone, unless they fulfil requirements set out regarding the degree of oil in parts per million measurements; location of the vessels; safety of navigation; and on whether it was an accidental spill.

Tankers and other vessels of certain measurements and power, including tugboats, are required to be equipped with preventive pollution instruments. There are also requirements for tanker construction with regard to, for instance, the foundation, the tank and pipes used, the ballast tank and the oil-water separator, including the obligation to have an oil book record. Plans for the installation of the required equipments should be submitted to the Director General for Sea Communication before being installed. Vessel captains and owners are liable for both public and private compensation in accordance with relevant legislations. Having met all the requirements and gone through inspections, a National Certificate for the Prevention of Pollution by Oil will be issued, which is subject to annual re-inspection.

4.3.6 Law No. 22 of 2001 on Oil and Gas

This Law replaces Law No. 44 of 1960. According to this Law, the principles and purposes of the oil and gas industry are based *inter alia* on the environmental outlook.[20] Article 40 of this Law further states that oil and gas companies are under an obligation to guarantee the management of the environment in the form of obligations to prevent and control pollution, and to reconstruction following environmental damage.[21]

4.3.7 Law No. 22 of 1999 on Regional Government

In 1998, through the People's Consultative Assembly (MPR) Decision No. XV/MPR/1998, the government was instructed to establish laws and regulations concerning autonomous regional administration. On 7 May 1999 the government enacted Law No. 22 of 1999 concerning Regional Government (Regional Autonomy).

Article 2, paragraph 1 stipulates that the Indonesian territory is divided into three autonomous regional territories of provinces, municipalities and regencies. Furthermore, Article 3 provides that the territory of a province consists of land areas and a sea area of 12 nautical miles measured from the 'coastline' towards the high seas or towards the archipelagic waters.[22] Meanwhile, the territorial jurisdiction of the municipalities and regencies is set to be one third of that assigned to the provinces, or a belt of four nautical miles, assumed to be measured in the same manner as that of the provinces.

This assignment of territories is accompanied by a corresponding right for the regional governments, among others, to enforce the regional and national laws and regulations, including those with regard to pollution of the marine environment.[23]

4.3.8 Other Laws and Regulations

Other laws and regulations that are directly or indirectly related to the protection of the marine environment from oil pollution, but that are outside the mandate of the Office for the State Minister of Environment, can be found in various laws and regulations containing relevant provisions, most of which are of a general nature. Examples of such provisions are cited below.

According to Law No. 1 of 1973 concerning the Indonesian Continental Shelf, exploration and exploitation of the resources of the Indonesian continental shelf must be carried out in accordance with the relevant laws and regulations containing provisions on the protection and preservation of the marine environment.[24]

---

[20] Art. 2.
[21] Art. 40 paragraphs (2) and (3).
[22] The use of the term 'coastline' has been challenged by many as an indication that the drafter of this Law was ignorant of the existence of methods used to measure maritime zones according to the 1982 UNCLOS, which Indonesia ratified in 1985.
[23] Art. 10 paragraph 2.
[24] Art. 4, Law no. 1 of 1973 on the Indonesian Continental Shelf.

Being one of the old laws within present Indonesian laws and regulations, Law No. 11 of 1974 on Waterworks already envisages the need to protect and secure water sources including maintaining its preservation.[25]

Article 8 of Law No. 5 of 1983 on the Indonesian Exclusive Economic Zone stipulates that whoever conducts any activity within the Indonesian EEZ has the duty to take steps towards preventing, minimising, controlling and surmounting the pollution of the environment. It further states that discharge of waste in the Indonesian EEZ may be effected only after having obtained the permission of the government.

Preservation of the environment is one of the basic principles governing Law No. 5 of 1984 on Industries. In addition to that, this Law also sets out as one of the purposes of industrial development the equal development of the people's prosperity and welfare through the utilization of resources, balanced with the preservation of the environment.

Law No. 5 of 1992 on Cultural Heritage Objects prohibits any destruction of cultural heritage objects and their sites, including their environment.[26]

Under Articles 6 and 7 of Law No. 9 of 1985 on Fisheries, it is prohibited to engage in fishing and aquaculture using any destructive method that may endanger fishery resources or their environment. Furthermore, as a general rule, it also prohibited to carry out any activities that may endanger fishery resources and/or their environment.

Other laws and regulations that also contain environmental provisions are:

1. Law No. 5 of 1990 on Conservation of Living Resources and Its Ecosystems;
2. Law No. 9 of 1990 on Tourism;
3. Law No. 4 of 1992 on Housing and Settlements;
4. Law No. 23 of 1992 on Health and Law No. 24 of 1992 concerning Spatial Use Management;
5. Law No. 11 of 1967 on Basic Mining Law and Law No. 41 of 1999 on Forestry; and
6. Government Regulation No. 20 of 1990 on Water Pollution Control.

5 PROBLEMS IN IMPLEMENTING ENVIRONMENTAL LEGISLATION

As a result of the establishment of the various laws and regulations described above, one major problem lies in the division of competences among the following bodies:

1. The Office of the State Minister for the Environment;
2. *BAPEDAL*;
3. Other national sectoral agencies, especially the Ministry of Forestry and Industry, the Ministry of Trade and Industry, the Ministry of Energy and Mineral Resources, the Ministry of Agriculture, the Ministry of Marine Affairs and Fisheries and the Ministry of Home Affairs;
4. The local arms of national agencies; provincial and municipal governments; and local bodies like the police, the army and the prosecutors.

---

[25] See art. 13 of Chapter VIII.
[26] Art.15 paragraph (1).

At the central level, there are significant jurisdictional issues vis-à-vis the other sectoral ministries which had traditionally regulated the areas now governed by the Office of the State Minister/*BAPEDAL*. It must be noted that neither the Office of the State Minister nor *BAPEDAL* are powerful departmental agencies with full-fledged provincial competence, whereas most other sectoral ministries are. Given these intersections in jurisdiction, there has been a proliferation of legislation enacted by a host of national and provincial agencies, all of which seems to have conflicting competences in different spheres of environmental protection. The challenge remains to ascertain which agency has particular competence in a given issue, a task which is by no means straightforward.

There is a high degree of complexity and overlapping interests on maritime issues among the governmental sectoral agencies. The laws establishing these agencies regulate the degree of competence for each agency. This has resulted in overlapping and conflicting jurisdictions over coastal and offshore areas, and also between central and local governments.

5.1 Government Agencies

Previously, there were about fourteen ministerial agencies involved in the management, development and administration of maritime activities. These include the following: Agriculture (Directorate General of Fisheries); Forestry (Directorate General of Forest Protection and Nature Conservation); Mining and Energy; Communication (Directorate General of Sea Communication); Security and Defence; Trade and Industry; Public Works; Tourism, Internal/Home Affairs; Education and Culture; Justice; Foreign Affairs; Finance and State Ministers for the Environment and for Science and Technology.

Aside from these ministerial agencies, there are six non-ministerial or inter-departmental agencies with roles affecting offshore zones or the oceans, the National Development Planning Board (BAPPENAS); the National Coordinating Agency for Surveys and Mapping (BAKOSURTANAL); the Coordinating Committee for National Territory (PANKORWILNAS); the Co-ordinating Board for Maritime Security (BAKORKAMLA); the Indonesian Institute of Science (LIPI); and the National Council on Maritime Affairs (Dewan Kelautan Nasional), which has now been replaced by a new Maritime Council (Dewan Maritim), and which is under the coordination of the Ministry of Maritime Affairs and Fisheries.

5.2 Department of Marine Affairs and Fisheries

By virtue of Presidential Decree No. 355/M/1999, the government established a new ministerial agency called the Department of Sea Exploration and Fisheries, which was later changed into the Department of Marine Affairs and Fisheries. Organs of this department consist of the offices of the Secretariat General, the Inspector General and five Directorate Generals for: Capture Fisheries; Aquaculture Fisheries; Capacity Building and Marketing; Coastal, Beaches and Small Island Affairs; Sea Surveillance and Marine Protection; and an Agency for Ocean Research and Exploration.

6 REGIONAL INITIATIVES

Due to the many incidents causing oil spills in the Straits of Malacca, in 1981 a Revolving Fund was established to combat pollution in the Straits of Malacca and Singapore with a 400 million yen (approx. US $3.8 million) donation from the Malacca Straits Council. The Fund allows any of the three littoral states to take an advance for use in combating an oil spill from a ship. When compensation is later received, the amount is paid back into the Fund. A Revolving Fund Committee (RFC), which is made up of one representative from each littoral state, controls the Fund. The littoral states take turns managing the Fund, each for five years. The RFC also conducts joint exercises to improve coordination and preparedness among the three countries to combat oil pollution.

In 1993, six ASEAN countries, namely Brunei Darussalam, Indonesia, Malaysia, the Philippines, Thailand and Singapore, established the ASEAN Oil Spill Response Action Plan or ASEAN-OSRAP. The objective of the Plan is to enhance those countries' ability to respond to oil spills exceeding the capacity of any individual country. It provides a cooperative plan for mutual assistance from member States in oil spill response. The area of responsibility for the ASEAN-OSRAP includes all waters within the exclusive economic zone of the ASEAN countries and the territorial waters surrounding Singapore. When an oil spill is confirmed, the lead Agency of the country concerned will initiate a Pollution Reporting System (POLREP) within the ASEAN countries. The lead Agency of the country whose waters are affected is responsible for initiating any action within his area of responsibility and in accordance with the country's National Oil Spill Contingency Plan. In the case of a large spill occurring near a common territorial border, one country may be in a better position to respond to the spill due to its geographical location than the country actually affected. The ASEAN-OSRAP includes a Protocol for the Equal Right of Access between member countries outlining the procedures for doing this.

Japan also sponsored the project on Oil Spill Preparedness and Response (OSPAR) in the ASEAN sea area in 1994. The aim of OSPAR is to promote cooperation between Japan and the ASEAN countries to combat oil spills in the region. Under the OSPAR Project, Japan donated to the ASEAN countries a sum of 1 billion Yen (approx US$9.4 million) to purchase oil spill equipment to reinforce national stockpiles in the region. The OSPAR project has contributed to the development of an ASEAN Oil Spill Information Network System and enhancement of equipment stockpile bases in Muara (Brunei Darussalam), Balikpapan (Indonesia), Port Klang, Johor, Penang and Labuan (Malaysia), Manila, Cebu and Davao (Philippines), Thailand and Singapore.

7 DRAFT NATIONAL CONTINGENCY PLAN

The drafting of a National Contingency Plan for oil pollution was started a long time ago. However, it failed to gain a significant priority status and was set aside. After the establishment of the Ministry of Marine Affairs and Fisheries, the then Minister, Sarwono Kusumaatmadja, put the issue back onto the drafter's table for further finalization. Since then, through the initiative taken by the Office of the State Minister for the Environment, and through several inter-agency meetings, the draft has been revised and finalized.

However, since there are still some inter-sectoral issues that need further discussion, the draft has not yet been enacted as an official regulation.

Notwithstanding the fact that the draft has not yet officially been enacted as an official regulation, this paper will try to touch upon some important issues, particularly with regard to the division of responsibilities.

The Draft divides responsibilities to various agencies and persons involved. The responsibility is divided, as follows:

1. The Minister of Communication: for emergency response to pollution of the sea by oil;
2. The Head of the Environmental Impact Management Agency will be responsible for combating pollution in and rehabilitation of the polluted environment located in cross border areas of the provinces or States.
3. The responsibilities of the Director General for Sea Communications include:
   - the establishment of SOP, and inventorying of available manpower and physical equipment;
   - the coordination with other government agencies and with non-government institutions at both central and regional levels
4. Port Authority (central & regional): has similar responsibilities to that of Director General of Sea-Communication at a Tier 1 level;
5. Vessel's Captain or Owner and Company's Director are responsible for emergency response to oil pollution of the sea originating from their activities;
6. A Vessel's Owner, Operator and Company's Director: for the oil spill and any impact originating from their activities;
7. Mission Commander: for providing guidelines, coordination of personnel and equipment; appointment of On Scene Commander; reporting of the development of activities and their termination; drafting the claim to be submitted through the Dir. Gen. Sea Communication.
8. On-Scene Commander, for announcement of his appointment; execution of Standard Operating Procedure for oil response; lead the activities; organization and control of personnel and equipment; and reports to the Mission Commander.

The draft also includes provisions regarding procedures for reporting, for the response activities, calculation of damages and submission of claims. Reports of any oil spill should be submitted to the Port Administrator or Harbourmaster. Response at Tier 1 shall be carried out by the Port Administrator or Harbourmaster acting as Mission Commander, while for Tier 2 they shall be carried out by the Head of the Sea and Coastguard (*Kepala Pangkalan Armada Penjagaan Laut dan Pantai*) as Mission Co-ordinator.

The Minister for Communication leads the calculation of damages and response costs, while the Head of the Environmental Impact Management Agency, through coordination with relevant agencies at central or municipal level, will calculate the cost of the impact of the oil spill to be used as the basis for a claim. Anyone who suffered from the impact of the oil spill may submit claims for damages directly or through the Head of the Environmental Impact Management Agency of the Minister for Communication, in accordance with the relevant laws and regulations.

This Draft of the National Contingency Plan to combat oil pollution is aimed at an integrated effort at the central, regional and municipal levels. However, this contingency plan will require further Standard Operating Procedures at all levels. Even though it is still a draft, as indicated below this draft has been put into practice several times, even in response to an actual oil spill, namely, the MT Nagasaki Spirit marine pollution casualty.

8 CASES INVOLVING POLLUTION OF THE SEA BY OIL

As already indicated, the Indonesian marine environment is very prone to oil pollution, and several major disasters have occurred, despite the efforts that have been taken by the government. Below are some examples of oil spill incidents that affected the Indonesian marine environment. Awareness of the danger of the oil spill into the marine environment first emerged when the Japanese tanker *Showa Maru* ran aground in the Strait of Malacca in 1975. The incident pushed the government to establish Standard Operation Procedures (SOP) for oil spill response in the Straits of Malacca and Singapore, through a Joint Decree between the Director General of Sea Communication and the Director General of Oil and Gas, signed in 1981,[27] whereby the Directorate General of Sea Communication has the role of Co-ordinator and Pertamina, the state-owned oil and gas company, has the role as oil pollution equipment operator.

8.1 THE NAGASAKI SPIRIT MARINE POLLUTION CASUALTY[28]

MT. Nagasaki Spirit (DWT 95,000), flying the Liberian flag, was carrying 40,000 tons of Khafji crude oil on her voyage from the Persian Gulf when it collided with the container ship MV. Ocean Blessing (DWT 25,000), flying the Panamanian flag, on 19 September 1992 at 16:32 hours local time, 40 miles from Port of Belawan in the Sumatra Island of Indonesia. The initial impact resulted in the fracturing of the Nagasaki Spirit's hull and the release of its crude oil, some burning, from her fractured tanks. The Ocean Blessing caught fire and was engulfed in flame. Several pictures taken from the air showed patches of black oil slicks and debris spread within a five-mile radius of the casualty ship.

The USS Niagara Falls was the ship in the closest vicinity at the moment of the accident. Her Commanding Officer took on the role of On Scene Co-ordinator for the search and rescue operation. The draft oil spill response plan was activated once news of the disaster was received. The National Co-ordination was conducted from the National Operation Centre for Combating Oil Pollution (NOCOP) at the head office of the Directorate General of Sea Communication. It used the command control, communication, and information system of the Command and Control Post for the maritime safety guard and rescue system, which operates 24 hours a day.

---

[27] Joint Decree no. DKP/1 /1 /27/Kpts/DM/MIGAS/81.
[28] T. Soentoro and H.M.J. Lumenta, 'Prevention and Abatement of Vessel-Borne Pollution: A Case Study of Marine Pollution Casualties: The Nagasaki Spirit', in M. Kusuma-Atmadja, T.A. Mensah and B.H. Oxman (eds), *Sustainable Development and Preservation of the Oceans: The Challenges of UNCLOS and Agenda 21*, (Honolulu, The Law of the Sea Institute, William S. Richardson School of Law, University of Hawaii, 1999), pp. 380-381.

The Port Administrator/Harbourmaster of Belawan, North Sumatra Province, was appointed as On-Scene Coordinator (OSC), and coordinated all related agencies in the area, alerted them and asked them to either stand by, or attend the scene. The Indonesian Navy sent its aircraft and surface craft; the Police and Customs sent their surface craft, and Pertamina sent its surface craft and some of their oil pollution control equipment. The OSC also coordinated the salvage ships from Singapore, which rendered assistance during the incident and later towed the ships for repair.

NOCOP coordinated the related agencies at the national level, including the Navy, Police, Customs, and Pertamina. The Environment Impact Management Agency (*BAPEDAL*) supported the NOCOP with other experts during the activities. At the international level, coordination was conducted with the Directorate General of Environment of Malaysia, the owners of the ship, and the salvage company. There was intensive coordination with the Directorate General of Environment of Malaysia, especially when the disabled ship entered Malaysian waters and, later, when it returned to Indonesian waters while salvage ships conducted the necessary operations.

This was the first accident to take place at the north entrance of the strait. Coordination with related parties was considered to be very good, and all related parties were very responsive. This showed that, although the National Oil Spill Response Plan was still a draft, it could be exercised and put into practice, and proved to be useful during the response operation. Coordination with Malaysia was also very good; it was perhaps because both countries had experience through the exercises conducted by littoral states under the scheme of the Revolving Fund.

Since Indonesia had ratified the CLC and Fund Conventions at that time, the damage compensation claim process was much easier compared to during the MT Showa Maru incident in 1975, before Indonesia was a party to those conventions.

8.2 THE MT NATUNA SEA OIL SPILL

On 3 October 2000, at approximately 06:15 Western Indonesia Time, the MT Natuna Sea, a 92,313 metric ton DWT, carrying 523,088 barrels of crude oil, struck a reef in Batam waters. The amount of oil spilt was about 20% of the total cargo. The location of the incident was at $01°11'29'$ NL, $103°53'05'$ EL, north of the Belakang Padang and Sambu Islands in the Indonesian territory. The MT Natuna Sea tanker was operated by Tanker Pacific Pte. Ltd., Singapore and insured by the London Steam Ship Owners Mutual Insurance Association, Ltd.

Claims for environmental damage were carried out at both the local and national level. The collection of the necessary information and the initial environmental impact evaluation were carried out by the Environmental Impact Assessment Agency (*BAPEDAL*). The Batam local government signed a Memorandum of Understanding with the London insurance company mentioned above, under which clearance for MT Natuna Sea to sail out of Batam waters to Singapore for dry-docking was carried out with a guarantee bond. Following this MOU, *BAPEDAL* took action on the environmental damage claim.

Basically, the oil spill damaged the coastal community, among others the fishermen, ship and boat operators, and marine culture business community. Using the 'economic valuation method' and applying it to various ecosystems, that is, those of the coral reefs, mangroves, sea grasses, sands and fisheries, a valuation with the addition of assessment costs was that the total cost of the oil spill was estimated at US$29,365,667.73. Damage costs for the various ecosystems were valued as follows: Coral Reef: US$550,188.44; Mangrove: US$9,916,193.60; Seagrass: US$2,598,171.07; Sand clean up: US$3,851,948.00; Fisheries: US$12,266,286.62; and Assessment costs of US$382,880.00.

The Office of the State Minister for the Environment, in cooperation with the local governments, prepared evidence to bring a civil lawsuit and to proceed with the criminal investigation. However, neither the lawsuit nor criminal investigations have been continued for unknown reasons.

However, after negotiations with London Steam Ship Owners Mutual Insurance Association, Ltd., the insurance company only agreed to provide compensation of only US$3,500,000.00. Payment to the affected fishermen, so far, has only reached US$1,500,000.00.

8.3 THE MT KING FISHER INCIDENT

The MT King Fisher, a Malta tanker, entered the port of Cilacap located in the southern coast of Central Java on 1 April 2000 at approximately 14:30 Western Indonesian Time, carrying 600 MB of crude oil originating from Dumai, and Attaka Crude Oil/Badak originating from Tanjung Santan, East Borneo, and bound for Pertamina IV at Cilacap. The tanker was owned by King Fisher Navigation Co. Ltd., Valetta, Malta, and operated by Estoril Navigation Co. Ltd. of Greece.

Upon its entry to the port of Tanjung Intan, Cilacap, the tanker ran aground and hit a reef located between Buoy 2 and 3, causing the ripping of its right side flank in two places measuring respectively of 340 x 120 cm and 160 x 60 cm. As a result of this impact, there was a leak and oil spilled into the area.

Procedures similar to those used in the previous incidents were taken through coordinated efforts between relevant agencies, both at the central and municipal levels. The result of the survey was an indication of six places where there was oil spilt, covering the whole coast, which was comprised mostly of recreational or tourism areas, fisheries, landing and auction ports, and sand mining sites. Several vessels of different purposes and measurements, including fishing devices, also suffered from the impact of the oil spill. This of course resulted in the loss of income not only for the fishermen, but also for the fishing industry, the local government's income derived from taxes, and the mariculture industry.

The claim was submitted for both short and long term compensation. Included in the short-term compensation claim were the clean-up costs and fishing devices. Long-term compensation covered the environmental destruction or degradation, and the loss suffered by the fishermen because of the destruction of the fishery habitat. The total claim for short-term compensation was estimated at around 682 billion rupiah, and the total estimate for long-term compensation is not yet available. The Indonesian Fishermen Association (HSNI), acting on behalf of the fishermen, claimed 272.769 billion rupiah for damages suffered by 21,263 traditional fishermen, 113 fishing boats and 120 fishing nets. After negotiations, the insurance company only agreed to pay the amount of 18 billion rupiah, excluding compensation for ecological damages.

9 CONCLUSION

The above description has tried to give a picture of where Indonesia has placed its efforts in order to combat pollution of the sea by oil. However, since most of the international instruments in this regard are very technical and require implementation of a technical nature, Indonesia has not yet been able to meet the ideal standards for preventing and combating of oil pollution. The sectoral approach in the establishment of laws and regulations has also become an obstacle in this effort.

With regard to legal remedies for compensation for damages caused by oil pollution, the final analysis is perhaps that an absence of procedures for submitting claims for damages and of clear methodology for calculating losses has created difficulties for the affected parties, including the government in its role as custodian of the ecosystem, and in this particular case, the marine environment.

LIST OF REFERENCES

T. Soentoro and H.M.J. Lumenta, 'Prevention and Abatement of Vessel-Borne Pollution: A Case Study of Marine Pollution Casualties: The Nagasaki Spirit', in M. Kusuma-Atmadja, T.A. Mensah and B.H. Oxman (eds), *Sustainable Development and Preservation of the Oceans: The Challenges of UNCLOS and Agenda 21,* (Honolulu, The Law of the Sea Institute, William S. Richardson School of Law, University of Hawaii, 1999), pp. 380-381.

PROTECTION OF THE MARINE ENVIRONMENT UNDER BELGIAN LAW

Marc A. Huybrechts
Katrien N. Van Damme

1 INTRODUCTION

The legal basis for the protection of the 'Belgian' part of the North Sea against sea-related pollution incidents and for the conservation, restoration and development of nature and the natural environment is provided for by the Statute of 20 January 1999 (henceforth: Belgian Statute Marine Environment).[1] To illustrate and explain the origin of the Belgian Statute on the protection of the marine environment, a short overview of the European and International rules by which this Belgian Statute was inspired is necessary.

The Belgian Statute on the protection of the marine environment implements the following European Directives:

- Council Directive 79/409/EEC of 2 April 1979 on the conservation of wild birds as amended by the Directive 91/244/EC.[2]
- Council Directive 85/337/EEC of 27 June 1985 on the assessment of the effects of certain public and private projects on the environment as amended by the Council Directive 97/11/EEG of 3 March 1997.[3]
- Council Directive 92/43/EEC of 21 May 1992 on the conservation of natural habitats and of wild fauna and flora as amended by Council Directive 97/62/EC of 27 October 1997.[4]

Also, the Belgian Statute on the protection of the marine environment complies, *inter alia*, with the following international agreements:[5]

- The Rio de Janeiro Convention on biological diversity of 5 June 1992.
- The Paris Convention for the protection of the Marine Environment of the North-East Atlantic of 22 September 1992.
- The Bern Convention on the conservation of European Wild Life and Natural Habitats of 19 September 1979.
- The Ramsar Convention on international water areas (habitat, water birds) of 2 February 1971.

---

[1] *B.S.* of 12 March 1999 (Belgian OJ).
[2] *OJ* L103, 25 April 1979.
[3] *OJ* L216, 3 August 1991.
[4] *OJ* L176, 20 July 1993; *OJ* L305, 8 November 1997.
[5] This list is not meant to be exhaustive.

Michael G. Faure and James Hu (eds), Prevention and Compensation of Marine Pollution Damage. Recent Developments in Europe, China and the US, 119-135.
© 2006 Kluwer Law International. Printed in the Netherlands.

However, the Belgian Statute on the protection of the marine environment not only reflects the international and European legislation. It also improves the current legislation and goes beyond what is provided for in European and international rules.[6]

2 CONTENTS OF THE BELGIAN STATUTE MARINE ENVIRONMENT

For a better understanding of the Belgian Statute on the protection of the marine environment, the table of contents is hereby provided:

- Chapter I: Definitions.
- Chapter II: General aims and principles.
- Chapter III: Protected marine reserves and protection of species.
- Chapter IV: Intervention and avoidance of damage to and disruption of the environment.
- Chapter V: The prevention and avoidance of pollution caused by vessels.
- Chapter VI: Permits and authorizations.
- Chapter VII: Report on environment effects and environmental assessment.
- Chapter VIII: Emergency measures to safeguard and protect the marine environment.
- Chapter IX: Compensation for damages and disruption. (which, for the purpose of our study, is the most important chapter)
- Chapter X: Supervision and control.
- Chapter XI: Penalty provisions.
- Chapter XII: Amendments.

The first part of the Belgian Statute on the protection of the marine environment provides for a legal basis for creating marine reserves and for protecting plants and animals. The second part provides certain measures for preventing and compensating damages to and the disruption of the marine environment.

Given the limited amount of space, this paper will be limited to the study of the protection of the marine environment related to the shipping industry and this with reference to the compensation for damages to and disruption of the marine environment.

---

[6] See *infra* section 4.2.1 the jurisdiction in the EEZ by Belgian authorities with reference to the purely operational discharges; and B. the concept 'disruption of the environment' together with 'damage to the environment'.

3 IMPORTANCE OF THE NORTH SEA AND THE RISK OF POLLUTION[7]

Irrespective of the rather limited surface area of the sea under Belgian jurisdiction,[8] an appropriate instrument was nonetheless necessary to provide a legal basis for protecting the marine environment, together with a solid international cooperation.

First, the Belgian waters include very valuable bird habitats where rare bird species live.[9] Part of the Belgian sandbanks are also been recognized under a special protection scheme within the European Habitat Directive.[10]

Second, the North Sea is an important route for ships having Antwerp, Rotterdam or Hamburg as port of call. The Belgian coast is adjacent to the (narrow) Strait of Dover or the Channel, which is one of the busiest shipping routes in the world. Moreover, there is considerable traffic in the Belgian territorial sea to and from the ports of Antwerp, Zeebrugge and Ostend.[11]

This intense degree of traffic in such narrow shipping lanes creates a serious risk for pollution, mainly resulting from possible collisions, as was illustrated by the recent collision between the Turkish Tanker 'Vicky' and the 'wrecked' pure car carrier 'Tricolor'.

Third, one also has to take into account the many illegal operational discharges that take place there. In the past year, the Belgian authority supervising the North Sea has recorded some five hundred cases of oily residue contaminations, which were probably caused by operational discharges.

For a good understanding of what is at stake in this paper, a clear distinction must be made between accidental discharges and illegal discharges. Accidental discharges can be the result of a collision at sea. Illegal operational discharges result from an intentional act by the vessel operator, namely, the cleaning of the tanks or ballast tanks at sea rather than a proper disposal of the waste or oily residue in a reception facility or otherwise. Thus, not only must we face the risk of pollution resulting from collisions at sea, but also from the practice of illegal spills. These illegal discharges (around five hundred a year in Belgian waters) occur daily and are therefore much more frequent than accidental spills (around twenty five year in Belgian waters). Only in 10 % of the cases of illegal discharges, can a vessel be found near the location of the spill.

An oil spill floats on the water for about eight to twelve hours and the illegal discharge itself takes about twenty minutes. When one takes into consideration that a vessel can traverse Belgian waters in only about four hours, the chances of actually catching a vessel during an act of illegal discharge seem small indeed. Therefore, international cooperation is necessary.[12] As a consequence, the ambition of this paper is,

---

[7] <http://www.mumm.ac.be>.
[8] More or less 40 miles coastline.
[9] The Ramsar Convention on international water areas (habitat, water birds), 2 February 1971.
[10] European Directive 92/43/EEC on the protection of natural habitats and wild flora and fauna, *OJ* L167/17, 21 May 1992.
[11] <http://www.mumm.ac.be>.
[12] International undertakings in which Belgium is a contracting party, or that Belgium approved by internal law: Convention for the Protection of the Marine Environment of the North-East Atlantic, Paris 2 September 1992 (OSPAR) approved by law of 11 May 1995, *B.S.*, 31.01.1998; Agreement for the cooperation in

among other things, to illustrate how the supervision of the North Sea is carried out by solid international cooperation, which is set up by the Bonn Agreement.[13]

4 THE BELGIAN STATUTE MARINE ENVIRONMENT

4.1 GENERAL APPROACH

4.1.1 Field of Application

The statute is the legal basis for the protection of the Belgian part of the North Sea against sea-related pollution incidents and for the conservation, restoration and development of nature and the natural environment. In other words, the statute is applicable in relation to the Belgian territorial sea, the exclusive economic zone (EEZ) and the continental shelf.[14]

The Belgian authorities do not only exercise jurisdiction in their territorial sea, but also within the EEZ, which means *inter alia* the exercise of jurisdiction with regard to the protection and conservation of the marine environment.[15] This jurisdiction relates to the protection and conservation of the marine environment in general, and as a consequence deals with both accidental and illegal operational discharges.

The enforcement of this form of jurisdiction in the EEZ by the Belgian authorities with reference to the purely operational discharges is innovative.[16]

The Statute also provides a very broad definition of 'the marine environment.'[17] As a consequence, the biotical, a-biotical, ecological components and the interfering components are included.[18]

---

dealing with pollution of the North Sea by oil and other harmful substances, Bonn 13 September 1993 (Bonn Agreement) approved by law of 16 June 1989, *B.S.*, 03.10.0996; Convention on the prevention of marine pollution by dumping of wastes and other matters, London 1972, (LC Convention) approved by law 20 December 1984, *B.S.*, 22.10.1985; International Convention for the prevention of the pollution from ships, London 1973, 1978, (MARPOL) approved by law 17 January 1984, *Belgisch Staatsbald,* 24.05.1984; Agreement on the conservation of small cetaceans of the Baltic and the North Sea, New York, 17 March 1992, (ASCOBANS) *B.S.*, 20 October 1993; The Scheldt and Meuse International Agreements 3 December 2002 to which Belgium is a contracting party; International Conferences for the protection of the North Sea (NSC), fifth reunion was held at Bergen, Norway, 20-21 March 2002; United Nations Open-Ended Informal Consultative Process established by Resolution 54/33 (General Assembly) to facilitate the annual review by the Assembly of developments in ocean affairs; overview can be found on <http://www.mumm.ac.be>.

[13] The agreement for cooperation in dealing with pollution of the North Sea by oil and other harmful substances, opened for signature at Bonn on 13 September 1983; <http://www.bonnagreement.org>; *infra* '5. Pollution in practice'.

[14] Section 2 Belgian Statute Marine Environment *juncto* EEZ Act, 22 April 1999, relating to Belgian's exclusive economic zone in the North Sea, *B.S.*, 10 July 1999.

[15] Section 4 of the Statute 22 April 1999 on the EEZ and Section 46 of the same Statute; also <http://www.mumm.ac.be>.

[16] The 1982 UN Convention on the Law of the Sea UNCLOS (entered into force 16 November 1994) provides jurisdiction for the Coastal States in the EEZ in addition to the sovereign rights in the territorial sea in Sections 211 §5 and 220, but not for purely illegal operational discharges.

[17] Section 2 of the Belgian Statute Marine Environment.

4.1.2 Aim and Basic Principles

Section 3 of the Belgian Statute on the protection of the Marine Environment describes the aim as being

> 'the conservation of it's own nature, the biodiversity and the unharmed character of the marine environment by measures aiming to protect it and by measures aiming to compensate damages and disruption of the environment.'

The general principles of environmental law are summarized in section 4 of this important Statute:[19]

- The *prevention principle* according to which prevention is better than curing the problem.
- The *precautionary principle* according to which preventive measures must be taken if there are grounds for concern regarding pollution.
- The *principle of sustainable management* according to which human activities must be managed in such a way that the marine ecosystem remains in a status which ensures the continued use of the sea.
- The *polluter pays principle* according to which the costs of the measures to prevent and fight pollution are to be borne by the polluter.
- The *principle of restoration* according to which in case the environment is damaged or disrupted, the marine environment must be restored to its original status in the first place as far as is humanly possible.

The *principle of strict liability* is also established.[20] It provides that in the event of any damage to, or disruption of the environment in sea areas as a result of an accident or an infringement of the law, the party having caused the damage to, or the disruption of the environment is obliged to remedy this, even if they are not at fault.

These principles apply on the 'users' of the sea areas *and* the government, and are of great importance for the interpretation and construction of the Belgian Statute on the protection of the Marine Environment, when applied in Court.

However, these principles (out of the principle of strict liability) are not applied *in absoluto*, seeing that section 4, § 1 of the Statute clearly mentions, *'these principles must be taken into consideration.'* Therefore, a certain discretionary appreciation is possible in the different situations in which actors, the government or even judges must decide.[21]

---

[18] F. Maes, 'De wet van 20 januari 1999 ter bescherming van het Mariene milieu in de zeegebieden onder de rechtsbevoegdheid van België Op weg naar een duurzaam gebruik van de zee' (1999) *Tijdschrift voor Milieurecht* (the Belgian journal on environmental law) 270-285.

[19] These principles can be found in the Paris Convention for the protection of the Marine environment of the North-East Atlantic of 22 September 1992.

[20] Section 37 of the Belgian Statute Marine Environment.

[21] F. Maes, *supra* note 18.

4.2 PREVENTION AND AVOIDANCE OF DAMAGE TO AND DISRUPTION OF THE MARINE ENVIRONMENT

4.2.1 Damages to and Disruption of the Marine Environment

The Belgian Statute on the protection of the Marine Environment is innovative in the sense that on top of the concept *'damage to the environment'* a new concept has been introduced being *'the disruption of the environment'*. Belgium together with Italy is, in comparison with the other European countries, innovative and shares this daring approach by adding the principle of 'disruption' of the environment.[22]

4.2.1.1 Damage to the Marine Environment

Damage to the marine environment is defined as: *every form of damage, loss or prejudice, suffered by an identifiable natural or legal person, caused by an impairment of the marine environment, irrespective its cause.*[23]

Damage to the environment can be assessed in terms of an 'economic' value (or loss of such value). Examples of damage to the environment are:

- damage to health;
- damage to an object;
- nuisance;
- loss of income, on account of preventive measures.[24]

On account of the definition it is clear that also 'pure economic loss' is included. This statute speaks of *any* damage, *any* prejudice.[25] This is a major departure of the damage concept in the CLC and Fund conventions that reject compensation for pure economic loss as is illustrated by the 'Sea Empress' decision.[26]

When does the concept 'damage' to the environment apply?

The first condition for damage to the marine environment is *'the impairment of the marine environment'*. Damage must be caused via the impairment of the marine environment.

The Belgian approach to the issue of causation or the causal link is different from other countries: every cause without which the damage would not have occurred as it occurred specifically, is taken into account and considered to be a cause; this is known as the theory of the equivalent causes which is the 'Belgian' causation theory.[27]

---

[22] In Italy, the Penal Code provides in its Section 185 the obligation to compensate 'patrimonial damage' and 'non-patrimonial damage' as a result of a criminal offence.
[23] Section 2, 6° of the Belgian Statute Marine Environment.
[24] F. Maes, *supra* note 18.
[25] A. Carette, 'De aansprakelijkheidsregeling uit de wet ter bescherming van het mariene milieu' (1999) 3 *Tijdschrift voor Milieurecht* 362-374.
[26] The Sea Empress Court of Appeal, 17 January 2003; *Algrete Shipping Co. Inc. and another v. International Oil Pollution Fund and others*, The Sea Empress, EWCA Civ 6 LLP 2003, vol. 1, 327-340.
[27] A. Carette, *supra* note 25, nos. 2, 4.

A second condition is that the damage must be *suffered personally* by the individual claiming compensation. In other words, this individual will have a subjective right or legitimate personal interest in relation to his personal integrity or personnel assets.[28]

4.2.1.2 Disruption of the Marine Environment

Disruption of the marine environment is defined as *a negative impact on the marine environment in as far as it is not covered by the traditional concept damages.*[29] In other words, when the adverse influence on the marine environment cannot be classified as 'damage to the environment,' it will qualify as a 'disruption of the marine environment.' The disruption of the marine environment is not assessable as an economic value; it deals with violations of non-appropriated components of the environment, in others' 'collective goods' or goods described as '*res nullius*' (goods belonging to no one), compounding the marine ecosystem.[30] One could also refer to the 'natural resources damage' concept.

As a consequence of this innovative attitude towards pollution introducing the concept of disruption to the marine environment, in case of an incident, not only the economical consequences will be considered but *all the consequences* will qualify for compensation under the Belgian statute.[31]

4.2.2 Compensation[32]

The *principle of objective (or faultless) liability* is established in section 37 §1 of the Belgian Statute on the protection of the Marine Environment. This principle provides that 'any' damage to or disruption of the marine environment as a result of an accident or an infringement of the law, must be made good by the party having caused the damage to or the disruption of the environment, even if this party is not at fault. In other words there is a duty to compensate irrespective of personal fault.

Every claimant must first establish his 'standing' in court. The party claiming compensation has to prove the damaging act, the loss that it suffered personally on account of this incident or impairment of the environment and the 'causal' link between the damaging act and the loss suffered.[33]

However, section 37 § 2 of the Belgian Statute on the protection of the Marine Environment provides some *'force majeur situations'* in which cases the party having caused the damage to or disruption of the environment will not be held liable. Therefore, the party having caused the damage or disruption can try to prove the application of one of the following situations:

- damage or disruption exclusively caused by war, civil war, terrorist acts, by a natural phenomenon that is unavoidable, and so on. (An ordinary storm is not an excuse!)

---

[28] A. Carette, *supra* note 25, no. 5.
[29] Section 2, 7° of the Belgian Statute Marine Environment.
[30] A. Carette, *supra* note 25, no. 6.
[31] *Ibid.*
[32] A. Carette, *supra* note 25, was relied upon on the occasion of the preparation of this contribution.
[33] A. Carette, *supra* note 25, no. 9.

- damage or disruption caused exclusively and solely by wilful, intentional misconduct or negligence by third parties in view of actually causing that damage or disruption. The third party (or person) must not be an agent or a representative of the (otherwise) liable person.
- damage or disruption *totally* caused by the *negligence or fault* of the public authorities or the governmental agents in relation to the maintenance and upkeep of the navigable waterways (navigation lights or other navigational equipment, necessary for navigation).

4.2.2.1 Who has the Right to claim for Compensation?

The right to claim *compensation for damage to the environment* is for the party having suffered the damage 'personally.'[34] According to Belgian law, the claimant who wants to establish his 'standing' in court needs to prove his lawful interest.[35]

In this respect, see the 'second' condition qualifying the damage concept.[36] The second condition spells out that the damage must be *suffered personally* by the individual claiming for compensation. In other words, this individual will raise a subjective right or legitimate personal interest in relation to his personal integrity or personal assets.

The right to claim *compensation for 'disruption' of the environment* is vested exclusively in the Belgian State.[37] By vesting the claim for compensation for disruption of the environment in the Belgian State, the need to prove any (further) lawful interest is overcome.[38]

The position of marine or environmental action groups is not dealt with in the Belgian Statute on the protection of the Marine Environment, nor does the statute give them a standing in court to entertain an action for damage that they did not themselves personally sustain.[39]

The fact that a specific right of suit is not given to these environmental action groups, does not mean that there is no possibility for them to claim the actual costs borne by these action groups when they intervene without the invitation of the government. Environmental action groups can make such a claim on the basis of section 40 § 2 of the Belgian Statute Marine Environment, which provides that compensation for disruption of the environment will be claimed by the government without violating the rights of other persons that claim *the actual costs borne by them.*[40]

4.2.2.2 Limitation of Liability

The Belgian Statute on the protection of the Marine Environment not only deals with the '*force majeure*' situations, but also does not infringe upon the right of the party having

---

[34] Section 37 §3 of the Belgian Statute Marine Environment; A. Carette, *supra* note 25, no. 25.
[35] Section 17 of the Belgian Code of Legal Proceedings.
[36] *Supra* 4.2.1.1 Damage to the marine environment.
[37] Section 37 §3 of the Belgian Statute Marine Environment.
[38] A. Carette, *supra* note 25, no. 30.
[39] A. Carette, *supra* note 25, no. 31.
[40] A. Carette, *supra* note 25, no. 32.

caused the damage to or the disruption of the environment, to limit his liability as to the total amount due.[41]

Therefore the Belgian Statute on Marine Environment is in line with the international treaties like the LLMC Convention London, 19 November 1976, and the CLC – Convention 1969.

4.2.2.3 Compensation: Restoration in Kind?

The Belgian Statute on the protection of the Marine Environment does not differ from the general law regarding compensation for damages as being a *restoration in kind*.[42] We can think of an oil spill, which has as a consequence the reduction of the amount of fish in a certain area. On account of the principle of the restoration in kind, new fish (a component of the harmed environment) will be introduced into the marine environment.

The general rule for compensation for damages to or disruption of the environment is limited by the principle of the *prohibition of the abuse of right*. Compensation for restoration in kind or the introduction of similar components must only be implemented insofar as the expenses are not 'unreasonable' in light of the aim in view.[43]

With regard to compensation for damages to the environment, expenditures must be compensated. This includes not only the cost of introducing new marine components (disruption of the environment), but also *expenses* born by the government or by private parties, that is, individual citizens acting on behalf of the government, in order to take measures to prevent or limit the damages.[44] The cost of these acts include not only *the costs actually borne in the specific event* but also *the fixed costs* that are directly related to the intervention by the public authorities and the costs borne beforehand to dispose of the necessary means.[45]

How did the legislator deal with damage or disruption which is not 'repairable' in kind?

The basic tenet of the 'restoration in kind' principle or the introduction of marine components is that this is the preferred remedy, but the principle should not operate as a limitation. In other words, the liability will not be limited in the event that a restoration in kind or introduction of similar marine components is not possible. Irreparable damage to, or disruption of the marine environment can be compensated by the imposition and payment of a monetary award.[46]

How must one assess compensation for disruption of the marine environment?

The non-economic damage can be based on a scientific theoretical model, which will determine the actual damage to the marine environment, instead of introducing similar

---

[41] Section 37 §4 of the Belgian Statute Marine Environment.
[42] Section 37 §5 of the Belgian Statute Marine Environment.
[43] *Ibid.*; A. Carette, *supra* note 25, no. 19.
[44] Section 38 of the Belgian Statute Marine Environment; A. Carette, *supra* note 25, no. 26.
[45] Section 39 of the Belgian Statute Marine Environment provides a Royal Decree that must determine how these fixed costs are to be assessed and claimed; up to now, this Royal Decree has not been promulgated; A. Carette, *supra* note 25, no. 27.
[46] A. Carette, *supra* note 25, nos. 34-36.

marine components in the marine environment.[47] Up until today, these scientific theoretical models have not yet been provided. Therefore, until such time that these scientific theoretical models are available, a compensation will be ordered on the basis of an '*ex aequo et bono*' approach.[48]

The fines for disruption of the environment will be received in a special fund for the marine environment and will be used for environmental purposes.[49]

In other words, even when the disruption of the environment cannot be restored in kind, the fines received to compensate the disruption will be effectively invested in the marine environment.

4.2.2.4 Bail or Guarantee Declaration

Section 42 of the Statute Marine Environment provides the possibility for the government to claim and impose for a 'bail' or guarantee declaration. The government can demand this bail or guarantee declaration as soon as there is a risk for damage to the marine environment. In other words, it does not apply to 'disruption' of the environment.[50] The amount of this bail or declaration must cover the foreseeable damage, taking in consideration not only the damage suffered but also future risks and consequences. The fines will be received in the special fund for the marine environment and will be used for environmental purposes.

5 POLLUTION IN PRACTICE

So much for the theory. In practice, the Belgian Statute on the protection of the marine environment is mostly used as a tool of 'deterrence.' With reference to the compensation for disruption of the marine environment, until today, the scientific theoretical models on the basis of which an economical assessment should be made possible, have not yet been provided. Therefore, until such time that these scientific theoretical models are made available, a compensation will be ordered on the basis of an '*ex aequo et bono*' approach.

The following section will illustrate the approach of 'deterrence' within a broader framework. Indeed, the Belgian Statute on the protection of the marine environment must also be seen within the international framework of combating marine pollution. Therefore, the Bonn Agreement, port reception facilities and port State control, the North Sea disaster plan and places of refuge, legal aspects and finally future developments to be expected, will all be briefly discussed.

---

[47] Section 40 §1 of the Belgian Statute Marine Environment foresees the possibility for a Royal Decree to assess in economic terms the disruption of the marine environment; until today this Royal Decree is not promulgated.
[48] A. Carette, *supra* note 25, nos. 36, 37.
[49] Section 40 §3 of the Belgian Statute Marine Environment.
[50] A. Carette, *supra* note 25, no. 29.

5.1 AIR SURVEILLANCE BY THE MANAGEMENT UNIT OF THE NORTH SEA MATHEMATICAL MODELS (MUMM)

During the day and in favourable weather conditions, the authorities of the North Sea often receive reports from various sources, like commercial aircraft, helicopters, yachtsmen, or other vessels concerning possible discharges by ships. However, these reports are often incomplete. For this reason, special aircraft will provide additional professional expert evidence.[51]

Air surveillance of the North Sea is organized in the context of the Bonn agreement.[52] The Bonn agreement is an overall strategy to reduce pollution of the seas from ships, and to enforce the MARPOL 73/78 Convention and the 1982 UNCLOS Convention effectively.[53] To this effect, the exchange of information on the detection and the prosecution of offenders between different countries is essential.[54] To this end, national authorities receive reports on oil pollution (POLREPs) and deal with questions concerning measures of mutual assistance.[55]

Each state party organizes its own surveillance program in accordance with the guidelines laid down in this agreement and joint international exercises are carried out several times a year.

In Belgium, the 'Management Unit of the North Sea Mathematical Models' (MUMM) monitors the sea areas for which Belgium is responsible.[56] A total of two hundred and fifty flying hours is scheduled every year. These flying hours do not only occur during office hours, but also occur in the evening or at weekends. Flight schedules are kept secret. Only one hour is needed to fly over the sea area under Belgian responsibility.

Air surveillance has a deterrent effect. Activity reports show that the number of total spills has gone down over the years.[57] Owing to this deterrent effect, aerial surveillance is an important instrument in the fight against illegal discharges.

Recent public announcements show that the Belgian Government wishes to supervise the Belgian area of the North Sea twenty four hours a day. By 2005, 'unmanned air vehicles' (UAV's) will be operational, and will be equipped with cameras weighing less than ninety kilos. Besides the unmanned air vehicles, the government will also use satellite and radar systems. The aim of introducing these new technologies is to enhance the chance of getting caught, or to increase the deterrent effect.

---

[51] Bonn Agreement, Manual Oil pollution at sea – part 2, 20.
[52] The agreement for cooperation in dealing with pollution of the North Sea by oil and other harmful substances, opened for signature at Bonn on 13 September 1983. The contracting parties are Belgium, Denmark, France, Germany, the Netherlands, Norway, Sweden, United Kingdom and the European Community.
[53] Both conventions describe the enforcement powers of the coastal State, the port State and the flag State. Within the Bonn Agreement, these different States cooperate.
[54] Bonn Agreement, Manual Oil pollution at sea – part 2, 5.
[55] Section 4(b) Bonn Agreement (*supra* note 51); <http://www.bonnagreement.org>.
[56] <http://www.mumm.ac.be>.
[57] <http://www.mumm.ac.be>; <http://www.bonnagreement.org for annual reports>.

5.2 RECEPTION FACILITIES AND PORT STATE CONTROL

To provide the marine environment with better protection against pollution by ships, the European Parliament and the Council enacted the Directive 2000/59/EC of the European Parliament and of the Council of 27 November 2000 on port reception facilities for ship generated waste and cargo residues.[58]

Reducing marine discharges of ship-generated waste and cargo residues enhances the protection of the marine environment. This can be achieved by improving the availability and use of reception facilities and by improving the enforcement regime.

The Directive pursues the same aim as the 73/78 MARPOL Convention for the prevention of pollution by ships.[59] This MARPOL Convention was signed by all European Member States.[60] The convention regulates discharges by ships at sea, while the Directive focuses on ship operations in community ports and addresses in detail the legal, financial and practical responsibilities of the different operators involved in delivery of waste and residues in ports.

Member states shall ensure that the costs of port reception facilities for ship-generated waste shall be covered through the collection of a fee from ships.[61] All ships calling at the port of a Member State shall contribute irrespective of the actual use of these facilities. This cost-recovery system encourages the delivery of waste on land and discourages them being dumped at sea.

To this effect, arrangements may include the incorporation of a fee in the port dues or a separate standard 'waste fee.' The commission interprets the incorporation of the fee in the port dues as meaning at least 30% of these port dues.

In the context of the port reception facilities, port State control can be carried out. For example, whenever a vessel does not properly dispose of its waste or oily residues, oil samples can be taken to prove the illegal discharge was effected by that vessel.

In Belgium, the reception facilities are provided for by private entities. As a consequence, the cost is not included in the port taxes. This system does not encourage the delivery of waste on land, and a change in it is worth considering. It is not only important to have reception facilities available. As well, the collection of a 'waste fee' from ships irrespective of the actual use of the facility would encourage the use of these facilities. In Belgium, these facilities are private projects.

5.3 NORTH SEA DISASTER PLAN AND PLACES OF REFUGE

When there is an incident, coastal States must be prepared. It is desirable to have a relevant emergency plan ready together with a designated place of refuge where the vessel

---

[58] *OJ* L332/81, 28.12.2000; see <http://www.europa.eu.int>.
[59] International Convention for the prevention of pollution from ships (London, 1973) and the Protocol of 1978 to the international convention for the prevention of pollution from ships (London, 17.02.1978) as approved in Belgium by the law 17.01.1984, *B.S.*, 24 May 1984.
[60] Before 01.05.2004 with the number of European Member States being 15.
[61] Section 8 Directive 2000/59/EC of the European Parliament and of the Council, 27 November 2000 on port reception facilities for ship generated waste and cargo residues, *OJ* L324/53, 29 November 2002.

can be brought, for example, to undergo repair operations.[62] It is in the interest of the coastal State to protect the coast line (and the marine environment) together with rendering assistance to vessels in distress.[63]

In Belgium, the responsibility for formulating marine environmental policies at the national level lies with the federal Minister for the Environment. A national emergency plan has been promulgated and designates the place(s) of refuge for Belgium.[64]

The national emergency plan or '*Rampenplan Noordzee*' consists in three phases, with those being pre-alarm, initial alarm and full alarm. Each phase clearly mentions what should be done, who is involved, and so on.[65] The vessel in distress could be admitted, according to this emergency plan, in a place of refuge, a port of refuge when a dry dockyard is not needed for repairs, or in a dry dockyard. The Belgian place of refuge is the anchor area known as 'Westhinder.'[66] This anchor area is a supervised area and provides protection to the vessel in distress against adverse weather conditions. Ironically, however, it provides no protection whatsoever against possible marine pollution. When the anchor area of Westhinder is unavailable due to specific circumstances, two other emergency anchor areas are available, but permission from the authority is required for their use. These emergency places are the anchor areas south the AZ buoy and north of the NE Akkaert buoy. In cases when neither of these areass can be used for anchorage, the governor of the Province must designate another place of refuge.

As other possible port(s) of refuge, the port of Ostend and the port of Zeebrugge are designated in the emergency plan. These options would be appropriate for a vessel in distress and needing assistance or repair without necessity of a dry dockyard. On these occasions, permission to enter the port must be given by the proper authorities.

The national emergency plan mentions dry dockyards. However, Belgium does not have its own dry dockyards, and therefore the dry dockyards of Vlissingen in the Netherlands and of Dunkirk in France are indicated as being the nearest dockyards for vessels in distress in Belgian territorial waters.

5.4 LEGAL ASPECTS: JUDICIAL PROCEEDINGS

When air surveillance or MUMM, observes an oil spill, the reporting agent will draw up a report (*process-verbal*) as to the facts of the case for the public authority.

In Belgium, evidence can be provided through various sources:

- written statements (*procés-verbaux*)
- photographs
- video footage and other information based on any of the following

---

[62] Directive 2002/59/EC section 20 provides coastal states must foresee places of refuge for vessels in distress.
[63] IMO Resolution A.949 (23) of 5 December 2003 Guidelines on places of refuge for ships in need of assistance; IMO Resolution A.950 (23) of 5 December 2003 Maritime assistance services, <http://www.imo.org/Safety>.
[64] 'Rampenplan Noordzee, Provinciaal Gouvernement van West-Vlaanderen', anno 2003.
[65] <http://www.bonnagreement.org> provides a full description of this emergency plan.
[66] The anchor area 'Westhinder' is adjacent to the Belgian territorial waters, and Belgium requests this area to be recognised as territorial Sea.

- SLAR (sight looking airborne radar being an instrument capable of detecting possible surface pollution),
- IR (Infra Red Sensor a sensor which provides the picture of the relative temperature defences),
- UV (Ultra Violet Scanner providing a picture of the total spill area),
- FLIR (Forward looking Infra Red Camera being similar to the IR sensor, but providing a different angle of view),
- Positioning systems (A system used to obtain the exact position of the observing platform and includes Decca, GPS or other navigation systems),
- Oil sample analyses by GC/MS (Gas Chromatography/Mass Spectrometry), which is a technique that can give a 'finger print' of the analysed oil)
- Use of colour code (the colour of the oil spill depends on the layer thickness or type of the oil. With the colour code, it is sometimes possible to estimate the amount of spilled oil.)
- Radio recordings (the conversation with the suspected violator recorded),
- Port inspection reports (a report from the port State Inspector with copies of all relevant documents. On request, the port State Officer can take oil samples to be analysed at a later stage),
- Computer modelling (being a digital calculation model used to backtrack from the position of a detected oil spill, with the objective of finding the original position of the discharge. The same model can also predict the drift of the slick).[67]

The collection of the evidence in Belgium is carried out by the Management Unit of the North Sea Mathematical Model (MUMM) by the air surveillance, the Ministry of Transport (Maritime Affairs) for port State control, and by COMPOSNAV (Naval Operations Command).[68] The MUMM and the associated office of the public prosecutor assess the evidence.[69] The court competent to assess the matter is the Criminal Court of Bruges or Antwerp.

In Belgium, the Belgian Statute on the protection of the marine environment is considered by the government as a tool of 'deterrence.' Up to now, compensation for disruption of the environment as such has not been prosecuted in court. Rather, a mere provisional claim of one EURO will be made in relation to the disruption.[70]

With reference to international legal assistance in criminal matters, the international police organization INTERPOL must be mentioned. Through its network, rapid information or requests can be exchanged between different States, for example, so as to allow a port State to control a vessel at the next port of call.[71]

---

[67] Bonn Agreement, Manual Oil pollution at sea – part 2, 18-19.
[68] Bonn Agreement, Manual Oil pollution at sea – part 2, 9-11.
[69] *Ibid.*
[70] The Royal Decree assessing the economical value of the disruption is still to be promulgated.
[71] Bonn Agreement, Manual Oil pollution at sea – part 2, 34.

5.5 FUTURE DEVELOPMENTS: ZERO TOLERANCE!

On 10 December 2003, the Belgian government announced its future approach concerning the marine environment and referred to a standard of 'zero tolerance.' To this effect, a five-point programme was presented.[72] This programme should come into full effect within the next four (2007).

5.5.1 Maximize the Chances of Apprehension: the Introduction of Unmanned Cameras

By introducing new technologies like 'unmanned air vehicles' (UAV's) and 'satellite and radar systems,' the Belgian area would be supervised twenty four hours a day. These UAV's will be equipped with unmanned cameras.

As a result, the chances of getting caught increase, and so too does the effectiveness of the policy of deterrence.

5.5.2 Referee Magistrate North Sea

To improve the policy and tools of prosecution, priority must be given to files regarding damage to and disruption of the marine environment, and the burden of proof must be well-organized by an appropriate scenario like centralizing the available expertise and coordination with a 'referee' magistrate for the North Sea. At present, this magistrate has been appointed. A referee magistrate is a sort of court rules on urgent proceedings.

There must be solid cooperation between the referee magistrate, the police authorities competent for navigational matters, the shipping inspection and the environmental inspection when gathering evidence through air surveillance, ship and port inspections (port State control). To this end, an appropriate framework will be set up by the middle of 2004.

5.5.3 Active International Cooperation

Cooperation between coast States, flag States and port States is of the utmost importance for dealing effectively with possible infringements. Belgium has to become actively present by announcing its policy of zero tolerance in international cooperation bodies, like the International Maritime Organization (IMO), the North Sea Network of Prosecutors and Investigators, and others.

5.5.4 Equipment

To assure the deployment and availability at sea of the equipment for fighting oil pollution, the Minister will give priority to the acquisition of multi-functional craft. Such vessels will be used for several purposes, like the fight against oil pollution, but also for

---

[72] Press release of 10 December 2003, presented by the Minister of North Sea affairs, Johan Van de Lanotte.

the fixing of buoys, and the upkeep of navigational tools, and so on. The investment will be a public/private project. In other words, the government will work closely with the private sector in view of implementing this target.

5.5.5 Compensation for Disruption of the Environment

As previously discussed in this paper, the Belgian Statute on the protection of the Marine Environment mentions that there must be a Royal Decree (i.e. an order in council) whose aim will be the assessment of the economic effects in monetary terms of a disruption of the marine environment. It is stated that priority will be given to the elaboration of this Royal Decree, but up to now it has yet been made available.

6 CONCLUSION

The Belgian Statute on the protection of the marine environment is a very strong instrument in the fight against pollution. It not only reflects international and European legislation, but also improves the current legislation and goes beyond what is provided for in European and international rules.

First, the Belgian authorities do not only exercise jurisdiction within their territorial waters, but also within the EEZ, and this in general for the protection of the marine environment. Purely operational discharges are therefore included in the scope of the Statute.

Second, a new principle is introduced, that being the disruption of the marine environment. Compensation is foreseen for damage to the environment and for the disruption of the marine environment. Not only will the economical consequences of pollution be considered. Rather, all of its consequences qualify for compensation under the Statute for the protection of the marine environment. Also, pure economic loss qualifies for compensation.

Practical experience has shown that, up to now, the Statute has been used mostly as a 'tool of deterrence.'

The economic assessment of the disruption of the marine environment is a problem. The Royal Decree must still be released to the effect of assessing the economic effect in money terms. It is made a priority on the list of future steps to be taken by the Belgian Government.

Until the release of a Royal Decree, dealing with the assessment of costs and expenses next to the damages borne in case of a 'pollution incident,' a merely provisional claim of 1 EURO will be filed in Court. Again, this attitude is merely an illustration of the policy of 'deterrence.'

LIST OF REFERENCES

A. Carette, 'De aansprakelijkheidsregeling uit de wet ter bescherming van het mariene milieu' (1999) *Tijdschrift voor Milieurecht* 362-374.

F. Maes, 'De wet van 20 januari 1999 ter bescherming van het Mariene milieu in de zeegebieden onder de rechtsbevoegdheid van België Op weg naar een duurzaam gebruik van de zee', (1999) *Tijdschrift voor Milieurecht* 270-285.

COMPENSATION FOR OIL POLLUTION DAMAGES: THE AMERICAN OIL POLLUTION ACT AS AN EXAMPLE FOR GLOBAL SOLUTIONS?

James Boyd

1 INTRODUCTION

US environmental regulation relies in large measure on polluter liability as both a deterrent and as a way to finance compensation for environmental damages. While most US environmental statutes also include *ex ante* operational, public process, and technological standards, the *ex post* imposition of liability is a distinctive feature of the American approach to regulation. Other nations employ liability, but no other nation as frequently or drastically imposes liability damages on the private sector. This paper describes the use of liability to deter and compensate for a particular kind of damage: that arising from marine and coastal oil and hazardous waste spills.

Liability for environmental damages is strict under both US common law and the major environmental statutes. Unlike fault-type liability rules, strict liability imposes the full burden of environmental costs on the pollution generator, independent of any precautions taken. In principle, strict liability leads to the internalization of otherwise externalized costs. Cost internalization is desirable for both distributive and normative reasons. Strict liability serves distributive goals by providing compensation to victims. It serves normative ones by creating financial incentives that lead to optimal levels of deterrence.[1]

The principal law governing US oil spills is the Oil Pollution Act, enacted in 1990. The law includes a suite of requirements designed to improve the safety of vessels and oil transport facilities, including technology and reporting requirements. The law also makes operators liable for three broad classes of costs: response and cleanup costs, damages to private property, and damages to public natural resources. Similar in impact are regulations derived from CERCLA, the Superfund law that govern vessel-related hazardous substance releases. Because CERCLA liability parallels oil pollution liability under OPA for marine spills, regulations deriving from both laws will be discussed.

This paper will focus on two aspects of OPA and CERCLA compensation law that are particularly distinctive. The first is the its creation of liability for damages to public natural resources. Natural resource damage (NRD) liability is the most distinctive and complex aspect of the compensation issue. Other kinds of compensation – cleanup and response costs and damages to private property – are more common in other legal systems

---

[1] See W.M. Landes and R.A. Posner, *The Economic Structure of Tort Law* (Harvard University Press, Harvard, 1987) for a history of the development and justifications for the theory of strict liability. Strict environmental liability is also mandated by statute, as in the Superfund amendments.

Michael G. Faure and James Hu (eds), Prevention and Compensation of Marine Pollution Damage. Recent Developments in Europe, China and the US, 137-163.
© 2006 Kluwer Law International. Printed in the Netherlands.

and more straightforward to calculate. Cleanup and response costs are the costs of minimizing an oil release and removing contamination from waters and shoreline. Property damages are damages to real estate, other property, or income.[2] These kinds of damages are relatively conventional and easy to calculate, since damages can be easily quantified in dollar terms. Natural resource damages are a more novel and challenging aspect of compensation law. In particular, NRDs require the government to calculate the social loss associated with damages to resources that are not privately owned or traded in markets. Without private ownership and trade there is no clear way to derive the social value of damages (by inferring them from prices, for example). For this reason, NRD liability is controversial and raises a host of legal and technical issues, as will be discussed in Section 3. The legal history of the NRD concept will be reviewed, as will the government's evolving approach to the calculation of such damages. The paper will also summarize the frequency and scale of NRD damages relative to other damages associated with marine oil spills.

The second distinctive aspect of OPA is its approach to funding cleanup and restoration. OPA requires financial assurance as a pre-condition for operation. These requirements are akin to mandatory insurance or minimum capitalization requirements and are designed to ensure that responsible parties have the funds necessary to pay for damages. OPA also includes an Oil Spill Liability Trust Fund designed to pay for costs that are not immediately recoverable from private, responsible parties. Section 5 discusses these financing issues.

Before describing OPA and its NRD provisions a description of the broad legal environment is necessary. First, vessels can be liable for damages, including NRDs, under several US statutes in addition to OPA. In particular, the Clean Water Act and the National Marine Sanctuaries Act provide legal authority for the collection of vessel-related NRDs. Second, NRDs are a part of other environmental statutes that do not apply to vessels. The Superfund law (CERCLA), for example, imposes NRDs for damages associated with hazardous waste sites, vessels carrying hazardous cargo (other than oil), and some releases to inland waterways. Because OPA and CERCLA are the principle statutes governing vessel liability, they will be the focus of this paper.

OPA is a potential model for other nations' approaches to marine oil and hazardous waste pollution. As will be argued, the imposition of NRDs is an important innovation that addresses a global need: restoration of damaged ecological services and acknowledgement that natural resources have significant economic value that should be included in the calculus of damages. Another important component of deterrence that is worthy of emulation is mandatory financial requirements. Financial assurance promotes deterrence and compensation by fostering polluter cost internalization. It also harnesses the expertise of financial intermediaries, like insurers, with capabilities in risk analysis and management. NRDs and financial responsibility requirements are what is most distinctive about US maritime spill law and are also what is most worthy of emulation by other countries.

---

[2] The government can recover lost government revenue (for example, lost fishing license revenue due to damage to a fishery).

2 OPA AND US NATURAL RESOURCE DAMAGES LAW

This section describes the legal basis for natural resource damage liability under US law, with particular emphasis on NRDs under OPA and CERCLA. Several US environmental statutes establish liability for injury to natural resources. The Deepwater Port Act of 1974 and the Clean Water Act amendments of 1977 introduced NRD liability to federal law.[3] Subsequent to, and in most ways superseding, those statutes, liability for NRDs was established under CERCLA,[4] OPA,[5] and the National Marine Sanctuaries Act.[6] These latter statutes significantly expanded the reach and amounts of potential NRD liability. OPA, for example, significantly increased prior liability limits and allowed for the recovery of income lost because of damage to public resources.[7] As noted above, liability for natural resource (and all other) damages under the statutes is strict in that liability is imposed irrespective of precautions, care, safety, or other measures undertaken to guard against injury. Among potentially responsible parties (PRPs), liability is joint and several.[8] There is a statute of limitations for the recovery of damages, and a causal link between the defendant's actions and a natural resource injury must be established.[9]

Also, many US states have their own natural resource damage laws.[10] In fact, more than half the states have independent statutory authority to pursue NRDs.[11] In some cases,

---

[3] For the CWA, see 33 USC 1251 et seq. Section 311 of the CWA regulates the discharge of oil and other hazardous substances into navigable waters, allows the government to remove the substance, and holds the responsible parties liable for that removal. The removal cost is defined to include 'costs for restoration or replacement of natural resources damaged or destroyed'. 33 USC 1321. The Deepwater Port Act of 1974, which preceded the CWA, established liability for damages to natural resources to be recovered by a federal trustee and used for restoration. 33 USC §1501-1524, 1982.

[4] Section 107 of the act establishes NRD liability and authorizes federal trustees to recover damages for assessing and correcting natural resource injuries. 42 USC 9607(f)(1).

[5] Section 1002 of the act establishes liability for 'injury to, destruction of, loss of, or loss of use of natural resources'. 33 USC 2702(b)(2)(A).

[6] The NMSA uses the same definition of natural resource damages as OPA, 16 USC 1432. Section 1443 establishes liability and authorizes civil actions to pursue cost recovery.

[7] The prior limitation on state-imposed liability was created by the Federal Limitation of Liability Act of 1851, 46 USC §183-189. OPA contains liability limits of its own, as discussed in section 5 *infra*, though these can be breached if the vessel operator is found to have been grossly negligent.

[8] 33 CFR §138.30(a). OPA and CERCLA do not explicitly provide for joint and several liability, though joint and several liability is strongly implied (CERCLA, 42 USC 9607; OPA). Court interpretations of the somewhat vague statutory language, however, have established joint and several liability as the rule when damages are indivisible. For a CERCLA case, see *United States v. Chem-Dyne Corp.*, 572 F. Supp. 802 (S.D. Ohio 1983) at 810-811. For an OPA case, see *Sun Pipe Line Co. v. Conewago Contractors, Inc.*, 1994 WL 539326, *8 (M.D. Pa. 1994), citing 136 Cong. Rec. H6933-02, H6936 (daily edition 3 August 1990).

[9] The statute of limitations for filing an NRD claim is three years from completion of cleanup at an NPL site, or three years after discovery of NRDs at a non-NPL hazardous waste site. CERCLA, 42 USC 9612(d)(2); OPA, 33 U.S.C 2712(h)(2).

[10] In some states – Pennsylvania, for example – even municipalities can sue for NRDs. 35 PA. Cons. Stat. Ann. 6020.507(a).

[11] As of 1995, one survey found that 28 states had passed laws authorizing NRD recovery. Environmental Law Institute, An Analysis of State Superfund Programs: 50-State Study, 1995 Update. ELI Project #941724, 46. For a more recent but somewhat less complete state survey, see Association of State and Territorial Solid Waste Management Officials, *Survey of State Remedial Program Activities in Natural Resource Damages*, February 1997, at <http://www.astswmo.org/Publications/ascii/nrdsur.txt>.

the liabilities arising under state law can significantly expand upon federal liability. For example, in contrast to federal NRD law, state NRD laws do not always cap liability.[12] In most cases, however, state claims are pursued through federal law,[13] and at least until the mid-Nineties, few state NRDs had been recovered.[14]

2.1 WHAT IS A NATURAL RESOURCE DAMAGE?

In physical terms, natural resource damages are damages to land, fish, wildlife, biota, air, water, groundwater, and other resources.[15] Physical injuries can take a variety of forms but typically relate to adverse changes in the health of a habitat or species population and in the underlying ecological processes on which they rely.[16] In legal terms, the definition of NRDs is restricted to resources that are owned, controlled, or managed by federal, state, or other governmental entities, including foreign governments.[17] Damages to pure private property interests are not considered natural resource damages under US law. However, the definition of natural resources is not limited to government-owned resources. What must be demonstrated is a 'substantial degree of government regulation, management, or other form of control over the property' that is injured.[18] Accordingly, injuries to natural resources on or associated with private property can lead to NRD claims.

Defendants found liable for NRDs face three primary damage components: first, the cost of resource restoration to baseline conditions; second, compensation for 'interim losses,' that is, the lost value of injured resources pending full restoration; and third, the reasonable cost of the damage assessments themselves.[19] In some cases, the acquisition of 'equivalent resources' can be used as a substitute for restoration.[20] NRDs can also arise from natural resource injuries that have not yet occurred. Under both OPA and CERCLA the government has authority to respond to, and recover costs for, 'threatened' releases of oil or hazardous substances that pose a danger to natural resources.[21]

In economic terms, the goal of federal NRD liability is to 'make the environment and public whole' following a pollution event.[22] In principle, this is straightforward and

---

[12] New Jersey is an example. NRD assessments are required under New Jersey law during any remedial investigation of a hazardous waste site. N.J.A.C. 7:26E-4.7. The Spill Compensation and Control Act establishes liability for NRDs, with no cap on liability. N.J.S.A. 58:10-23.11(u)(b)(4). In other respects the state's law is similar to the federal, such as the definition of natural resource injury (N.J.A.C. 7:26E-1.8).

[13] States may recover NRDs under federal authority (CERCLA) at non-NPL sites within their jurisdictions. See Association of State and Territorial Solid Waste Management Officials, *supra* note 11, which found that 30 of 38 states responding pursued NRD claims through federal law.

[14] According to the ELI report, *supra* note 11, as of 1995, only eight states had recovered NRD claims under their statutes. However, it is likely that this number has expanded significantly in recent years.

[15] OPA 33 USC §2701(20); CERCLA 42 USC §9601(16).

[16] 15 CFR 990.52. 'Potential categories of injury include, but are not limited to, adverse changes in: survival, growth, and reproduction; health, physiology, and biological condition; behavior; community composition; ecological processes and functions; physical and chemical habitat quality or structure; and public services'.

[17] CERCLA §101(16); OPA §1001(20).

[18] Qualifying resources are 'resources the government substantially regulates, manages, or controls'; see *Ohio v. United States Department of Interior*, 880 F.2d 432 (D.C. Cir. 1989), at 460-461.

[19] CERCLA §101(6); OPA §1001(5), §1002(b)(2).

[20] 33 USC 2706(d)(1)(Λ).

[21] 33 USC 2702(a); 42 USC 9606(a).

[22] 15 CFR 990.53.

consistent with legal and economic theories of deterrence that emphasize the desirable consequences of social cost internalization by polluters. In practice, the determination of compensating remedies can be quite difficult, as will be discussed in Section X.

To clarify further the nature of natural resource damages, it is useful to distinguish between remediation and restoration. Remedial activities are clean-up actions designed primarily to reduce threats to public health. Restoration activities are directed at the recovery of resources themselves. This distinction is clearly evident in CERCLA. Remedial actions are typically undertaken only if a site is placed on the National Priorities List (NPL).[23] The criteria for placement on this list emphasize threats to public health. In contrast, CERCLA restoration authority is the authority to restore or replace natural resources to the conditions that would have existed without the hazardous release.[24]

2.2 COMMON LAW FOUNDATIONS

The public trust doctrine is the common law foundation upon which liability for NRDs is based.[25] Well before the passage of the relevant federal statutes, natural resource damages were collected under common law.[26] In fact, the public trust doctrine is a legal concept stemming from Roman law, applied historically to navigation and fishing rights. The doctrine held that the public (usually commercial) interests in navigable waters superseded any private claim to them.[27] The doctrine has evolved in US common law to represent the more general notion that the public has an explicit legal interest in the nation's natural resources.[28] Under US law, these interests are now understood to include the preservation of non-commercial public interests, like recreation.[29] As a result, citizens can challenge private or government actions that threaten the public's interest in natural resources.[30] The same principle can be used to compel administrative actions by the government to protect resources under their care.[31] Natural resources are often considered to be 'held in trust' by the federal or state government. A government's responsibilities as

---

[23] The US Environmental Protection Agency (EPA) can mandate certain actions, such as temporary relocation of residents, even if a site does not appear on the NPL. Under the National Contingency Plan, however (which establishes rules governing the response to hazardous releases), remedial actions may be taken only at sites on the NPL. 40 CFR 300.68(a).

[24] CERCLA §107(f)(1); 40 CFR §300.615(c)(3),(4).

[25] See J. Power, 'Reinvigorating Natural Resource Damage Actions through the Public Trust Doctrine', (1995) 4 *NYU Environmental Law Journal* 418.

[26] *State v. Jersey Central Power & Light Co.*, 308 A.2d 671 (N.J. Super. Ct. Law Div 1972), reversed on other grounds, 351 A.2d 337 (N.J. 1976).

[27] J. Sax, 'The Public Trust Doctrine in Natural Resource Law: Effective Judicial Intervention', (1970) 68 *Michigan Law Review* 471, 475. See *Illinois Central Railroad v. Illinois*, 146 U.S. 387 (1892) for an early statement of the principle in US law.

[28] See *National Audubon Society v. Superior Court of Alpine County*, 658 P.2d 709, 719 (Cal. 1983).

[29] As in *Matthews v. Bay Head Improvement Association*, 471 A.2d 355, 363 (N.J.), cert. denied, 469 U.S. 821 (1984); *Marks v. Whitney* 491 P.2d 374 (Cal. 1971), concluding that it is in fact unnecessary to define precisely what public use is threatened by a natural resource damage.

[30] Citizens can sue 'for the purpose of vindicating the public trust', *State v. Deetz*, 66 Wis. 2d 1, 13, 224 N.W.2d 407 (1974).

[31] *Sierra Club v. U.S. Department of the Interior*, 376 F. Supp. 90 (N.D. Cal. 1974).

trustee can be challenged by private citizens.[32]

Common law tort and contract principles have also been used to gain private recovery for natural resource injury. When individuals have a significant interest in public resources, they can sue to recover loss of those resources. A common example is when water pollution closes a commercial fishery, as in *Louisiana v. M/V Testbank*.[33] A nuisance action can be brought, usually by a state, but in some cases also by private citizens who can show some individualized injury to themselves.[34]

2.3 ADMINISTRATIVE ISSUES

OPA and CERCLA govern NRDs arising from the release of oil and hazardous substances, respectively. When a release occurs, either the US Environmental Protection Agency (EPA) or the US Coast Guard (USCG) has authority over removal responses and other immediate actions. When oil or hazardous substances are released on land, CERCLA and OPA require the EPA to investigate and respond. The USCG is responsible for releases taking place in coastal waters, including the Great Lakes. Authority for response to inland waterway releases is divided between EPA and USCG.[35] Technical response actions and other procedures to be followed in the event of a release are governed by what is called the National Contingency Plan.[36]

A different set of agencies is responsible for longer-term damage assessment and restoration activities. Restoration, assessment, and settlement of NRD claims are undertaken by federal, state, and tribal trustees. Only appointed, governmental trustees can seek natural resource damages. Private persons do not have the right to assert claims for natural resource damage.[37] The principal federal trustees are the National Oceanic and Atmospheric Administration (NOAA) and the Department of Interior (DOI).[38] Generally speaking, NOAA is the federal trustee for claims arising under OPA, and DOI for claims arising under CERCLA. Two sets of rules guide the agencies' respective NRD assessment procedures.[39] These rules also act as a blueprint for the determination of appropriate

---

[32] For instance, Wisconsin holds the beds of its navigable waters in trust for the use and enjoyment of its citizens. *Muench v. Public Service Comm'n*, 261 Wis. 492, 501 53 N.W.2d 514 (1952).

[33] 524 F. Supp. 1170, 1173 (E.D.La. 1981), aff'd 752 F.2d 1019 (5th Cir. 1985), cert. denied, 477 U.S. 903 (1986). In legal parlance, the issue decided in *Testbank* was whether the 'contractual relational economic loss' suffered by the plaintiffs was recoverable in tort. The case involved the collision of two vessels with the release of toxic chemicals. The pollution led to the government's closure of an area used for commercial fishing and in turn to a commercial loss for the affected fishermen.

[34] Cases where private plaintiffs successfully brought claims of public injury include *International Paper Co. v. Ouellette*, 479 U.S. 481 (1987); *Middlesex County Sewerage Auth. v. National Sea Clammers Ass'n*, 453 U.S. 1 (1981).

[35] 40 CFR 300.5.

[36] 40 CFR 300.

[37] *Artesian Water Co. v. Government of New Castle County*, 851 F.2d 643 (3rd Cir. 1988).

[38] The Departments of Agriculture, Defense, and Energy can also be trustees, 40 C.F.R. §300.600. State and tribal trustees vary and are designated by the governor of each state or by tribal chairmen, 40 C.F.R. §300.605, 610. Under OPA, foreign officials can also act as natural resource trustees for foreign resources. OPA §1006(a)(4),(b)(5).

[39] 15 CFR 990 (the NOAA regulations); 43 CFR 11 (the DOI regulations).

restoration actions. Accordingly, the damage assessment rules, together with the analysis of a specific site, largely determine the nature and scale of NRD recoveries.[40]

There is no direct private right of action to recover NRDs under federal law.[41] However, common law principles and federal administrative law extend the scope of possible citizen action.[42] The Administrative Procedures Act (APA) allows citizens to challenge government administrative actions.[43] Governmental actions affecting aesthetic and environmental interests associated with the use and enjoyment of natural resources are well within APA's ambit.[44] A central issue in such cases is the degree to which a private plaintiff can show a concrete, individual form of harm to a legally protected interest. Injury to a natural resource alone is an insufficient basis for such a claim.[45] Nevertheless, because so many government actions can have natural resource consequences, the scope for citizen-initiated action is potentially broad.

Moreover, most federal environmental statutes explicitly authorize citizen suits that can lead to injunctive relief and, in some cases, civil damages.[46] For example, although

---

[40] This should not be taken to suggest that recoveries are easily calculated via reference to an objective, unambiguous schedule of damages. The legal and technical uncertainties surrounding NRD assessment imply that PRP-trustee bargaining, as much as objective criteria, will determine the ultimate scale of financial damages. The damage assessment rules are discussed in more detail in Section 4.

[41] G. Spyridon and S. LeBanc, 'The Overriding Public Interest in Privately Owned Natural Resources: Fashioning a Cause of Action', (1993) 6 *Tulane Environmental Law Journal* 287, citing *Artesian Water Co. v. New Castle County*, 659 F.Supp. 1269, 1288 (D.Del. 1987), aff'd, 851 F.2d 643 (3rd Cir. 1988).

[42] See B. Breen, 'Citizen Suits for Natural Resource Damages: Closing a Gap in Federal Environmental Law', (1989) 24 *Wake Forest Law Review* 851.

[43] APA covers the breadth of federal agency action. 'A person suffering legal wrong because of agency action, or adversely affected or aggrieved by agency action within the meaning of a relevant statute, is entitled to judicial review thereof'. 5 USC 702.

[44] As in *Sierra Club v. Morton*, 405 U.S. 727 (1972), in which the plaintiff sued under the APA alleging that the US Forest Service erred in granting a construction permit for a ski area adjacent to a national park. The Supreme Court's ruling held that 'Aesthetic and environmental well-being, like economic well-being, are important ingredients of the quality of life in our society, and the fact that particular environmental interests are shared by the many rather than the few does not make them less deserving of legal protection through the judicial process', at 734 (though plaintiff's claim failed because of an inability to adequately demonstrate an injury-in-fact).

[45] See *Lujan v. Defenders of Wildlife* 504 U.S. 555 (1992), which challenged several federal agencies' decisions relating to the use of federal lands. 'Respondents mistakenly rely on a number of other novel standing theories... [such as] that any person using any part of a contiguous ecosystem adversely affected by a funded activity has standing even if the activity is located far away from the area of their use' (at 556). However, see Justice Blackmun's dissent: 'As I understand it, environmental plaintiffs are under no special constitutional standing disabilities. Like other plaintiffs, they need show only that the action they challenge has injured them, without necessarily showing they happened to be physically near the location of the alleged wrong' (at 595).

[46] A recent Supreme Court case addressed the conditions under which a citizen suit can compel a defendant to comply with regulatory permit violations affecting recreational and aesthetic interests. Even though no environmental harm occurred, the permit violation itself was found to create an injury-in-fact subject to redress (the same issue confronted in Lujan, *supra* note 45). *Friends of the Earth, Inc. v. Laidlaw Environmental Services, Inc.* 120 S. Ct. 693 (2000). For a more detailed analysis, see M. Healy, 'Standing in Environmental Citizen Suits: Laidlaw's Clarification of the Injury-in-Fact and Redressability Requirements', (2000) 30 *Environmental Law Reporter* 10455, 10465. 'Laidlaw would thus appear to permit any person with a proper interest in a resource affected by pollution levels that are illegally high as a result of defendant's statutory violations to show injury-in-fact as long as the person feels injured by that higher level of pollution'.

NRDs are not compensable to individuals under OPA, citizens can file suit to compel a federal agency to fulfil its role as a natural resource trustee.[47] Typically, such suits are barred when a government agency is considered to be diligently prosecuting an enforcement action.

Numerous technical difficulties are associated with the calculation and application of natural resource damages, as is discussed in the next section. Nonetheless, these difficulties should not obscure the fact that legal authority for the collection of NRDs is well established in the United States.

3 CALCULATING DAMAGES

Natural resource damages are an established legal concept with a clear economic rationale. In practice, however, NRDs raise a host of difficult issues for regulators. By their nature, NRDs acknowledge that natural resources produce a collective social benefit. These benefits are distinct from benefits associated with private property interests. This means that the value of NRD-related ecological benefits is not typically revealed by market transactions. Instead, government trustees must calculate the lost social value using techniques that are technically challenging and sometimes socially controversial. This section describes the history of alternative NRD damage calculation methods, explores the technical issues involved with damage calculation, and reviews the current application of damage estimation methods.

As will be seen, there are two broad ways in which to calculate an NRD. The first is to measure the lost social benefits arising from the damage. Basing penalties on the lost social benefits is the correct and most precise way to 'make the public whole.' Unfortunately, it is also a difficult task. Measuring lost benefits requires extensive data and sophisticated ecological and economic assessment techniques. Moreover, many of the methods used to calculate lost benefits are subject to scientific debate. For a regulatory agency these characteristics are problematic, due both to cost and political acceptance. For these reasons, an alternative approach is increasingly employed: damages based on replacement cost. The focus of this approach is not on the determination of lost social benefits, but rather on the replacement of biophysical functions and the services they generate. For example, if an oil spill damages a bed of seagrass, the objective is to replace the seagrass and the services it provides, such as habitat for species that are commercially and recreationally valuable. The cost of replacement is the damage imposed. Ecological replacement costs should not be thought of as an 'easy' task for regulators. However, replacement costs are much easier to calculate than lost benefits.[48]

---

[47] 33 USC §2706(g). 'Review of actions by any Federal official where there is alleged to be a failure of that official to perform a duty under this section that is not discretionary with that official may be had by any person in the district court in which the person resides or in which the alleged damage to natural resources occurred. The court may award costs of litigation (including reasonable attorney and expert witness fees) to any prevailing or substantially prevailing party'.

[48] For government reports that describe the practical advantages of replacement cost-based damages over lost benefit-based damages see Texas General Land Office, Texas Parks and Wildlife Department, Texas Natural

3.1 THE HISTORY OF NRD ASSESSMENT

CERCLA and OPA directed two US agencies, the Department of Interior (DOI) and the National Oceanic and Atmospheric Administration (NOAA), respectively, to develop rules governing natural resource damage assessment. This section reviews the development of these procedures and outlines the basic determinants and methods used to value natural resource damage.

DOI first published damage assessment rules in 1986.[49] The rules established two basic procedures: Type A for small releases of oil and hazardous waste, and Type B for large and complex releases.[50] These original rules took a relatively narrow view of the types of injuries that were compensable, the scope of compensation, and the methods to be used in damage assessment. The rules strongly favoured a market-oriented approach to damages and established a hierarchy of assessment methodologies. If there was a competitive market for the resource, a diminution in the resource's value (the damage) was to be captured by a price change. If this was inappropriate or impossible, standard appraisal methods were to be used. Only when neither of these was determined by the trustee to be appropriate would 'non-market' procedures be used. In addition, the original DOI rules limited damage awards to the lesser of the resource's replacement cost or the diminution in use value associated with the injury.[51] Moreover, damages were not to include those associated with non-use values (e.g., option, existence, or bequest values).

Two 1989 cases, *Ohio* v. *Department of Interior* and *Colorado* v. *Department of Interior*, forced DOI to revise those rules.[52] The revised rules were published in 1994. In *Ohio* the court ruled that CERCLA does not in fact mandate a least-cost approach to damages, thus invalidating the 'lesser of' damage rule. Instead, the court strongly favoured the use of restoration as the basis for damages, even if restoration is more expensive than monetary estimates of lost use value.[53] Based on the ruling, the current regulations allow for damages based on diminution in value only in cases where restoration is infeasible or where restoration costs are judged 'grossly disproportionate' in

---

Resource Conservation Commission, NOAA, and the US Fish and Wildlife Service, *Damage Assessment and Restoration Plan and Environmental Assessment for the Point Comfort/Lavaca Bay NPL Site Recreational Fishing Service Losses*, 2001; and D. Chapman, N. Iadanza and T. Penn, *Calculating resource Compensation: An Application of the Service-to-Service Approach to the Blackbird Mine Hazardous Waste Site, Damage Assessment and Restoration Program* (NOAA, Washington, 1998).

[49] 61 FR 20609, 1986.
[50] CERCLA sec. 301(c)(2)(A-B). OPA, 33 USC 2706(d). Type A assessment procedures are used for small incidents with limited duration and cost.
[51] 43 CFR 11.35(b)(2-3), 1987. This 'lesser of' damage rule is reflective of the common law standard for determination of tort damages.
[52] *Ohio v. Department of Interior* (880 F.2d 432, 442 (D.C. Cir. 1989) and *Colorado v. Department of Interior* (880 F.2d 481 (1st Cir. 1989). For the purposes of this analysis, the Ohio case is more important, since it related to Type B assessment procedures. The Colorado case came to broadly similar conclusions regarding the original DOI rules but relates primarily to the more limited Type A procedures.
[53] The NRDA rules' current focus on the replacement cost of resources, rather than their estimated market value, is a direct outgrowth of these cases. H.R. Conf. Rep. no. 653, 101st Cong., 2d Sess. 108 (1990). Also, see discussion in R. Randle, 'The Oil Pollution Act of 1990: Its Provisions, Intent, and Effects', (1991) 21 *ELR News and Analysis* 10119.

relation to lost benefit measures, including lost non-use benefits. The *Ohio* court also invalidated the exclusive reliance on market-derived definitions of damage.

Under the current rules, non-use values like option, existence, and bequest values are compensable.[54] The revised rules acknowledge that 'the mere presence of a competitive market [for resources] does not ... ensure the price will 'capture fully' the value of the resource.'[55]

The statutes are careful not to limit damages to those that can be directly measured in markets or that are based on observable resource uses. Second, as long as the agencies' own damage assessment rules are adhered to, there is a 'rebuttable presumption' of the analyses' correctness and legal validity.[56] Accordingly, the presumption provides agencies with considerable latitude in their choice of damage assessment methods. It has been suggested that Congress included the agency procedural advantage in OPA and CERCLA in order to expand the scale and applicability of NRDs beyond common law damage rules.[57] A remarkable aspect of both OPA and CERCLA is that they have given DOI and NOAA significant latitude to resolve these difficult valuation issues.[58]

In 1996, NOAA followed the 1994 DOI rules with rules of its own, to be applied to assessments authorized under OPA.[59] The rules define the goals of compensation and establish procedures to assess injury, establish causality, and calculate damages. Damage settlement can be arrived at without precise adherence to the rules. Any settlement, however, requires adherence to the broader goals for compensation established by the rules.[60] Settlements that fail to adhere to these standards can be challenged.[61] As noted above, the current emphasis is on restoration rather than a monetized estimate of lost benefits as the measure of damages. Although the 'lesser of' rule has been abandoned,

---

[54] Compensable value includes 'all of the public economic values associated with an injured resource, including use values and nonuse values such as option, existence, and bequest values.'. 56 FR 19760, 29 April 1991.

[55] 56 FR 19759, 1991.

[56] CERCLA §107(f)(2)(C); OPA §1006(e)(2).

[57] See F. Anderson, 'Natural Resource Damages, Superfund, and the Courts', in R. Kopp and V.K. Smith, (eds), *Valuing Natural Assets: The Economics of Natural Resource Damage Assessment, Resources for the Future* (RFF Press, Washington, 1993). (['It is likely that], the rebuttable presumption was placed in the statute specifically in anticipation that the government would adopt regulations that would press well beyond traditional damage awards').

[58] It should be emphasized that the difficulties are not related to 'financial' or 'economic' issues alone. The physical determination of injury itself poses significant technical challenges. For example, it is difficult to establish baseline conditions given natural variability, and the possibility of preexisting contamination. Also, the biological impact of loss of a subset of a biological (species) on a larger community is highly uncertain. The incomplete loss of a community can allow for accelerated biological restoration. As a final example, toxicity measurements for a given release on a given population are themselves a source of uncertainty.

[59] The rules are codified at 15 CFR 990 (the NOAA rules for OPA damages) and 43 CFR 11 (the DOI rules for CERCLA damages).

[60] 15 CFR 990.25. 'Trustees may settle claims for natural resource damages. at any time, provided that the settlement is adequate in the judgment of the trustees to satisfy the goal of OPA and is fair, reasonable, and in the public interest, with particular consideration of the adequacy of the settlement to restore, replace, rehabilitate, or acquire the equivalent of the injured resources and services'.

[61] See *Kennecott Utah Copper Corp. v. U.S. Department of Interior*, 88 F.3d 1191 (D.C. Cir 1996), in which the court rejected an inadequate NRD settlement and defined minimal standards that settlements must meet. In particular, the settlement failed to involve restoration, a conclusion the court felt was unwarranted given the trustees' limited analysis of options.

cost remains relevant to the determination of remedies. Technical feasibility and cost-effectiveness must be considered in the choice of restoration projects.[62] In addition to restoration, however, the rules allow for compensatory damages, which relate to the loss in value experienced between the time of injury and full restoration. As the NOAA rules put it, the goal of the damage assessment is to 'make the environment and public whole ... [and is to be] achieved through the return of the injured natural resources and services to baseline and compensation for interim losses of such natural resources and services from the date of the incident until recovery.'[63]

The NOAA rules provide trustees with wide latitude to choose among alternative valuation methodologies, including market price–based valuation methods, appraisal methods, hedonic analysis, and travel cost methods.[64] The aforementioned methods are associated with the estimation of use values. As noted earlier, however, the rules explicitly allow trustees to recover lost non-use values.[65] The estimation of non-use values raises significant methodological concerns and is viewed with particular alarm by potentially liable parties. For this reason, an independent panel was convened in 1993 to assess the validity of the so-called contingent valuation methodology to measure non-use values. The NOAA Panel established a set of guidelines for the use of CV methods and concluded that CV can provide a valid economic measure of value associated with resources people do not actually use but whose existence they may nevertheless value.[66] The rules now permit CV for estimating use and non-use values but only when 'no use values can be determined.'[67]

3.2 THE CHALLENGE OF ECOSYSTEM SERVICE ASSESSMENT

For more than twenty years, economists have been experimenting with methods to estimate the monetary value of non-marketed ecological services. For an even longer period of time, ecologists and other natural scientists have been grappling with ways to describe biophysical processes and the impact of human activity on biophysical outcomes. NRD assessment requires both biophysical analysis and economic analysis of the services generated by biophysical functions. Because ecological and economic systems are complex, scientifically rigorous assessment is a challenge.

Economic valuation of natural resource damages is built on a foundation of biophysical assessment. Ecosystem structure and functions, as described and evaluated by

---

[62] 43 CFR 11.82(d).
[63] 15 CFR 990.10. For an example, see Chapman, D., Iadanza, N. and Penn, T., Calculating Resource Compensation: An Application of the Service-to-Service Approach to the Blackbird Mine Hazardous Waste Site, NOAA Technical Paper 97-1, 16 October 1998.
[64] 43 CFR 11.83.
[65] 43 CFR 11.83; 15 CFR 990.30. 'The total value of a natural resource or service includes the value individuals derive from direct use of the natural resource, for example, swimming, boating, hunting, or birdwatching, as well as the value individuals derive from knowing a natural resource will be available for future generations'.
[66] The panel concluded that '(contingent valuation) produces estimates reliable enough to be the starting point of a judicial process of damage assessment, including passive-use values (i.e., nonuse values)'. Report of the NOAA Panel on Contingent Valuation, 58 FR 4601, 15 January 1993, at 4610.
[67] 43 CFR 11.83(c)(2)(vii).

ecological science, generate services valued by people. It is the services created by biophysical functions that yield social benefits. Ecological analysis, of course, is concerned with the biological, chemical, and hydrological relationships that determine biological production.[68] An ecosystem's structure, like size, vegetation, boundaries, and its functional aspects, such as the ability to absorb floodwater or remove contaminants from surface water are biophysical contributors – as inputs – to the services the habitat generates. While economic and biological systems are clearly different in important respects, both economics and ecology seek to understand the activity or productivity of such systems by understanding the systems' basic components and the functional relationships between those components.

The distinction between biophysical functions and ecosystem services can be defined in the following way: ecosystem services are the outcomes of ecosystem functions that yield value to people. For example, the ability of wetlands to mediate extremes of flood and drought at a downstream location is a biophysical function. A service is created if the absorbed floodwater yields less damage to buildings, roads, and agriculture or if the higher flows in limited rainfall years support a recreational fishery. Even if an ecosystem rates highly in terms of a functional characteristic, that function may not provide a socially valued service.[69] As another example, any consideration of the value of lost commercial and recreational fishing opportunities due to an oil spill in coastal wetlands must understand the role of the wetlands in fishery population dynamics. Leaving economic issues aside, ecological science is itself an evolving field where biological and physical interactions are often poorly understood, particularly where spatial phenomena are concerned.[70]

For conventional, marketed goods economics prescribes market-based methods for valuing goods and services (and losses to those goods and services). Market-based methods use prices and behaviour directly observed in the real world to calculate benefits lost or gained. The key to these conventional methods, and their relative simplicity, is that conventional goods and services are privately owned and traded in functioning markets. Accordingly, prices and demand for such products are a good guide to the products' social value. As described earlier, however, natural resources held in public trust are not owned, nor are they traded in markets. Accordingly, there are no direct price signals of their social value. Therein lies the problem for economic assessment. Instead, practitioners must employ relatively unconventional 'non-market' valuation techniques.

---

[68] There is also a long history of integrated economic and biological production function analysis in agricultural and natural resource economics. Among other things, agricultural studies show how substitution of one farm input for another (e.g. land for fertilizer, tractors for man-hours) affects production levels, or how landscape characteristics affect yields. For a general overview see J. Penson, O. Capps and C. Rossen (eds), *Introduction to Agricultural Economics* (2nd edn, Pearson Education, Prentice Hall, Harlow, 1999).

[69] Another way to make the point is to consider functionally identical ecosystems. Functional equivalence does not imply equivalent social value. Wetlands with an equivalent ability to remove nitrates from groundwater or absorb floodwater pulses will nevertheless differ in their social value. This follows since the number of people whose drinking water is purified and the number of homes protected from flooding will not be identical.

[70] See D. Tilman and P. Kareiva, *Spatial Ecology: The Role of Space in Population Dynamics and Interspecific Interactions* (Princeton University Press, Princeton, 1997).

Non-market valuation approaches fall into three broad categories: revealed preference, expressed preference, and derived willingness to pay.[71] Revealed preference studies look at the price people pay for marketed goods that have an environmental component. From those prices, inferences about the environmental benefits associated with the good can be made. Example: the value of habitat to commercial fishing. Another example is when people purchase a home near an aesthetically pleasing habitat with available access for recreation, home prices reflect the value of the aesthetic and recreation services realized by the homeowners.[72] Alternatively, for people who do not live near the site, recreational and aesthetic services can be valued by the time and money spent travelling to the area. These direct travel cost expenditures and imputed costs of travel time reveal a willingness to pay for the recreational services.[73] Differences in quality attributes can be valued if there are perceptible differences in the number, length, or cost of trips taken to sites of different quality.[74] The quality change may be from an alteration to the site. Consider another example: sediment, nutrient or pathogen trapping that affects swimming beach quality at a remote location can be valued if the relationship between the habitat of the remote site and the recreational quality of the beach can be estimated.

An expressed willingness-to-pay study asks people, in a highly structured way, what they would be willing to pay for a set of environmental improvements. Contingent valuation studies and contingent ranking are examples of this.[75] Surveys of expressed willingness-to-pay are expensive, controversial, and answers to questions may be affected by the specific context. The more complex the habitat change and its consequences, the more difficult the challenge of survey design and interpretation may be. A principal drawback to this approach is the risk that people may misunderstand the precise service being valued when undisciplined by the need to spend their own money. Also, respondents may not have fully formed preferences for the service. For both of these reasons, they may misstate their willingness to pay.[76]

In a valuation survey, the questions are structured in such a way that a particular economic calculation can be extracted from the results-willingness to pay in terms of a hypothetical amount of your income for a change in the state of the environment. This

---

[71] For a good overview of these methods see A.M. Freeman, *Measurement of Environmental and Resource Values: Theory and Methods* (Resources for the Future Press, Washington, 1993).

[72] Hedonic analysis is used in this type of study. See, e.g., B. Mahan, S. Polasky and R. Adams, 'Valuing Urban Wetlands: A Property Price Approach' (2000) 76 *Land Economics* 100.

[73] There is a substantial literature on this subject. See, e.g., K. McConnell, 'On-Site Time in the Demand for Recreation' (1992) 74 *American Journal of Agricultural Economics* 918.

[74] An important issue in travel cost studies, for example, is the definition of relevant substitutes for the sites in question. Northeast-Midwest Institute and National Oceanic and Atmospheric Administration, *Revealing the Economic Value of the Great Lakes* (publisher, place, 2001) ('omitting the prices and qualities of relevant substitutes will bias the resource valuations'), at p. 73.

[75] See R. Carson, N. Flores and N. Meade, 'Contingent Valuation: Controversies and Evidence', (2001) 19 *Environmental and Resource Economics* 173-210 for a review and defense of contingent valuation's role in the evaluation of environmental goods and services.

[76] See generally R.J. Kopp, W.W. Pommerehne and N. Schwarz (eds), *Determining the Value of Non-Marketed Goods* (Kluwer Academic, Boston, 1997) (presenting a good collection of articles relating to the contingent valuation method).

calculation is expected to represent the nature of an individual's preferences and is a way to aggregate those preferences over individuals in a population.[77]

For many years, researchers have also derived benefits via simulation studies based on engineering analysis.[78] For instance, if we want to know the value of having a wetland reduce flood damages, we can, in principle, estimate the dollar value of real property damage due to a flood event, and estimate the greater likelihood such an event will occur if the wetland is destroyed.

An important characteristic of benefit monetization studies is that they need not and typically do not monetize benefits arising from the entire suite of services generated by a site. This is true because different services typically require different assessment procedures. Recreational services will require one kind of study, eliciting existence values requires another approach, and understanding an individual's valuation of flood prevention yet another. Contingent valuation surveys can be designed to value a wider suite of benefits, but this complicates the administration and design of the survey. In many situations, it is not feasible to implement a survey to determine the value associated with each parcel of habitat of open space, preservation of an acre of wetland habitat, or increase in catch per unit of effort for a recreational fishery. Therefore, economists have long argued for the use of benefit transfer methods as way to avoid site-specific monetization exercises and minimize the need for costly new data collection.[79] Benefit transfer methods essentially take the benefits estimated at well-studied reference sites and relate those benefits to the benefits likely to be found at a site of interest for preservation, mitigation, or exchange. The 'transfer' of the benefits is made a function of differences in the reference site and site of interest. Benefit transfer still requires data collection and careful econometric analysis, but it reduces somewhat the burden of new data collection.

3.3 THE POLITICS OF NRD

The previous section describes the scientific and technical challenge of non-market benefit estimation. Although mainstream economics now accepts the validity of non-use values at a conceptual level, the methods used to calculate those values remain controversial.[80] In particular, the science behind both the ecology and economics of this kind of assessment has been the subject of much debate.

---

[77] If a measure of public opinion is desired to support decision making, other public opinion polling approaches and calculations also might be considered. For example a survey that asked about levels of agreement with statements about options and tradeoffs may be seen as a kind of 'valuation' effort.

[78] O. Herfindahl and A. Kneese, *Economic Theory of Natural Resources* (Charles E. Merill Publishing Co., Ohio, 1974).

[79] For an overview of benefit transfer methodologies, see the special issue of (1992) 28 *Water Resources Research* devoted to it. Also see S. Kirchhoff, B. Colby and T. LaFrance, 'Evaluating the Performance of Benefit Transfer: An Empirical Inquiry', (1997) 35 *Journal of Environmental Economics and Management* 75-93, and R. Kopp and V.K. Smith, *supra* note 57, p. 329.

[80] Northeast-Midwest Institute and National Oceanic and Atmospheric Administration, *Revealing the Economic Value of the Great Lakes* (published by the Northeast Mid West Institute and NOAA, US Department of Commerce, Washington, 2001), ('The development of natural resource damage assessment regulations was controversial because stakeholders disagreed over what damages would be assessed, how damages would be calculated, and how damages to environmental goods and services not valued in traditional markets would be calculated'), at p. 22.

Fear of NRD liability has prompted a set of legislative reform proposals, associated primarily with the reauthorization of CERCLA. Although no reforms have been enacted and the possibility of reforms remains in doubt politically, the reauthorization debate provides evidence of the concerns being raised by the private sector regarding financial responsibility for natural resource damages. Criticisms highlight the discomfort with which the regulated community approaches the valuation of NRD claims. The law's requirement that the public be 'made whole' following a natural resource injury, combined with the inherent difficulties of natural resource valuation have been a source of political controversy and calls for reform.[81]

A common feature of reform bills is the restriction or elimination of recovery for lost non-use values.[82] Some reform packages prohibit the use of the CV method to value NRDs.[83] Note, however, that none of these reform bills have passed, and there has been significant opposition to weakened NRD provisions, including opposition from the Clinton administration and from other administrative agencies.[84] Some states limit the methodologies that can be used by states to calculate the scale of NRDs. An example is Michigan's prohibition against use of the CV method to determine damages.[85] It should also be noted that the most controversial valuation methods, like CV, have in actuality been employed infrequently.[86]

3.4 DAMAGES: LOST BENEFITS VERSUS REPLACEMENT COST

Most NRD cases demand significant amounts of data regarding both biophysical conditions and demand for the services damaged. The cases tend to be complex and have

---

[81] Consider the following representative comment in the US legislature: 'The liabilities imposed on vessel owners under OPA are subject to dramatic inflation because of the methodologies embraced by agencies of the United States in calculating natural resource damages.'. Testimony of Svein Ringbakken, International Association of Independent Tanker Owners, Hearing before the Subcommittee on Coast Guard and Maritime Transportation of the Committee on Transportation and Infrastructure, House of Representatives, 26 June 1996.

[82] See S.8, Part 403, 105th Cong., A Bill to Reauthorize and Amend the Comprehensive Environmental Response, Liability, and Compensation Act of 1980, introduced 21 January 1997 (there will be 'no recovery under this Act for any impairment of nonuse values'), and S.1537, 106th Cong., Superfund Amendments and Reauthorization Act of 1999, introduced in the Senate 5 August 1999.

[83] See S.8, *supra* note 82, which states that a trustee's claim for recoverable assessment costs 'may not include the costs of conducting any type of study relying on the contingent valuation methodology.'

[84] See National Association of Attorneys General, Resolution, Superfund Reauthorization, Adopted 22-26 June 1997, available at <http://www.senate.gov/~epw/105th/joh_9-04.htm> (accessed 6 July 2000). Also, see C. Openchowski, 'Superfund in the 106th Congress' (2000) 30 *The Environmental Law Reporter*, 10648, 10659.

[85] Michigan law establishes liability for natural resource injury (Act 451, MCL 324.20126a). However, contingent valuation methods cannot be used for damage calculations 'unless a determination is made by the department that such a method satisfies principles of scientific and economic validity and reliability and rules for utilizing a contingent nonuse valuation methods or a similar nonuse valuation methods are subsequently promulgated'. MCL 324.20104(3). The state of Texas lists a set of acceptable valuation methods. The list includes contingent valuation but requires that contingent valuation studies be undertaken only in accordance with guidelines established by NOAA. 31 TAC §20.32(f).

[86] See testimony of Douglas Hall, NOAA, Subcommittee on Water Resources and Environment, House of Representatives, 11 July 1995. 'There have only been six contingent valuation studies completed to date, and only one in which the Federal Government was involved in litigation'.

not always been successful.[87] Whatever the precise approach, non-market benefit studies are complex, time-consuming, and costly. Methods seeking monetary estimates of ecosystem benefits are technically challenging, fraught with dangers that may not be obvious to non-practitioners, and require significant amounts of data collection. In other words, economic valuation methods are expensive and difficult to execute, particularly by non-economists. A professionally credible study costs, at a minimum, hundreds of thousands of dollars and can easily cost orders of magnitude more.[88]

It should not come as a surprise that NRD assessments have shifted from an emphasis on monetary benefit estimation to non-monetary techniques – in particular, habitat equivalency analysis.[89] This is understandable given the practical difficulty of ecological benefit studies. It is important to note, however, that in principled economic terms replacement cost is not the preferred method for damage assessment. The reason is that the costs of replacing a lost biophysical function may exceed the social benefit of the function. In effect, replacement cost damages can lead to wasteful expenditure on relatively unimportant resource improvements.

Consider a more conventional example to make the point. If a window is broken in an abandoned building, we would not necessarily require its breaker to replace the window, since the building may be unused or near demolition. Replacing the window would in such a situation be wasteful. This same argument can hold in some cases of damage to natural assets. As argued above, the alternative, explicit benefit estimation, can be costly and difficult. Also, the underlying uncertainties associated with biological systems makes it difficult to say that a particular biophysical function is as unimportant as a broken window in an abandoned building. The protection against particularly inefficient outcomes is the 'grossly disproportionate' standard, whereby benefit-based damages can be required only in cases where restoration is infeasible or where restoration costs are 'grossly disproportionate.'[90]

The NOAA assessment rules favour restoration over monetary measures largely because restoration costs are easier to estimate.[91] Similarly, the DOI is contemplating revisions to its NRD assessment rule, including a reduced role for methodologies that economically quantify damages, in favour of resource-based measures.[92] A restoration cost estimate relies on easily computable capital and labour costs (e.g., the costs of dredging, species reintroduction, or contaminant neutralization). These costs are easier to

---

[87] For an excellent description of the NRD process in court see D. Chapman and W.M. Hanemann, 'Environmental Damages in Court: the American Trader Case', in A. Heyes (ed.), *The Law and Economics of the Environment* (Edward Elgar, Cheltenham, 2001).

[88] See B. Conner and R. Gouget, 'Getting to Restoration', (2004) *The Environmental Forum* 19-29. In the case of the Exxon Valdez, government studies alone cost more than $100 million.

[89] See C.A. Jones and K.A. Pease, 'Restoration-Based Measures in Natural Resoiurces Liability Statutes', (1997) 15(4) *Comtemporary Economic Policy*; also see, M.P. Medina, 'Just Do It', (2001) *The Environmental Forum*.

[90] See *supra* note 53.

[91] For a discussion of damage assessment challenges, see R. Renner, 'Calculating the Cost of Natural Resource Damage', (1998) 32(3) *Environmental Science and Technology* 86.

[92] Interior Department Revises NRDA Rule to Lessen Focus on Economic Impact, *Environmental Policy Alert*, 6 September 2000, 7-8.

predict, rely on fewer economic valuation methodologies, and are verifiable *ex post*.[93] According to a NOAA director,

> Earlier damage assessment procedures emphasized determining a monetary value for the loss of use of the injured resources. Criticisms of this approach led NOAA to ... develop regulations that focus on damages measured by the actual cost of restoration... Focusing on determining the appropriate scale of restoration projects is preferable to focusing on the monetary amount of damages.... Instead of collecting damages, then determining how to spend that money on restoration, the goal of assessment is now focused on timely, cost-effective restoration of the natural resources that have been injured.[94]

In general, there remains a pronounced desire to avoid monetization of losses and gains. Monetization is not prohibited but is rarely favored.[95]

How do replacement cost damages get assessed in practice? The easy part of this damage calculation is the cost of restoration to the damaged resource itself. The capital, labour, and other costs necessary to the restoration are the cost of replacement and form the basis of damages. There is a more difficult aspect of the damage calculation, however. This challenge arises because it is necessary to compensate for losses arising in the period between an incident and full restoration – a period that can span decades.[96]

These are called 'interim losses' and, by definition, cannot be achieved via on-site restoration. Accordingly, interim losses require a search for comparable restoration actions. This requires a comparison of natural resource services provided by different types of natural or constructed assets and across different sites. Numerous challenges are associated with this kind of comparison. A first step is the identification of comparable restoration alternatives. Comparable alternatives mean restoration actions that provide the same type of resources or that provide the same kind of services as the injured resources. Once comparable actions have been identified, the government must then determine 'the

---

[93] The private sector prefers restoration cost to monetization, largely because of the former's greater predictability. See Guide to P&I Cover, *The Standard* (<http://www.standard-club.com/pubs/g2pi/pub_a_24.htm>; accessed 28 July 2000), responding to the NOAA rules: 'The final rules are clearly much better than the initial draft rules because they concentrate on restoration. However, the rules still create the very real likelihood that trustees will produce large and speculative claims in the United States in respect of damages to natural resources'.

[94] Testimony of David M. Kennedy, Office of Response and Restoration, NOAA, before a joint House Hearing, Subcommittees on Coast Guard and Maritime Transportation and Water Resources and Environment, 24 March 1999, <http://www.house.gov/transportation/cgmt/03-24-99/03-24-99memo.htm> (accessed 13 July 2000).

[95] See *supra* note 48. Also, 15 CFR 990.53(d)(3)(ii). 'If valuation of the replacement natural resources and/or services cannot be performed within a reasonable timeframe or at a reasonable cost ... trustees may estimate the dollar value of the lost services and select the scale of the restoration action that has a cost equivalent to the lost value'.

[96] There are two reasons that off-site restoration is typically needed to achieve full social compensation. First, complete physical restoration of the injured resource may be impractical. For instance, complete restoration of a damaged site is often cost prohibitive. If so, some other form of compensating restoration will be pursued, usually involving the enhancement of another comparable, but not identical, resource. Second, interim natural resource service losses must be compensated. Restoration of a site to prerelease conditions does not compensate for these interim losses. Supplemental restoration actions, either on site or off site, must be undertaken to compensate for those types of losses.

scale of those actions that will make the environment and public whole.'[97] Scaling often relies on the concept of ecosystem services.[98] If a resource's services are identified and the scale of those services estimated, the same can be done for a second, comparable resource. Protection, restoration, or enhancement of the second resource can then be defended as a comparable action. The NOAA rules, for instance, refer to both 'resource to resource' and 'service to service' valuation methods.[99] One particular replacement cost method is called Habitat Equivalency Analysis, which scales restoration to compensate for interim losses.[100] The method involves adding up and discounting the stream of lost benefits and comparing that to the level of services created by a restoration project.[101]

Habitat equivalency analysis and other replacement cost methods are used to avoid the challenges of monetary, benefit-based damage calculations. They are a pragmatic alternative, though not without their own challenges. What should not be missed is the importance of the US government's ability as a trustee to collect damages for injury to public natural resources, irrespective of the method used to calculate the damage. Other countries can learn from the US experience in tackling the difficult issues raised by resource value methods. However, the more important message is that natural resource damages are real in economic terms. Governments concerned with deterring and compensating for these real losses can look to the US NRD experience as a successful first step.

4 NRD CLAIMS – SCALE AND FREQUENCY

The widespread application of NRD claims is relatively recent.[102] For this reason, and because there is no central repository of data on NRD claims, it is difficult to accurately

---

[97] 15 CFR 990.53(d). For the CERCLA rules, see 43 CFR 11.80. 'Damages may also include, at the discretion of the authorized official, the compensable value of all or a portion of the services lost to the public for the time period from the discharge or release until the attainment of the restoration, rehabilitation, replacement, and/or acquisition of equivalent of the resources and their services to baseline'.

[98] 43 CFR 11.71. 'Services include provision of habitat, food and other needs of biological resources, recreation, other products or services used by humans, flood control, ground water recharge, waste assimilation, and other such functions that may be provided by natural resources'. 43 CFR 11.70: 'Upon completing the Injury Determination phase, the authorized official shall quantify for each resource determined to be injured and for which damages will be sought, the effect of the discharge or release in terms of the reduction from the baseline condition in the quantity and quality of services provided by the injured resource ...'.

[99] 15 CFR 990.53(d)(2).

[100] See *Habitat Equivalency Analysis: An Overview, Damage Assessment and Restoration Program*, NOAA, 2000 ('The public can be compensated for past losses of habitat resources through habitat replacement projects providing additional resources of the same type', 1).

[101] *Ibid.*, at 3 ('the process of scaling a project involves adjusting the size of a restoration action to ensure that the present discounted value of project gains equals the present discounted value of interim losses'). For a concrete example, see D. Chapman, N. Iadanza, N. and T. Penn, 'Calculating resource Compensation: An Application of the Service-to-Service Approach to the Blackbird Mine Hazardous Waste Site', *Damage Assessment and Restoration Program*, NOAA, 1998.

[102] The first NRD case filed under CERCLA (New Bedford) was in 1987. Also, see D. Woodward and M. Hope, 'Natural Resource Damage under the Comprehensive Environmental Response, Compensation, and Liability Act', (1990) 14 *Harvard Environmental Law Review* 189 (citing lack of NRD cases as of 1990).

summarize the range of NRD awards collected in the United States. By far the largest NRD case was the Exxon *Valdez* recovery. The *Valdez* case involved USD 2.1 billion in cleanup costs, approximately USD 1 billion in natural resource damages,[103] and a USD 5 billion punitive damage award.[104] Although a particularly visible case, *Valdez* was unique in the scale of injury caused, damages awarded, and methodologies used, including extensive use of the CV methodology.[105]

The government's pursuit of NRDs continues to evolve as it gains legal and economic experience with this kind of case. Meanwhile, several government and research studies provide at least a rough guide to the scale and likelihood of NRD claims. Two US General Accounting Office (GAO) reports have addressed the scale of NRD claims under CERCLA (note that these numbers reflect NRDs collected both inland and in the coastal zone. The first report found that as of April 1995, federal trustees had come to monetary NRD settlements in fifty cases[106] for a total of USD 106 million, with recoveries ranging from USD 4,000 to USD 24 million.[107] As of that time, the five largest NRD settlements (not necessarily recoveries) ranged from USD 12 million to USD 24 million.[108] Another fifty cases had settled with no NRD-specific payments, usually because site remediation activities were judged to have sufficiently addressed natural resource damages. The numbers were updated in a second report later in the year.[109] As of July 1996, total NRD settlement amounts had increased to $109 million from cases at sixty-two sites.[110]

More up to date numbers are available for NRD recoveries involving NOAA's Damage Assessment and Restoration Program. These numbers reflect coastal NRD recoveries brought under OPA, CERCLA, NMSA, or the CWA. They do not reflect 'inland' NRD recoveries.[111] Between 1991 and 2002 total NRD awards total USD 290 million from fifty-six cases. The average NRD recovery for these coastal cases is USD 4 million per case. The average NRD costs of oil-spill related damages are lower than for hazardous waste-related damages. For cases authorized under CERCLA, the average NRD award is USD 7 million, while for cases authorized under the OPA the figure is USD 3.2 million.[112] There is no clear time trend in the scale of NRD claims.

---

[103] 4 Oil Spill U.S. Law Rep., November 1994, at 13-14.
[104] The punitive damages have been appealed several times and have not been collected. In re Exxon Valdez, no. A89-095-CV (HRH) (D. Ct. Alaska), 24 September 1996.
[105] See R.T. Carson et al., *A Contingent Valuation Study of Lost Passive Use Values Resulting from the Exxon Valdez Oil Spill* (Report to the Attorney General of Alaska), 10 November 1992.
[106] At that time there were 1,290 sites on the National Priorities List.
[107] US General Accounting Office, Outlook for and Experience with Natural Resource Damage Settlements, GAO/RCED-96-71, April 1996, 4-5.
[108] As of July 1995, only about 40% of the US$83.8 million had been collected.
[109] US General Accounting Office, Superfund: Status of Selected Federal Natural Resource Damage Settlements, GAO/RCED-97-10, November 1996.
[110] *Ibid.* As of July 1996, about 80% of the settlements had been collected.
[111] These data are based on NOAA, Damage Assessment and Restoration Program, Natural Resource Damage Settlements and Judgments, 2002. The numbers reported here are based on statistics that exclude two inland cases (Blackbird Mine and Iron Mountain Mine) from the NOAA data.
[112] As noted earlier, cases can also be brought under the CWA and the NMSA.

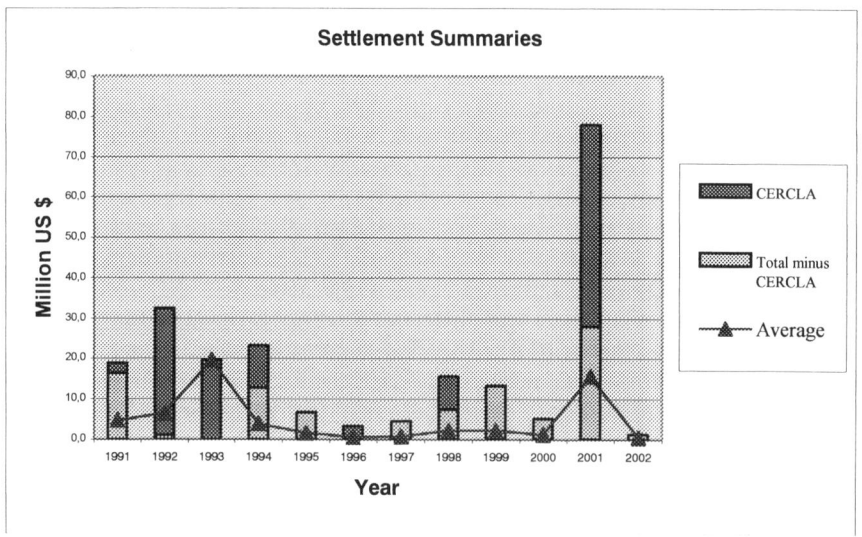

Figure 1 shows total 'coastal' NRD claims, by year.[113] The CERCLA claims roughly correspond to damages caused by non-oil hazardous substances. The remainder are predominately oil-related damages.

In one sense, NRD costs are a relatively small fraction of total oil spill liability. This is due to the fact that NRD claims are brought in only 10 % of the coastal oil spills reported each year.[114] NRD costs can be a large percentage of a particular case, however. A study of spills between 1984 and 1997 found that NRD damages – when they are claimed – account for 26% of the total damages, which include response costs, third-party claims, and government penalties.[115]

The significance of NRD liabilities remains a subject of debate. In the eyes of NRD reform advocates (typically representing the regulated community), future NRD liabilities pose a huge potential cost.[116] In the eyes of government trustees, NRD claims represent a

---

[113] See *supra* note 111.
[114] See B. Conner and R. Gouget, *supra* note 88. This percentage is in part a reflection of the significant costs associated with damage assessment and the pursuit of NRD claims. See US General Accounting Office, Outlook for and Experience with Natural Resource Damage Settlements, GAO/RCED-96-71, April 1996, at 4. 'Department of Justice officials state that the level of appropriations to fund federal natural resource damage programs is the single most important factor in determining how many sites can be assessed for damages'.
[115] D. Helton and T. Penn, *Putting Response and Natural Resource Damage Costs in Perspective*, mimeo, paper ID #114, International Oil Spill Conference, 1999, <http://www.darp.noaa.gov/pdf/costsofs.pdf> (accessed 17 July 2000). The study looked at 48 oil spill incidents occurring between 1984 and 1997.
[116] See statement of George Mannina, Director, Coalition for NRD Reform, Committee on Environment and Public Works, US Senate, 4 September 1997. 'When the NRD coalition formed two years ago, we were told NRD was a small problem involving only a few sites. A scant two years later, federal trustees state that they want to use their NRD authority at half the NPL sites and at 80,000 surface lagoons, 14% of all US lake acreage and 4% of all US river miles'.

small and manageable fraction of the environmental costs generated by polluting vessels and facilities.[117]

5 FINANCIAL ASSURANCE'S ROLE IN COMPENSATION AND DETERRENCE

In combination with liability law, financial responsibility rules foster the internalization of social costs by polluters by ensuring that firms possess the resources needed to compensate society for environmental costs. Insolvency can truncate the penalties borne by strictly liable tort defendants and thereby reintroduce the possibility of externalized social costs.[118] Any such externalization means that potential defendants will not be sufficiently motivated to take precautions against risk.[119] Bankruptcy and corporate dissolution defeat the law's ability to force polluter cost internalization by allowing many firms to abandon environmental responsibilities after reaping short-term financial gains. Non-recoverable environmental obligations are more than a theoretical possibility. The US landscape is littered with environmentally damaging operations that were either abandoned entirely or left unreclaimed due to bankruptcy.[120]

---

[117] See testimony of Douglas Hall, NOAA, 1995 House Hearing, *supra* note 86. 'With regards to the claim that NRD is a sleeping giant that is going to bankrupt industry, this is simply not the case. It amounts to nothing more than speculation that is unsupported by the record. We have had less than 5% of the sites on the national priorities list that have required compensation for natural resource injuries in addition to remediation. The compensation for natural resource damage at all NPL remedial sites has been less than 1% of the cost of remediation'.

[118] Bankruptcy proceedings do not always limit the ability to recover costs, though they certainly do not foster cost recovery. Generally speaking, debtors are protected from creditors (which can include tort victims) by the 'automatic stay' provision of the US bankruptcy code, 11 U.S.C. §362(a). There is, however, a 'police and regulatory power exception' to the automatic stay. The exception states that the automatic stay does not apply to the 'commencement or continuation of an action or proceeding by a governmental unit to enforce such governmental unit's police or regulatory power', 11 U.S.C. §362(b)(4). Courts have generally held that a CERCLA action is an exercise of the government's police or regulatory power and is not subject to the automatic stay. See generally, R.L. Epling, 'Impact of Environmental Law on Bankruptcy Cases', (1991) 26 *Wake Forest Law Review* 69.

[119] For analyses which have explored or employ this reasoning see A. Schwartz, 'Product Liability, Corporate Structure and Bankruptcy: Toxic Substances and the Remote – Risk Relationship', (1985) *Journal of Legal Studies* 689-736; S. Shavell, 'Liability for Harm versus Regulation of Safety', (1984) *Journal of Legal Studies* 357-374 and S. Shavell, 'The Judgement Proof Problem', (1986) *International Review of Law and Economics* 43-58; W.M. Landes and R.A. Posner, *The Economic Structure of Tort Law* (Harvard University Press, Harvard, 1987); L. Kornhauser and R. Revesz, 'Apportioning Damages among Potentially Insolvent Actors', (1990) *Journal of Legal Studies* 617-651; A.M. Polinsky and S. Shavell, 'A Note on Optimal Fines when Wealth varies among Individuals', (1991) 81 The American Economic Review 618 and J. Boyd and D. Ingberman, 'Noncompensatory Damages and Potential Insolvency', (1994) *Journal of Legal Studies* 895-910. Note that the mere possibility of bankruptcy is sufficient to weaken liability's ability to deter. The corollary to this statement is that bankruptcy need not already be observed for it to have an effect on firm incentives. From the standpoint of *ex ante* decision-making, whenever bankruptcy occurs with a positive probability the incentive to make costly investments in risk reduction is reduced.

[120] See J. Boyd, 'Financial Responsibility for Environmental Obligations: Are Bonding and Assurance Rules Fulfilling Their Promise?', (2002) 20 *Research in Law and Economics* (also available at <http://www.rff.org/Documents/RFF-DP-01-42.pdf>).

Two financial assurance rules govern marine oil and hazardous waste operations. The first rule, authorized by both OPA and CERCLA, governs water-borne vessels that carry oil or hazardous substances.[121] The second rule, authorized by OPA, governs offshore facilities used for oil exploration, drilling, production, or transport.[122] The vessel and offshore facility FR rules describe, among other things, implementation schedules, types of facilities to which the rules apply, financial instruments with which compliance can be achieved, minimum coverage amounts, and enforcement procedures. Individual coastal states may also impose financial responsibility requirements on vessels using their waters. For example, in 2004, Massachusetts passed a law requiring up to USD 1 billion in coverage for certain vessel operators.[123]

The vessel rule applies to tank vessels of any size, foreign-flag vessels of any size, and mobile offshore oil and gas drilling units.[124] Some smaller commercial vessels, like barges not carrying oil or hazardous substances, are excluded from the regulations. The offshore facility rule applies to facilities 'in, on, or under' navigable waters. Covered facilities include platforms, terminals, refineries, and pipelines used for oil exploration, drilling, and production.[125]

Vessel liability limits are a function of the vessel's size and type of cargo (oil vs. hazardous substances).[126] For large vessels, the limit (and coverage requirement) is USD 1,200 per gross ton for oil cargo and USD 600 per gross ton for hazardous substance cargo. Offshore facility liability limits are based on calculations of the volume of a 'worst-case' oil spill discharge. There are four types of 'allowable mechanism' that can be used by firms to demonstrate the existence of coverage: insurance, surety bond, self-insurance, and financial guaranty.[127] All four mechanisms are designed to ensure that liabilities can be satisfied, up to the statutory coverage requirements.

Insurance and surety bonds are financial commitments, purchased from third parties, guaranteeing payment of claims arising from liabilities of the purchaser. Generally speaking, insurance contracts 'pay out' when a liability claim arises. Surety bonds are somewhat different in that the risk of loss remains with the principal (rather than being transferred to an insurer). The surety pays out only when the principal defaults.[128] In most other respects, however, these two instruments are substantially similar. Not all insurers and sureties are acceptable coverage providers. In the case of surety companies, they must be one of those certified by the US Treasury Department.[129]

---

[121] 33 U.S.C. §2702; 42 U.S.C. §9607(a)(1). The vessel financial responsibility rules are codified at 33 C.F.R. part 138.

[122] OPA §1016. The offshore facility financial responsibility rules are codified at 30 C.F.R. part 253.

[123] State of Massachusetts, Chapter 251 of the Acts of 2004: An Act Relative to Oil Spill Prevention and Response in Buzzards Bay and Other Harbors and Bays of the Commonwealth.

[124] 33 CFR 138.12.

[125] 30 CFR 253.3.

[126] See 33 CFR §138.80(f)(3).

[127] The mechanisms are described at 33 CFR 138.80 (for vessels) and 30 CFR 253.28-31 (for offshore facilities).

[128] *Schmitt v. Insurance Co. of North America* (1991) 230 Cal.App.3d 245, 257. Typically, though, either the principal or surety may be sued on a bond, and the entire liability may be collected from either the principal or the surety. This characteristic of surety bonds is also tempered by the FR rules' 'direct action' requirements, described below.

[129] 30 CFR 253.31; 33 CFR 138.80(b)(2).

Self insurance allows relatively deep-pocketed companies to satisfy the coverage requirement by demonstrating sufficient financial strength. The vessel rule requires that two measures of financial strength, 'working capital' and 'net worth,' both be greater than the coverage requirement. The tests require not only financial strength, but financial strength based on domestic assets, in order to foster cost recovery. When using the financial test, firms must make annual reports that are independently audited according to generally accepted accounting practices.

A financial guaranty, or indemnity, agreement allows another firm, like a parent corporation, to satisfy the coverage requirement. Financial guarantors must themselves pass the corporate financial test and agree to guarantee the liabilities of the potentially liable firm. The requirements are identical to those for self-insurers, including the domestic assets requirement.

An important policy question is whether financial assurance for environmental liabilities should be made mandatory. An alternative, of course, is to rely on the voluntary provision of environmental coverage by private markets. Clearly, if left to themselves, private markets will demand and supply environmental insurance coverage. We know this, because we can observe such markets in active operation. The question, of course, is whether markets will voluntarily provide adequate insurance coverage from the standpoint of public policy. When purchased voluntarily, insurance benefits the purchaser by reducing risk, which is valuable in the presence of risk aversion. Note, however, that insurance is not voluntarily purchased as a means to internalize costs. In fact, cost internalization is directly at odds with the profit motive. When a firm expects to externalize some fraction of large liability claims against it (the judgement-proof problem) the desire for insurance in response to risk aversion will be countered by the desire to avoid internalizing otherwise externalized costs.

Why does insurance imply greater cost-internalization by an insured party? In the US, any insurer providing coverage to a potentially judgement-proof customer must build into the insurance cost its own exposure to the insured's liabilities. Insurers have been exposed to such costs under US law. Demand for environmental insurance thus usually involves this trade-off: risk-spreading versus greater cost internalization. When a firm's potential liabilities exceed its capital value the dis-incentive to purchase insurance (greater cost internalization) is likely to outweigh the benefits of insurance (reduced uncertainty).

Finally, political opposition to mandatory financial responsibility serves as a confirmation that insurance is unlikely to be provided voluntarily. While some opposition is due to complaints regarding arguable imperfections in the mandatory mechanisms themselves (e.g., the P&I Clubs' opposition to the lack of policy defences), many of the complaints are due simply to the obvious costs of compliance.[130] These costs, of course,

---

[130] A focus of lobbying efforts in the US is for the relaxation of the conditions under which firms can self-insure (i.e., not have to purchase insurance from third parties). It is clearly in a firm's economic self-interest to qualify. According to one committee reviewing such proposals 'additional mechanisms for qualifying as a self-insurer are needed to ensure that the costs of demonstrating OSFR do not cause serious economic harm to responsible parties'. Minerals Management Service, OCS Policy Committee Passes Recommendations on Oil Pollution Act Financial Responsibility Requirements (#50033), 4 May 1995 (viewed at <http://www.mms.gov/ooc/press/1995/50035.txt>). The public policy implications of relaxed qualifying conditions are negative, since relaxed conditions thwart cost internalization. The unsurprising message to be

are exactly what will inhibit the spontaneous voluntary development of an adequate cost-internalizing environmental insurance market. However, private sector coverage mechanisms are available, and at rates that can be absorbed by most of the firms to which the requirements apply. Moreover, the trends in coverage availability and premium affordability are positive.[131]

Even with OPA and CERCLA's financial responsibility provisions, a significant fraction of spill damages go uncollected due to bankruptcy or dissolution. One illustrative statistic is that between 1990 and 2002, more than 17% of payments from the Oil Spill Liability Trust Fund were unrecoverable due to bankruptcy or dissolution.[132]

6 CONCLUSION

Two distinctive features of the Oil Pollution Act and related aspects of CERCLA are worthy of international emulation. The first is the creation of liability for natural resource damages. For both economic and environmental reasons, NRD liability is an important component of statutes designed to deter environmentally damaging behaviour and provide compensation necessary to restore injured resources. Society in the broadest sense derives significant benefits from ecological assets in their natural state. While difficult to estimate with precision, these social benefits should be included in a regulatory calculus that seeks to internalize costs and make the polluter pay. NRDs are the principal mechanism to foster internalization of costs imposed on collective, public environmental resources.

The second distinctive feature of compensation law is the inclusion of mandatory asset and insurance requirements, or financial responsibility rules. These rules address a practical problem associated with the use of liability: the possibility that a defendant will dissolve or lack the wealth necessary to pay its liabilities after an accident. Financial responsibility rules are thus an important compliment to liability law. Particularly because marine accidents can in an instant create multi-million dollar liabilities, regulations should be in place to ensure that such liabilities will in fact be internalized by a polluter.

While significant implementation issues must be confronted, particularly as regards the calculation of natural resource damages, NRD liability and financial responsibility rules foster an environmentally sound and economically efficient approach to the regulation of marine and coastal spills.

---

taken from such a statement, however, is that greater cost internalization is costly. It is something firms wish to avoid, rather than something they will pursue voluntarily.

[131] See J. Boyd, *supra* note 120.
[132] The trust fund is a public revenue fund used to finance responses to spills. The fund then seeks collection of response and other costs from the responsible party. National Pollution Funds Center, *Year in Review* (<http://www.uscg.mil>, 2001) p. 10.

LIST OF REFERENCES

F. Anderson, 'Natural Resource Damages, Superfund, and the Courts', in R. Kopp and V.K. Smith, (eds), *Valuing Natural Assets: The Economics of Natural Resource Damage Assessment, Resources for the Future* (RFF Press, Washington, 1993), 26-61.

J. Boyd, 'Financial Responsibility for Environmental Obligations: Are Bonding and Assurance Rules Fulfilling Their Promise?', (2002) 20 *Research in Law and Economics* (also available at <http://www.rff.org/Documents/RFF-DP-01-42.pdf>).

J. Boyd and D. Ingberman, 'Noncompensatory Damages and Potential Insolvency', (1994) *Journal of Legal Studies* 895-910.

B. Breen, 'Citizen Suits for Natural Resource Damages: Closing a Gap in Federal Environmental Law', (1989) 24 *Wake Forest Law Review* 851.

R. Carson, N. Flores and N. Meade, 'Contingent Valuation: Controversies and Evidence', (2001) 19 *Environmental and Resource Economics* 173-210.

R.T. Carson et al., *A Contingent Valuation Study of Lost Passive Use Values Resulting from the Exxon Valdez Oil Spill* (Report to the Attorney General of Alaska), 10 November 1992.

D. Chapman and W.M. Hanemann, 'Environmental Damages in Court: the American Trader Case', in A. Heyes (ed.), *The Law and Economics of the Environment* (Edward Elgar, Cheltenham, 2001).

D. Chapman, N. Iadanza and T. Penn, *Calculating resource Compensation: An Application of the Service-to-Service Approach to the Blackbird Mine Hazardous Waste Site, Damage Assessment and Restoration Program* (NOAA, US Department of Commerce, Washington, 1998).

B. Conner and R. Gouget, 'Getting to Restoration', (2004) *The Environmental Forum* 19-29.

R.L. Epling, 'Impact of Environmental Law on Bankruptcy Cases', (1991) 26 *Wake Forest Law Review* 69.

A.M. Freeman, *Measurement of Environmental and Resource Values: Theory and Methods* (Resources for the Future Press, Washington, 1993).

M. Healy, 'Standing in Environmental Citizen Suits: Laidlaw's Clarification of the Injury-in-Fact and Redressability Requirements', (2000) 30 *Environmental Law Reporter* 10455-10465.

D. Helton and T. Penn, *Putting Response and Natural Resource Damage Costs in Perspective*, mimeo, paper ID #114, International Oil Spill Conference, 1999.

O. Herfindahl and A. Kneese, *Economic Theory of Natural Resources* (Charles E. Merill Publishing Co., Ohio, 1974).

C.A. Jones and K.A. Pease, 'Restoration-Based Measures in Natural Resoiurces Liability Statutes', (1997) 15(4) *Comtemporary Economic Policy* 111-122.

S. Kirchhoff, B. Colby and T. LaFrance, 'Evaluating the Performance of Benefit Transfer: An Empirical Inquiry', (1997) 35 *Journal of Environmental Economics and Management* 75-93.

R.J. Kopp, W.W. Pommerehne and N. Schwarz (eds), *Determining the Value of Non-Marketed Goods* (Kluwer Academic, Boston, 1997).

L. Kornhauser and R. Revesz, 'Apportioning Damages among Potentially Insolvent Actors', (1990) *Journal of Legal Studies* 617-651.

W.M. Landes and R.A. Posner, *The Economic Structure of Tort Law* (Harvard University Press, Harvard, 1987).

B. Mahan, S. Polasky and R. Adams, 'Valuing Urban Wetlands: A Property Price Approach' (2000) 76 *Land Economics* 100.

K. McConnell, 'On-Site Time in the Demand for Recreation' (1992) 74 *American Journal of Agricultural Economics* 918.

M.P. Medina, 'Just Do It', (2001) *The Environmental Forum* 3-5.

C. Openchowski, 'Superfund in the 106th Congress, (2000) 30 *The Environmental Law Reporter* 10648-10659.

J. Penson, O. Capps and C. Rossen (eds), *Introduction to Agricultural Economics* (Prentice Hall, Harlow, 1999).

A.M. Polinsky and S. Shavell, 'A Note on Optimal Fines when Wealth varies among Individuals', (1991) 81 *The American Economic Review* 618.

J. Power, 'Reinvigorating Natural Resource Damage Actions through the Public Trust Doctrine', (1995) 4 *NYU Environmental Law Journal* 418.

R. Renner, 'Calculating the Cost of Natural Resource Damage', (1998) 32(3) *Environmental Science and Technology* 86.

J. Sax, 'The Public Trust Doctrine in Natural Resource Law: Effective Judicial Intervention', (1970) 68 *Michigan Law Review* 471, 475.

A. Schwartz, 'Products Liability, Corporate Structure, and Bankruptcy: Toxic Substances and the Remote – Risk Relationship', (1985) *Journal of Legal Studies* 689-736.

S. Shavell, 'Liability for Harm versus Regulation of Safety', (1984) *Journal of Legal Studies* 357-374.

S. Shavell, 'The Judgment Proof Problem', (1986) *International Review of Law and Economics* 43-58.

G. Spyridon and S. LeBanc, 'The Overriding Public Interest in Privately Owned Natural Resources: Fashioning a Cause of Action', (1993) 6 *Tulane Environmental Law Journal* 287.

D. Tilman and P. Kareiva, *Spatial Ecology: The Role of Space in Population Dynamics and Interspecific Interactions* (Princeton University Press, Princeton, 1997).

D. Woodward and M. Hope, 'Natural Resource Damage under the Comprehensive Environmental Response, Compensation, and Liability Act', (1990) 14 *Harvard Environmental Law Review* 189.

THE PRESTIGE CASE. INTERNATIONAL AND SPANISH LEGAL REGIME FOR COMPENSATING DAMAGE*

María Paz García Rubio

1 RAISING THE PROBLEM

On 13 November 2002, the Bahamas registered tanker *Prestige*, laden with 77,000 tonnes of heavy fuel oil, broke in two off the Galician coast (Spain) spilling an unknown but substantial quantity of its cargo. The bow and stern sections, which are lying 3,500 metres below water, are estimated to contain 13,300 tonnes and 900 tonnes of oil respectively. In this contribution, the problem related to compensation for the damage caused to individual and collective rights by the disaster of *Prestige* will be considered.

The possible ways of achieving this compensation are various. They can consist of a public legal system based on the logic of social assistance, paid for by all citizens and following principles of distributive justice. They can also consist of a system of civil liability, of which the logic is one of commutative justice between the author of the damage and the victims. The person who has to pay for the damage is the person who has caused it. In all European legal systems, the most usual mechanism for obtaining compensation for injury caused by an accident like the *Prestige* continues to be, principally, civil liability. It is probable, however, that in such a complex case the mechanism of civil liability is insufficient. However, it continues to be the one preferred in many legal systems, which is inspired by the 'polluter pays' principle.

In this contribution, the system of norms and problems which arise from torts in the case of the *Prestige* will be shown. It should be noted from the outset that this is not a simple question. In this topic there are norms of various origins and natures, which are at times contradictory. There are at a minimum two groups of rules. On the one hand, there are international norms which are included in international conventions promoted by the International Maritime Organization (IMO) and which have been ratified by Spain. On the other hand, there are Spanish norms, which, equally, are diverse in nature. For the moment there are no norms on this topic in the European Union because Directive 2004/35/CE of the European Parliament and of the Council of 21 April 2004 on environmental liability with regard to the prevention and remedying of environmental

---

* Proyecto de investigación 'Daños y responsabilidad resultantes del vertido de hidrocarburos en el mar' (Ref.: BJU2003-01161), financed by the Ministerio de Ciencia y Tecnología; and Proyecto de investigación 'Indemnización de daños en el caso Prestige. Estado de la cuestión y propuestas de modificación legislativa' (Código: PGIDIT03CSO20201PR), financed by the Consellería de Innovación Industria y Comercio de la Xunta de Galicia.

damage[1] excludes from its field of application the damage in the sea caused by Oil Pollution.[2]

Prior to analyzing this set of norms, a general overview of the system of liability for damage to natural resources will be outlined.

2 GENERAL OVERVIEW OF THE SYSTEM OF CIVIL LIABILITY AS A MECHANISM FOR COMPENSATING ENVIRONMENTAL DAMAGE

The main function of the Spanish system of civil liability is to compensate the injured person for the damage which has been caused to them. It tries to re-establish the situation prior to the occurrence of the damage. However, it does not do so by making the injury disappear but by paying the injured, normally, with money. Only because this compensation obliges the party responsible to pay does the civil liability also have an indirect preventive function as it makes appealing the practice of actions that could give rise to the obligation to compensate. It never attempts to punish the defendant, although having to pay significant sums of money for serious damages can have, in the eyes of the author, a punitive character.

In the Spanish system there are two reasons for which a person can be liable for the damage caused: firstly, fault or negligence (subjective liability), and second, the simple activity of creating a risk (strict liability). In the first case, a party who is at fault or who has acted negligently must pay, but a party who has acted correctly does not have to pay. In the second case, the party who has carried out an activity which creates a risk must pay, even if the party has acted with due diligence. From a practical point of view, the importance of the difference is that if the liability is strict, the injured party does not have the burden of proving that the defendant was at fault, but only that his action or lack of action caused the plaintiff damage. The difference is relevant in cases like the *Prestige* because, for example, the owner of the ship that caused the oil pollution is strictly liable independent of whether he was the most diligent ship owner.

The most important elements in civil liability are: 1) establishing the liable persons or entities; and 2) determining the damage and identity of the owners of the damaged goods or interests, that is, the victims. In a problem of oil pollution at sea as in the case of the Prestige there exist both pecuniary and non-pecuniary losses. The former comprises the harm to the victims' assets, which can be given a monetary value. These interests can be either private or public (for example, the costs of the sanitary services of people who had to be assisted during the process of cleaning beaches). These pecuniary losses cover both gains and losses (*damnum emergens* and *lucrum cesans*); as an example, in the case of Eagan Sea, part of these damages was that suffered by the fishermen, fish farms, cleaning operations, wholesale fish dealers, transport companies, tourism, and so forth. The non-pecuniary losses affect elements or interests which are difficult or impossible to assess economically, because such elements or interests cannot be replaced by money. Examples of these are emotional injury even when no physical harm of any kind has been done.

---

[1] *OJ*, L143/56, 30 April 2004.
[2] Article 4.2 Directive 2004/35.

Without doubt, these harms have been very intense in the case of the Prestige, although it is very unusual for compensation to be awarded for these harms.

The assessment of damage in any case of civil liability requires a valuation of the injuries. This valuation is not easy in the pecuniary losses caused by oil pollution. A great deal of the damage affects a wide range of people, who are at times difficult to identify. The damage also lasts a long time and may only appear for the first time many years after the event. The difficulty is even greater in the case of non-pecuniary losses. Who has not suffered watching the images broadcast on TV showing the catastrophe?

However, without doubt, the element, which introduces the greatest degree of complexity to this topic is 'environmental damage' (of which damage to biodiversity is a component[3]). This damage affects unowned elements like the air, water, migratory birds and wild animals. Also in the Spanish legal system there are no specific norms to assess these types of damage nor is it a straightforward issue to identify the plaintiff who could claim the compensation. Initially, this could be the State as a trustee of public natural resources. In Spanish Law as it stands, there exist legal technical difficulties which do not permit ecological associations or other similar organizations to claim compensation.

It is easier to identify the tort feasors. Normally, all the possible authors who might have contributed to the pollution through their actions or lack of actions are included. However, what is not so straightforward is identifying the precise responsibility of each one as regards the plaintiff, and the relationships between the various tort feasors.

3 INTERNATIONAL FRAMEWORK. INTERNATIONAL CONVENTION ON CIVIL LIABILITY FOR OIL POLLUTION DAMAGE AND INTERNATIONAL CONVENTION ON THE ESTABLISHMENT OF AN INTERNATIONAL FUND FOR COMPENSATION FOR OIL POLLUTION DAMAGE

The characteristics of sea transport of petrol, the requirements of the petrol industry and the attempt to guarantee direct compensation for oil pollution damage are factors which have contributed to the construction of an international regime of compensation for this type of damage. This regime is promoted by the International Maritime Organization (IMO) and includes several Conventions of which Spain is a Member.

The system established in these International Conventions consists of the installation of a mechanism of liability together with another for the direct compensation for damage which will come from two previously established Funds (International Oil Pollution Compensation Funds). According to the first mechanism the entity liable for oil pollution damage caused by oil spills from tankers will only be the owner of the tanker[4] (in the *Prestige* case the owner was identified as a Liberian company called Mare Shipping Inc., owned by a Greek family). This liability can be limited to a lump sum that depends on the

---

[3] About this concept E.H.P. Brans, 'EC Proposal for an Environmental Liability Directive: Standing and Assessment of Damages', (2002) 2 *Environmental Liability* 135-146.

[4] According to article I.3 International Convention on Civil Liability for Oil Pollution damage (1992 version) 'Owner means the person or persons registered as the owner of the ship or, in the absence or registration, the person or persons owning the ship. However in the case of a ship owned by a State and operated by a company which in that State is registered as the ship's operator 'owner' shall mean such company'.

size of the ship (in the *Prestige* case the maximum sum will be of approximately 75 million EUR). The limit will also benefit the owner's insurer.

The liability of the owner of the ship is strict, that is, without fault, even though there are some exceptions. Article III.2 of the International Convention on Civil Liability for Oil Pollution Damage, 1992, states that 'No liability for pollution damage shall attach to the owner if he proves that the damage: (a) resulted from an act of war, hostilities, civil war, insurrection or a natural phenomenon of an exceptional, inevitable and irresistible character, or (2) was wholly caused by an act or omission done with intent to cause damage by a third party, or (3) was wholly caused by the negligence or other wrongful act of any Government or other authority responsible for the maintenance of lights or other navigational aids in the exercise of that function'. None of these circumstances exist in the *Prestige* case. However, it would not be unusual that in this incident the owner of the tanker might attempt to lessen or exclude his liability using as a justification article III.3 of the Convention according to which 'If the owner proves that the pollution damage resulted wholly or partially either from an act or omission done with intent to cause damage by the person who suffered the damage or from the negligence of that person, the owner may be exonerated wholly or partially from his liability to such person'. It is easy to imagine that in the *Prestige* case the person mentioned previously may be the Spanish State.

The regime of civil liability described above is accompanied by another system, which established a compensation to be paid by the Fund for Compensation for Oil Pollution Damage (IOPC). The Fund[5] is financed by levies on certain types of oil carried by sea. The levies are paid by entities which receive oil after sea transport and normally not by States. It is a mechanism for giving compensation without fault and creates a formula for a relative 'sharing of damages among society'. The compensation paid by this Fund is limited (in the *Prestige* case the maximum total limit including the amount actually paid by the owner of the ship would be approximately 180 million Euros). The money from the Fund is used to pay the victims who have not received equitable reparation according to the International Convention on Civil Liability. To sum up, these types of funds were created to complement the system of limitation of ship owner liability and, if the damage occurs in a State which is a Member of the Fund, this Fund will pay an amount for the damage which exceeds the limit of the owner's liability but only up to the pre-established limits. These Funds are also obliged to compensate, but only up to the pre-established limits, when the International Convention on Liability does not impose any responsibility; for example, because there exist some circumstances which exclude the ship owner's liability or because the owner or its insurance company are insolvent. (Article 4 International Convention on the Establishment of an International Fund). Also in this case, the Fund can lessen or exclude its obligation to pay if '[...] the Fund proves that the pollution damage resulted wholly or partially either from an act or omission done with the

---

[5] There are at present two IOPC Funds: the 1971 Fund and the 1992 Fund. These two intergovernmental organizations were established at different times (1971 and 1992), have different maximum amounts of compensation and had different Member States. The membership of the 1992 Fund is increasing. Due to a number of denunciations of the 1971 Fund Convention, this Convention ceased to be in force on 24 May 2002 and the 1971 Fund therefore has no Member States.

intent to cause damage by the person who suffered the damage from the negligence of that person [...]'.

In many of the incidents which have happened to date, the limits established in the two Conventions mentioned above have been seen to be completely insufficient to cover the integrity of the real damage. This insufficiency is a disgrace in the *Prestige* case because the Spanish authorities and mass media assess the damage to be between 600 and 1,000 million EUR. The insufficiency derives, on the one hand, from the low quantity established in the aforementioned Conventions, and on the other, from the restriction of the type of damage which can be compensated according to article. I.6 of the International Convention on Civil Liability for Oil Pollution Damage.[6] According to the majority of authors both non-pecuniary losses and environmental damage except for the cost of preventive measures are excluded from the system. This means omitting very important types of damage in the *Prestige* case. Personal injuries or damage to health, even when they could have had a direct effect on the patrimony of the victims, would also not be covered.

Nevertheless, under the International Convention on Civil Liability for Oil Pollution Damage, article III.4, 'No claim for compensation for pollution damage may be made against the owner otherwise than in accordance with this Convention...' Consequently it would not be possible to claim from the owner of the ship more compensation by using the national Spanish laws. Also the Convention on Civil Liability, following the principle of 'channelling the liability', which restricts the claim to a single person, prohibits claims against other persons, like the servants or agents of the owner or the members of the crew, the pilot or any other person who performs services for the ship, any charterer, manager or operator of the ship, any person performing salvage operations with the consent of the owner or on the instructions of a competent public authority, any person taking preventive measures and all servants or agents of these persons. This immunity does not apply if the damage is the result of their personal act or omission '... committed with the intent to cause such damage, or recklessly and with knowledge that such damage would probably result'. To sum up, the system of the Conventions does not fully close the door on national claims of liability but it does limit them greatly.

Article VIII of International Convention on Civil Liability for Oil Pollution and Article 6 of the International Convention on the Establishment of an International Fund for Compensation for Oil Pollution Damage provide that compensation rights shall be extinguished unless an action is brought thereunder within three years of the date when the damage occurred. However, in no case shall an action be brought after six years from the date of the incident which caused the damage.

According to Article IX of Convention on Civil Liability, where an incident has caused pollution damage in the territory, including the territorial sea or an area referred to in Article II, or one or more Contracting States or measures have been taken to prevent or minimize pollution damage in such territory including the territorial sea or area, actions

---

[6] In this Convention 'pollution damage' means: '(a) loss or damage caused outside the ship by contamination resulting from the escape or discharge of oil from the ship, wherever such escape or discharge may occur', provided that compensation for impairment of the environment other than loss of profit from such impairment shall be limited to costs of reasonable measures of reinstatement actually undertaken or to be undertaken (b) 'the cost of preventive measures and further loss or damage by preventive measures'.

for compensation may only be brought in the Courts of any such Contracting State or States. Reasonable notice of any such action shall be given to the defendant.

In sum, the explanation above reveals that although on some occasions the damage caused by oil pollution can be covered by the rules of the Conventions of the IMO, a case like that of the *Prestige* demonstrates its structural deficiency for various reasons. Firstly, there are restrictions on the type of damage which can be compensated. Second, there are unreasonably low limits of liability. Third, there are difficulties involved in putting the system into practice. It should not be forgotten that, in the case of the Eagan Sea, compensation for the damage was paid ten years after the incident, by means of an agreement between the Spanish Tax Office and the International Oil Pollution Compensation Fund and the corresponding agreements between the injured parties and the Spanish State.

It is precisely the insufficiency of the IMO Conventions system in the case of the Exxon Valdez off the coast of Alaska that can explain why the US is not a member of this system. In this country, the Oil Pollution Act of 1990 was promulgated. The Act contained preventive rules and an internal federal system, which is much stricter than the international one.[7]

4 THE SPANISH SYSTEM OF CIVIL LIABILITY AND ITS APPLICATION IN THE *PRESTIGE* CASE

Despite the fact that Spain is obliged to comply with the rules of the IMO Conventions, the claims placed to date in Spanish Courts for the oil spill caused by the Prestige follow the norms of internal Spanish Law. It is not known whether this has been a conscious or unconscious attempt to avoid the international system. Due to this, the basic lines of the Spanish domestic system of liability will be described. At the same time, an attempt will be made to identify whether such an internal system can cover some of the gaps that have been detected in the international system of the IMO.

Under Spanish Law, the claim for compensation in cases of civil liability must be lodged within one year of the injured gaining knowledge of the damage. This claim can be placed before a Criminal Court, a Civil Court or an administrative Court. The case will be decided by a Criminal Court if the plaintiff claims both the criminal responsibility of the persons who have committed crimes and the civil liability derived from the damage caused by such crimes. In this situation the Criminal Court that considers the criminal responsibility can also consider the civil liability of the defendant. The Court will therefore apply (if it takes into consideration only Spanish Domestic Law) the norms of civil liability included in the Spanish Criminal Code.[8] However, the injured has the right to file only a criminal suit and reserve the civil claim for a later civil process. In this

---

[7] See a comparative perspective of the international oil pollution regime and the US system after 1990, P. Del Olmo García and I. Pintos Ager, 'Responsabilidad civil por vertidos de hidrocarburos. ¿Quienes han de pagar los daños causados por el Prestige', (2003) *InDret* 1, <http://www.indret.com>.
[8] Arts. 109-126 Spanish Criminal Code.

situation the Civil Court should consider only the questions derived from civil liability according to the same norms included in the Criminal Code.

However, the civil procedure can also be applied using the norms of the Civil Code when the Criminal Court decides that despite the enormous economic and ecological disaster there is no one is criminally responsibility. In all cases it is necessary to wait for the decision of the Criminal Court to start the civil proceedings.

The criminal process has been used by some injured to request compensation for damage in the *Prestige* case. The accused was the captain of the ship. The problems that can arise if the Spanish State were to apply '*ius punendi*' against the captain shall not be considered here. It should, nevertheless, not be forgotten, as was mentioned previously, that the Conventions on Liability and Compensation for Oil Pollution Damage are Spanish Law, and consequently must be used by Spanish Courts. This is because they are international treaties and they have priority over internal rules. Therefore, the norms of civil liability, which must be applied, are those contained in the Conventions. It is again pointed out that the International Convention on Civil Liability for Oil Pollution only allows the injured to claim compensation from the owner of the ship and according to the norms of the Convention. No claim for compensation for pollution damage may be made against other persons unless the damage resulted from their personal act or omission, committed with the intent to cause such damage, or recklessly and with knowledge that such damage would probably result (article III). To sum up, the accusation made against captain Mangouras of having caused enormous damage will need to be supported by evidence that his acts or omissions were intentional. Nevertheless, a claim could be made in a civil process against some other persons, for example, the owner of the merchandise or the ship classification society, who are not included in the Conventions. It would constitute liability with fault, which would oblige the plaintiff to prove the negligence of these persons.

The problem is made even more complex if the victims decide to sue the State because they consider that the Spanish State has also contributed in causing the damage. That is that, in addition to the persons liable for the Prestige, for its navigation and its merchandise, the Spanish State has also caused the damage. In this case if a claim is made for civil liability in a criminal process the criminal judge will have to decide if any civil servant has committed a crime which contributed to causing the damage. If the judge holds that such a crime exists, the State can be declared in the criminal process to be subsidiarily civilly liable. In this case the judge should apply the norms of civil responsibility included in the Criminal Code. If, however, the criminal judge considers that there is no criminal liability of the civil servants, the victims can claim direct liability of the State in the administrative Courts. In this case Articles 139 and following of the Legal System of the State Administration Act (Ley de Régimen Juridico de la Administración del Estado y del Procedimiento Administrivo Común) are applied. These norms establish a system of strict liability.

In the cases in which there is civil liability of any natural or legal person not included in the International Convention on Civil Liability for Oil Pollution or of the Spanish Administration, all types of damage, including personal injuries and non-pecuniary losses, will be compensated. It is more difficult to decide whether compensation may be claimed for purely environmental damage under Spanish Domestic Law. This difficulty does not arise because such types of damage are not recognized by Spanish Law, as in fact the

Spanish system of liability includes a very broad concept of damage.[9] The difficulty is due to the problems in identifying who can be the plaintiff in the claim for this type of environmental damage. Considering again the examples of damage to the biodiversity (some species have been lost due to the pollution), water damage, damage to the landscape, and so on, who can make a claim? At first sight the person who should be called on to defend these interests, which are purely collective, is the State. This is because, according to article 45(2) of the Spanish Constitution, the State is obliged to defend and restore the environment with the help of the general public. However, in the case of the *Prestige* the assumption of the role of the plaintiff by the Spanish State provokes a certain perplexity. This is because the State can also be liable and it is impossible to be both plaintiff and defendant. It is possible to say that the victim has contributed with its acts to increase the original harm. However, this situation only produces the effect of exonerating wholly or partially the liability of the primary causer of the damage. It would be plausible for the local or autonomous governments to make a claim against the Spanish Central State although this would give rise to many legal and political difficulties. It is also difficult to answer the question of whether the claim for purely environmental damage or the collective non-pecuniary losses can be made by ecological associations. The Spanish norms on the judicial defences of collective interests are not very clear. On the one hand, it seems that the possibility of a direct action of the ecologist organizations has its justification in Article 7(3) of the Judiciary Act (Ley Orgánica del Poder Judicial). This Article provides: 'The Courts will protect the rights and legitimate interests, both individual and collective. For the defence of the latter, recognition is given to the right of the corporations, associations and groups which can be affected, or which are legally constituted for collective defence and protection.' A similar norm is contained in Article 19(1)(b) of the Contentious Administrative Act for cases of claims made in this type of Courts. However, in the civil processes a similar norm does not exist. What ecological associations can do is exercise the popular claim ('acción popular') in the criminal process. In this case, according to Article 339 of Criminal Code, they can ask the judge to order the adoption, at the expense of the author of the polluting activity, to re-establish the ecological balance that has been disturbed.

5 THE ROYAL DECREE-LAW 4/2003, OF 23 JUNE 2003

The Spanish Government published the Royal Decree-Law 4/2003[10] on the payment of compensation in relation to the harm caused by the accident of the Prestige.

According to the introduction of this Royal Decree-Law its objective is to ensure the victims of the catastrophe obtain the most immediate compensation for the damage. Below the most important norms of this statute will be analyzed.

---

[9] See on strict liability in favour of the environment in some Spanish cases, M. Martin Casals, J. Ribot and J. Solé, 'Spain', in B.A. Koch and H. Koziol (eds.), *Unification of Tort Law: strict liability* (Kluwer Law International, The Hague/London/Boston, 2002) p. 302-321.

[10] The Royal Decree-Law is a statute made by the Goverment in case of urgent need (art. 86 Spanish Constitution).

5.1 Legal Framework of the Agreement between State and Victims regulated in the Royal Decree-Law

The first article of this statute authorizes the Official Institute of Credit (ICO) to pay, to a maximum of 160 millions of euros, for the damage caused in Spain by the Prestige. This payment will be given to the victims who voluntarily accept settlements signed/concluded with the Tax Office and always in cases where these are natural or legal persons who have suffered damage for pollution which can be compensated according to the Conventions on Liability and Compensation for Oil Pollution Damage.

Furthermore, under the first additional article, the autonomous communities and local corporations affected by the accident of the prestige could conclude agreements with the Central Spanish Government in order to obtain compensation for the damage caused by the incident.

In the two above-mentioned articles there are two types of agreements. The first is an agreement between private individuals and the Tax Office of the Spanish State and the second is an agreement between two public institutions. In both cases the legal framework is in article 3(d) of Public Contracts Act and in article 6(1). of Legal System of the State Administration Act.

On the surface, these are agreements intended to solve quickly the issue of compensation for the damage caused by the Prestige. It may be argued that these agreements are a type of settlement because they allow the parties to resolve the conflict of interest generated between the State and the victims as a consequence of the catastrophe. These agreements do not mean that the Spanish State accepts responsibility for the incident either partially or wholly, something which is affirmed repeatedly in different articles of the Royal Decree-Law.

This agreement, like in any other settlement, must presuppose a mutual advantage for both parties. This advantage consists, in particular, of saving in the cost of litigation through settlement of the two parties '...by giving, promising or retaining something avoid the start of a legal process or put an end to proceedings that have already commenced' (article 1809 Civil Code). The settlement is thereby a contract in which there are reciprocal concessions. However, these concessions do not always cover, nor must they, pretensions of justice. Among other things, justice can cede for reasons of opportunity; for example, one of the parties may prefer to be awarded less money and have it awarded earlier than if he were to he wait for a judgement. Likewise, the other party may prefer to pay a smaller quantity than he might have to pay if he were to wait for the trial.

5.2 Subjective Framework. The Victims

Under Articles 1 and 3 of Royal Decree-Law, the persons who can receive the money are those persons or entities who have suffered damages as a result of the accident of the Prestige, according to Conventions on Liability and Compensation for Oil Pollution Damage. It is also necessary that the public organism communicate to the ICO the condition of their being/status as victims.

Consequently, the only natural or legal persons who may benefit from the Royal Decree-Law are those persons or entities that have suffered damages according to both

IMO Conventions. Nevertheless, the persons or entities who would be protected by domestic Spanish Law but are not included in the Convention systems mentioned above may not be party to this settlement. It should once again be stressed that, pursuant to article I.6 of International Convention on Civil Liability for Oil Pollution Damage, this treaty only covers damage caused by pollution, damage derived from preventive measures and further loss or damage caused by preventive measures. Therefore, the non-pecuniary losses and purely environmental damage that does not consist of preventive measures in the sense of this Convention[11] are excluded from the settlement. This means that the Royal Decree-Law does not cover some very significant types of damage caused by the Prestige, like non-pecuniary losses. Personal injuries would also not be covered even if they had affected the victims' assets. It is possible to award compensation for each of these types of damage according to the Spanish civil liability system, but the types are not included in the Royal Decree-Law.

Article 6 of Royal Decree-Law provides that the victims could sign settlements either individually or through associations or groups of victims and through public corporations (this refers chiefly to fishermen's guilds) who act on their behalf. Reference to victims' groups appears to include entities without personality, and this demonstrates an element of incoherence with article 1 of the Royal Decree-Law, which requires personality in order to sign the settlement.

Many of the victims considered in the Royal Decree-Law of the Government have already received several types of compensation from the authorities pursuant to two previous Decrees.[12]

The Royal Decree-Law includes fiscal benefits, unemployment benefits, subsidies to the victims for cessation of their productive activities and special loans for the repairs to and replacement of the productive infrastructures damaged by the catastrophe.

With relation to the loans granted to the victims, the Royal Decree-Law provides that 'persons who sign the settlements will receive from the State an official receipt for capital and interests of the loan in the amount corresponding to the compensation to which they have the right'. That is, the amount of the loans plus interest will be considered to have been paid as an advance on the compensation that they are entitled to receive. Consequently, it should be understood that if the victim has the right to compensation which is more than the loan which he was granted, he will have the right to be paid the difference. Similarly, it seems that if the beneficiary of the loan has received more, he will have to pay back the excess.

The Royal Decree-Law does not clarify the legal situation of other subsidies, in particular, the subsidies for cessation of activities, that have already been received by the victims.

Furthermore, as mentioned above, the Royal Decree-Law refers exclusively to the Conventions on Liability and Compensation for Oil Pollution Damage both in relation to the types of damage included within its framework and to the calculation of the amounts that must be paid. This implies that the Spanish Government considers the following types

---

[11] Under art. I.7 of Convention on Civil Liability for Oil Pollution 'preventive measures' means any reasonable measures taken by any person, after an incident has ocurred, to prevent or minimize pollution damage.
[12] Royal Decrees 7/2002, 22 November and 8/2002, 13 December.

of damage will not be compensated: (1) personal injuries (these have not been considerable but they could increase significantly in the future), (2) the non-pecuniary losses caused by the psychological suffering derived from the incident, and (3) purely environmental damage. The Royal Decree-Law does not recognize all these types of damage despite the fact that they should be compensated under the Spanish Civil Liability system. That is, their value will not be considered in the amount paid by the State in the settlement although it will be part of the claim of the victim.

To calculate damages, Article 1 of the Royal Decree-Law provides the signing of a protocol to the International Oil Pollution Compensation Fund, which establishes the methods of technical cooperation to determine the damages and identify the victims. However, in subsequent Decrees, the Spanish Government contemplates the possibility that this protocol will not be signed. Finally, the total amount of compensation offered by the Spanish Government to pay the damages contemplated in the Conventions on Liability and Compensation for Oil Pollution Damage has a ceiling of 160 millions of euros. Clearly this sum will not cover all harms even if they are limited to those included in these Conventions.

5.3 Waiving to the Claims by the Victims

In exchange for a payment in advance of the compensation, the signatories of settlement 'waive any claim, appeal and any other complaint which has not been resolved irrevocably and unconditionally and also waive the right to start in the future any judicial or extra-judicial claim, complaint or demand of any character related to the accident of the Prestige'.

This waiving is the basic concession made by the victims who accept settlement and which to some extent is normal in this type of agreement. In exchange for a rapid payment of the damage, the victims waive irrevocably their rights to any type of judicial or extra-judicial claim.

The expressions used in the norm seem to mean that waiving affects all areas, and includes any defendant and any place. Nevertheless, the aforementioned article gives rise to various problems of interpretation. Firstly, as regards its temporal framework the Royal Decree-Law provides that those who waive can affect not only the unresolved proceedings but also future claims. In the first case the waiving will produce a definitive end to the proceeding.

From a material point of view and although there is an attempt for the text of the norm to cover everything, it is clear that the waiving cannot affect those claims which cannot be waived under Spanish Law, as is the case of criminal claims.[13]

Nevertheless, since it is an issue that can clearly be waived, such a waiving will affect any claim for compensation for the damage caused by the disaster. Consequently, victims who accept the settlement also waive the right to claim against anybody, including the Spanish State. They also waive the right to claim damages that will not be compensated, like non-pecuniary losses, personal injures and purely environmental damage (all of which could be compensated under Spanish Civil Liability System). Furthermore, they also

---

[13] Art. 106 Judicial Criminal Act.

waive the right to claim future damages, that is, damages for harm that has not yet appeared but that could appear in the future. It also seems that they waive the right to file other suits different from these, which compensate damage.

There have apparently been attempts by the Spanish Government to prevent liability claims against the Spanish State in Spanish Courts. It has also tried to monopolize the claims that could be made in any other national or international court or authority. These circumstances must be taken into account because, while the Spanish Government offers victims the compensation provided for in the IMO Conventions on Liability and Compensation for Oil Pollution Damage, at the same time it reserves for itself any claim in any court of another State or international organization and against natural or legal persons who are not included in the Conventions. Indeed, Spanish Government has already filed a lawsuit in New York against the American Bureau of Shipping (ABS), which was the classification society that certified that the ship was in sailing condition.

The Royal Decree-Law attempts in this way efficiently to grant a settlement between the State and the victims against third parties, which is essential under Spanish Law. We have our doubts that such a level of efficiency is produced. Nevertheless, there remains a question about the compatibility between the anticipatory waivers and basic principles of Spanish Constitutional Law, like the right to free access to the courts to defend legitimate rights and interests (Spanish Constitution, Article 24). As a consequence of this, it is possible that the Royal Decree-Law is not consistent with the Spanish Constitution.

In any settlement there must be a balance between the two sides. In this case, the balance does not exist as the Spanish State buys the claims of the victims at a cost that is much lower than the real value. For the Spanish State, this settlement is cheap. Despite this, the uncertainty regarding the final result of possible judicial claims will mean that only optimistic victims will prefer to instigate legal proceedings, while the rest will agree to settle.[14] In this way the Spanish State will have taken advantage of its position to protect itself against its real responsibility.

However, it must be pointed out that following the change of government after the 14 March 2004 elections, the new administration has declared that it wishes to repeal or, at least, modify Royal-Decree Law 4/2003, so that damage which might appear in the future is not included in the waiver made by the victims. This modification seems to be imminent, although at the time of writing (21. June 2004), it had not yet been made.

6 FINAL NOTE

It is not possible to analyze in depth the questions deriving from possible claims in international courts. It is known that the oil spill has reached the French coast and to a lesser extent the coasts of Portugal and the United Kingdom. Furthermore, natural persons and the network of companies related to the Prestige have different nationalities and domiciles in different States; moreover, the contracts related to the ship, its cargo and its

---

[14] A. Rubí i Puig and J. Piñeiro Salguero, 'El blindaje del Gobierno en la crisis del Prestige. Comentario al Real Decreto Ley 4/2003, de 20 de junio, de ayudas a los damnificados', (2003) *InDret* <http://www.indret.com>.

transport have been performed or carried out in various places. To sum up, the factors which make it possible for the problem to be international are so numerous that they make it very probable that claims will be made in the courts of various States, as is evidenced by the suit filed by the Spanish State against the classification company. Some States will be party to the International Conventions of the IMO of 1992. All the States will have their own domestic system for compensating damage. In each case it will be necessary to analyze the coordination between the different systems. Finally the difficulties, the uncertainties, and the problems can multiply.

LIST OF REFERENCES

E.H.P. Brans, 'EC Proposal for an Environmental Liability Directive: Standing and Assessment of Damages', (2002) 2 *Environmental Liability* 135-146.

P. Del Olmo García and J. Pintos Ager, 'Responsabilidad civil por vertidos de hidrocarburos. ¿Quíenes han de pagar los daños causados por el Prestige', (2003) *InDret* 1, <http://www.indret.com>.

M. Martin Casals, J. Ribot and J. Solé, 'Spain', in B.A. Koch and H. Koziol (eds.), *Unification of Tort Law: strict liability* (Kluwer Law International, The Hague/London/Boston, 2002), pp. 302-321.

A. Rubí i Puig and J. Piñeiro Salguero, 'El blindaje del Gobierno en la crisis del Prestige. Comentario al Real Decreto Ley 4/2003, de 20 de junio, de ayudas a los damnificados', (2003) *InDret* <http://www.indret.com>.

PART III: COMPENSATING MARINE POLLUTION DAMAGE IN CHINA

THE ENFORCEMENT OF INTERNATIONAL CONVENTIONS FOR THE PREVENTION OF POLLUTION FROM SHIPS AND COMPENSATION FOR POLLUTION DAMAGE IN CHINA

Han Lixin and Guan Zhengyi

1 INTRODUCTION

It is widely accepted that the 21st century will be a century of the oceans. The marine economy plays an important role in wider national economies. Thus, protecting the marine environment has become a problem that attracts the attention of both international communities and national governments. Beginning in the early 20th century, the United Nations, the International Maritime Organization (IMO), the Comité Maritime International (CMI), and others drew up a series of international conventions for the prevention of pollution and for the compensation for pollution damage, all of which have had a significant impact on marine environmental protection. According to conditions in China, the Chinese government has concluded or acceded to some of these international conventions. Accession, though, is not the final aim of these conventions. Rather, they should promote the protection of the marine environment and a sustainable development of society through an appropriate enforcement of these international conventions, and by balancing the rights and obligations between the injuring parties and the victims.

On the basis of an introduction to such measures enforcing international conventions throughout the world, this paper expounds the relevant legislation and the current state of application of the conventions in China, and finally puts forward suggestions as to how to enforce the conventions in China.

2 BRIEF INTRODUCTION TO THE INTERNATIONAL CONVENTIONS FOR THE PREVENTION OF POLLUTION FROM SHIPS AND COMPENSATION FOR POLLUTION DAMAGE TO WHICH CHINA HAS ACCEDED

The international community has formulated a series of conventions for the prevention of pollution from ships and for compensation for pollution damage, which are as follows:

Michael G. Faure and James Hu (eds), Prevention and Compensation of Marine Pollution Damage. Recent Developments in Europe, China and the US, 181-191.
© 2006 Kluwer Law International. Printed in the Netherlands.

1   International Convention for the Prevention of Pollution of the Sea by Oil, 1954
2   The International Convention Relating to Intervention on the High Seas in Cases of Oil Pollution Casualties, 1969*
3   The Convention on the Prevention of Marine Pollution by Dumping of Wastes and other Matters from Ships or Airlines, 1972*
4   Protocol to the International Convention Relating to Intervention on the High Seas in Cases of Marine Pollution by Substances other than Oil, 1973*
5   The International Convention for the Prevention of Pollution from Ships 1973, and the Protocol of 1978 relating thereto (73/78 MARPOL Convention); and the Amendment of 1995 relating thereto (Supplementary V)*
6   The International Convention for the Safety of Life at Sea, 1974 (SOLAS Convention) and its Amendments thereafter*
7   The International Convention on Standards of Training, Certification and Watchkeeping for Seafarers, 1978 (STCW Convention. A revised STCW Convention came into force in 1997)*
8   The United Nations Convention on the Law of the Sea, 1982*
9   The International Convention on Oil Pollution Preparedness, Response and Co-operation, 1990 (OPRC 1990)*
10 The International Convention on Civil Liability for Oil Pollution Damage, 1969 and the Protocols of 1984 and 1992 relating thereto (CLC)[1]*
11 The International Convention on the Establishment of an International Fund for Compensation for Oil Pollution Damage, 1971 (Fund Convention) and Protocols of 1984 and 1992 relating thereto[2]*
12 The Convention on the Liability of Operators of Nuclear Ships, 1962
13 Vienna Convention on Civil Liability for Nuclear Pollution, 1963 and the Protocol of 1997 Relating thereto (Vienna Convention)
14 The Convention Relating to Civil Liability in the Field of Maritime Carriage of Nuclear Material, 1971
15 The International Convention on Liability and Compensation for Damage in connection with the Carriage of Hazardous and Noxious Substances by Sea, 1996 (HNS Convention)
16 The International Safety Management Code, 1998*
17 The International Convention on Liability and Compensation for Bunker Oil Pollution Damage, 2001, etc.

These conventions are all important international legal documents about protecting the marine environment. Among these conventions, Nos. 10, 11, 12, 13, 15 and 17 are those conventions that deal with compensation for damage caused by ships' oil pollution. The

---

[1]   Protocol 1992 of CLC 1969 is also called CLC 1992. The Chinese government acceded to CLC 1992 on 1 January 1999. It has came into force on 5 January 2000. CLC 1969 has been invalid in China at the same time.
[2]   Fund Convention 1992 is only applicable in Hong Kong SAR of China.
*   With regard to the above-mentioned international conventions, China has acceded to those conventions with the mark of '*'.

others are technical conventions concerning prevention of oil pollution from ships. This paper mainly discusses the application of the substantial conventions dealing with the compensation for damage caused by ships' oil pollution in China.

3 THE MAIN MEASURES OF ENFORCING INTERNATIONAL CONVENTIONS

For those treaties that come into force internationally to be implemented in any one country, the precondition is that the national laws of that country must 'admit' the treaties. With regard to the application of international conventions, this involves the relationship between international and national laws. Theoretically, there are the doctrines of the Monistic (Dependent) and the Dualistic (Independent) application. In enforcement practices for international conventions around the world, there are three main methods, which are as follows:

3.1 ADOPTION

This method is Monistic. That is, it accepts that international rules (including international conventions) are laws, and that they, together with the national laws, belong to the same legal system. They can be applied directly in the country after they are adopted into the national laws of that country. The following countries mainly adopt this method: France, Switzerland, the Netherlands, Japan, and some other European continental countries.

3.2 TRANSFORMATION

This method is Dualistic. That is, it accepts that international rules (including international conventions) are laws, but classifies them as belonging to a different legal system from their internal laws. The way to enforce international laws is to transform the treaties into national laws through national acts. This transformation refers to *domestic legislation which implements the conventions*. In Great Britain, it is also referred to as *statutory law carrying out international conventions*. 'In UK, the provisions in one convention would not be laws before they are carried out as a decree or legal document. There are several legislative techniques when granting international conventions legal effect: the decreed may be enacted on the basis of the treaty (the treaty itself not be incorporated into the decreed); the decreed may point out the legal force to grant the convention either in its title or preface or not. The decreed could list the main body in English (or French or other languages) into its catalog, or it also could list the main body in English or French(or other languages) into different part from its contents...'[3] Besides the UK, the following countries, *inter alia*, also take the Dualistic approach: Australia, Canada, Italy and Germany, and so on.

---

[3] J. Collins et al (eds), *Dicey and Morris on the Conflict of Laws* (13th edn, I, Sweet & Maxwell, London, 2001).

3.3 MIXTURE OF ADOPTION AND TRANSFORMATION

Some countries use the methods of adoption and transformation simultaneously when they apply conventions in their own country. Based on the treaties' varied characteristics and contents, they apply some treaties directly through adoption, while transforming some into domestic laws with various legislative measures. The United States is a typical example of a country that uses both ways. American judicial practice divides treaties into self-executing and non-self-executing types. Only self-executing treaties can be applied directly in the United States, and thus non-self-executing treaties require some legislative act (so they can usually be applied in the United States after legislation is made to carry them out). As to the distinction between self-executing and non-self-executing treaties, it is generally considered that the former are those that have clear and definite provisions, which can be applied directly by domestic courts and administrative organisations. The latter, on the other hand, are those with general obligations, which cannot be applied directly. However, in practice, the interpretation of this distinction and the affirmation of these two kinds of treaties in different countries is varied and largely arbitrary.[4]

4 THE CURRENT STATUS QUO FOR THE APPLICATION OF INTERNATIONAL CONVENTIONS ON THE PREVENTION OF POLLUTION FROM SHIPS AND FOR COMPENSATION FOR POLLUTION DAMAGE IN CHINA

4.1 THE EXISTING REGULATIONS FOR THE APPLICATION OF CONVENTIONS IN CHINA

There is no regulation specifically about the relationship between international conventions and national laws and how to apply international conventions in China in The Constitution of the P.R.C. Furthermore, The Procedure Law for Treaty Conclusion of the People's Republic of China only provides for the three stages for concluding a treaty: it is concluded by the State Council; it is ratified by the Standing Committee of the National People's Congress; it is ratified by the Chairman of the People's Republic of China according to the decision made by the Standing Committee of the National People's Congress.[5] In relation to the application of conventions in China, there are some clauses giving priority to the application of international conventions only in certain legal sectors. For instance, Article 142(2) of the General Principles of the Civil Law of the P.R.C. 1987 provides: 'If any international treaty concluded or acceded to by the People's Republic of China contains provisions differing from those in the civil laws of the People's Republic of China, the provisions of the international treaty shall apply, unless the provisions are ones on which the People's Republic of China has announced reservations'. Similarly, Article 97 of Marine Environment Protection Law of the P.R.C., which was implemented on 1 April 2000 stipulates: 'If any international treaty concluded or acceded to by the People's Republic of China contains provisions differing from those in the law, the

---

[4] G. Jiang, <http://www.Chinalawedu.com/news/2003_12/5/1000536033.htm>.
[5] Art. 3 of The Procedure Law for Treaty Conclusion of the People's Republic of China.

provisions of the international treaty shall apply, unless the provisions are ones on which the People's Republic of China has announced reservations'.[6]

In analysing the above-mentioned provisions, it is the belief of some scholars that this is a direct application of conventions in principle and that it is unnecessary to transform the conventions into Chinese national laws.

4.2 CHINA'S PRACTICE IN IMPLEMENTING THE CONVENTIONS

In China, the practice of the application of international conventions does not completely correspond to the doctrines of either Monism or Dualism. As to the conventions, China has accepted, China applies these conventions directly or indirectly in different ways depending on their different characteristics, and fulfils its obligations as a state party to the conventions. The following are the main methods employed:[7]

i. To adopt or partly adopt the same rules as the international conventions. For example, Chapter 8 of the Collision of Ships of the China Maritime Code was drafted on the basis of the Collision Convention 1910; Chapter 9 of the Salvage at Sea is based on the Salvage Convention 1989 and most of the provisions of Chapter 5 of the Contract of Carriage of Passengers by Sea come from the Athens Convention 1974 and its protocol of 1976.

ii. For some highly technical international conventions, the competent authorities of the State Council may provide the scope of application of the international conventions by way of distributing a notice. For example, it indicates clearly in the Notice of Certain Problems in Relation to the Enforcement of the Convention on the International Regulation of Collision Prevention at Sea 1972 (6 April 1981) that 'all ships sailing or berthing at sea and ports, their manoeuvre and signal display shall comply with the Regulation' (meaning here the International Regulation of Collision Prevention at Sea 1972). That is to say, be they Chinese or foreign ships, as long as they are sailing or berthing at sea or ports, they must abide by the Regulation of Collision Prevention 1972.

iii. When China accepts conventions or when conventions accepted by China come into force, the competent authorities of the State Council usually notify the relevant departments by way of distributing a notice, indicating also the time at which the international convention comes into force in China. The notice is a request for the departments to carry out conventions at that time, but there are not any specific laws and regulations to define the scope of the application of the conventions. For example, the Notice in relation to the coming into force of Amendment (Supplementary V) of 1995 to the 73/78 MARPOL Convention reads that '...the Notification of IMO was received recently, during the period provided in the implied acceptance procedure (till 1st January 1997), no opposition opinion of any state parties was received, so... the

---

[6] In addition, please see art. 189 of Civil Procedure Law of the P.R.C. and art. 268 of Chinese Maritime Code (CMC).
[7] See Y. Si et al., 'Suggestion on the Legislation of the Relationship between International Maritime Conventions and National Laws', (1999) *Annual of Chinese Maritime Law*.

Amendment has been deemed to have been accepted from 1st January 1997, and it will come into force on the same day. China is a state party to the 73/78 MARPOL Convention, and has accepted Supplementary V, the Amendment of 1995 will come into force to China. Now distributing the body of the Amendment to you, please be ready to enforce it at the appointed time'.

It can be seen from the above-mentioned methods that no method defines how to apply the international conventions that China has acceded to or concluded with. Moreover, there is no method that defines whether cases should be divided into domestic and foreign-related ones. The relationship between international conventions that China accepted and national laws therefore remains open.

4.3 THE CURRENT STATUS QUO FOR THE APPLICATION OF INTERNATIONAL CONVENTIONS ON THE PREVENTION OF POLLUTION FROM SHIPS AND FOR COMPENSATION FOR POLLUTION DAMAGE IN CHINA IN PRACTICE

In practice, conventions with strong technical characters and general provisions like the International Convention Relating to Intervention on the High Seas in cases of Oil Pollution Casualties, 1969; 73/78 MARPOL Convention and its Amendment of 1995; 1974 SOLAS Convention and its Amendments thereafter; 1978 STCW convention and its Amendments; United Nations Convention on the Law of the Sea,1982;, OPRC 1990 and the 1998 ISM Code, accepted by China, tend to be applied strictly to all Chinese and foreign ships bound by the provisions.

However, there are considerable disputes in China about how to apply substantive conventions like the CLC 1969 and its protocols of 1984 and 1992. The controversy focuses on the scope of application of conventions that China has entered into, and namely, whether they only applied in oil pollution cases with foreign factors, or apply in any oil pollution cases, no matter whether there are foreign factors or not. Theoretically, then, it is unanimously held that in oil pollution cases with foreign factors the CLC shall be applied. However, in oil pollution cases without foreign factors, there are different opinions. Some scholars believe that, firstly, according to Article 2 of CLC 1969, 'This Convention shall apply exclusively to pollution damage caused on the territory including the territorial sea of a Contracting State and to preventive measures taken to prevent or minimise such damage', when China acceded to this convention, no reservation was made to any provisions of the convention. Secondly, Article 97 of the Marine Environmental Protection Law of the P.R.C. provides that if any international treaty of ocean environmental protection concluded or acceded to by the P.R.C. contains provisions differing from those contained in this Law, the provisions of the relevant international treaty shall apply. As the CLC is a convention on protecting the ocean environment, and has been acceded to by China, therefore, as long as the oil pollution damage happens in Chinese territory, it is argued that the CLC should be applied in any such case, no matter whether it has foreign-related elements or not. However, other scholars hold that only cases of compensation for oil pollution from ships with foreign factors shall be given the priority in applying CLC, for the reason that Article 142 of General Principles of the Civil Law is provided in Chapter VIII 'Application of Law in Relation to Foreign-related Civil

Matters' and Article 268 of Chinese Maritime Code (CMC) is regulated in Chapter XIV 'Application of Law in Relation to Foreign-related Matters'. These two chapters contain provisions dealing with foreign-related matters, such that China adopts the principle of keeping international conventions to handle civil legal relationships concerning foreign factors. With respect to oil pollution cases without foreign particulars, it is therefore contended, national laws shall be applied.

In Chinese marine judicial practice, there are contradictory cases about whether the CLC should be applied to oil pollution damage cases without foreign factors. For example, in the case of M/V 'Yan Jiu You 2', the Qingdao Maritime Court did not apply the CLC 1969. The facts of the case occurred on 16 August 1994, when M/V 'Yan Jiu You 2' was taking shelter from a typhoon. It was pulled to shore by the storm and the bottom of the ship was broken. Most of the 995-tonne oil cargo was spilled, causing serious damage to aquatic plants. The ship-owner of M/V 'Yan Jiu You 2' applied to the Qingdao Maritime Court for a limitation of his liability for maritime damage claims in accordance with the CLC 1969. The Qingdao Maritime Court, however, held that the CLC 1969 should not be applied to ships of less than 2000-tons sailing in Chinese coastal waters, and so the Court rejected their application.[8]

In the case of M/V 'Min Ran Gong 2', the Guangzhou Maritime Court applied the CLC 1969. The facts of this case were that at 2:26 a.m., on t 24 March 1999, M/V 'Dong Hai 209', belonging to Taizhou East Ocean Co. Ltd. (Taizhou Co. in short), collided with M/V 'Min Ran Gong 2', belonging to the China Shipping Fuel Providing Fujian Co. Ltd. (Fujian Co. in short), when the two ships were sailing in the sea area between Lingding Island and Qiao Island. The M/V 'Min Ran Gong 2' was seriously damaged and most of 1032-tonne oil carried on board was leaked into the sea, causing serious degradation of the water quality and ecological environment in the western sea area of the Zhujiang River delta. The damaged area extended for 380 kilometres. In order to claim compensation for the environmental and fishing resources damage, the Zhuhai Environmental Protection Administration and the Guangdong Marine and Aquatic Department, as the plaintiffs in the case, brought a suit against Taizhou Co. and Fujian Co. The Zhuhai Environmental Protection Administration considered that the oil pollution was caused by the collision between the two ships. The collision caused serious damage to the marine ecological environment and its resources, and so it was requested that the two accused ships compensate for economic losses of RMB 7,079,724 Yuan as well as the costs of the lawsuit. The Guangdong Marine and Aquatic Department requested that the two accused ships jointly and severally compensate the plaintiffs for pollution damage to the total value of RMB 11,002,600 Yuan and costs.

Owing to the large amount of the compensation, one of the defendants, the Fujian Co., applied for a limitation of liability for compensation on 8 October 1999, in accordance with the CLC 1969. At present, in China, only Chapter 11 of the China Maritime Code (CMC) refers to a limitation of liability for maritime claims, and there are still no special provisions to regulate civil liability for compensation for oil pollution damage occurring during coastal oil transportation. Therefore, in this case, there existed several opinions on

---

[8] G. Xu and C. Zhou, *On Legal Problems of Limitation of Liability for Compensation for Oil Pollution Damage Caused by Ships*, <http://www.ccmt.org.cn>.

the applicable law and on the potential for the limitation of liability of the defendants. The first opinion held that the General Principles of the Civil Law of the P.R.C., rather than the CMC, should be applied, and that no right of limitation of liability should be given to the defendant (Fujian Co.) because there was no special provisions for oil pollution damage compensation in the CMC, and so the amount of compensation must be determined according to General Principles of the Civil Law of the P.R.C. The second advocated that the Marine Environment Protection Law of P.R.C. should be applied in this case. The third argued that the oil pollution damage compensation was also a kind of maritime compensation, and so the right of limitation of liability of the defendant should be governed by Chapter 11 of CMC. Finally, the fourth opinion held that the CLC 1969 should be applied, since while there are no special provisions for oil pollution damage compensation in China, China has acceded to the CLC 1969, and it came into force on 29 April 1982.

The defendant's application for a limitation of liability was objected to by the two plaintiffs, who argued that the M/V 'Min Ran Gong 2' was not covered by the CLC 1969.[9] In addition, there were no foreign factors in this case. According to the Regulation for Prevention of Pollution at Sea from Ships of the P.R.C (Regulation for Prevention of Pollution in short),[10] this case should have been decided only under Chinese national laws, and not by Chapter 11 of CMC or the Regulation for Limitation of Liability for Ships with a Gross Tonnage not Exceeding 300 Tonnes and Those Engaging in Transport Service or Other Coastal Works between the Ports of the P.R.C. Therefore, the defendant should take all responsibility for the indemnification of the marine environmental pollution damage. The Guangzhou Maritime Court held that the relative national laws like the Marine Environment Protection Law of the P.R.C., and the Regulation for Prevention of Pollution did not exclude the ships carrying oil cargo in bulk less than 2000 tons and engaging in costal transportation from applying the CLC 1969. It found that the Environment Protection Law of the P.R.C. provided that, in relation to environmental protection, international conventions should be applied with priority. The CLC 1969, as an international convention concerned with environmental protection to which China did not make a reservation when she acceded, was then held to apply to ships carrying oil cargo in bulk less than 2000 tonnes and engaged in coastal transportation in China. Therefore, because in this case the applicants did not have actual and private fault, they were entitled to limit their liability according to the law. According to the provisions of the CLC 1969 and its protocol of 1976, the Guangzhou Maritime Court issued a verdict permitting the application of the limitation of liability for oil pollution damage compensation, and ordered that the defendants pay into a limitation fund of 52,934 SDR.[11]

In the above two cases, the oil-leaking ships, the M/V 'Min Ran Gong 2' and 'Yan Jiu You 2', were both engaged in coastal transportation and carried a bulk weight of oil less than 2000 tonnes. The pollution in both cases occurred in China, and in fact, the two cases are almost the same, but the verdicts concerning the limitation of liability for oil

---

[9] M/V 'Min Ran Gong 2' is a steel tanker, G/T497, N/T325.
[10] Art. 13 of this Regulation provides that not only this Regulation but also CLC 1969 are applicable to ships for international service carrying oil in bulk as cargo exceeding 2000 tons. At present, the Communication Ministry of the P.R.C. is revising this Regulation.
[11] *Annual of China Maritime Trial*, People Communication Press, 2000, 515-516.

pollution damage compensation and on the application of the CLC were quite different. There is a great dispute over the two contradictory verdicts. Meanwhile, problems have been revealed regarding the application of these substantive international conventions to which China has acceded.

5 SUGGESTIONS ON THE APPLICATION OF INTERNATIONAL CONVENTIONS

We believe that, with regard to the application of the technical conventions for the prevention of oil pollution from ships, all ships governed by these conventions, be they foreign-related or not, should meet the requirements provided by them, including those as to the crews on board. However, the substantive international conventions, including the CLC, to which China is a party, should be applied only to cases with foreign factors, and not to purely domestic cases. The reasons for this are as follows:

i. Art. 142 of the General Principles of the Civil Law of the P.R.C. and Art. 268 of *the* CMC are provided in chapters enacted especially for the application of law in relation to foreign particulars. From the system of legislation, we can find that the intention of Chinese law-makers, when dealing with foreign particulars, is to hold that if there are international conventions which China has concluded or acceded to that are in conflict with Chinese national laws, those international conventions shall be given priority.
ii. Often, the provisions for the scope of application contained in conventions is clearly limited. An example of this reads as follows: 'Where all the persons interested belong to the same State as the court trying the case, the provisions of the national law and not of the Convention are applicable'.[12] Although there is no such provision in the CLC 1969, and its scope of application stresses that this Convention applies exclusively to pollution damage caused on the territory including the territorial sea of a Contracting State, and to preventive measures taken to prevent or minimise such damage, it is unreasonable to apply it to domestic cases (where all the persons interested belong to the same State as the court trying the cases). Moreover, Article 13 of Regulation for Prevention of Pollution provides that not only this Regulation but also the CLC 1969 are applicable to ships for international service carrying oil as cargo in bulk exceeding 2000 tonnes. This provision shows that ships sailing in Chinese coastal waters and ships for international service carrying oil in bulk as cargo but not less than 2000 tonnes are not governed by the CLC 1969. At present, the Communication Ministry of the P.R.C. is revising this Regulation, but the draft revision still provides that only Chinese ships for international service and foreign ships entering the waters governed by China should have mandatory insurance or other financial security certificates according to the provisions of CLC.
iii. In order to protect the interests of parties suffering from oil pollution damage, the CLC not only provides for the limitation of liability of the ship owners, but also has provisions on mandatory insurance, or other similar financial security, and for direct

---

[12] See art. 12(2) of Collision Convention 1910 and art. 15(2) of Salvage Convention 1910.

action against such insurers or the person who provides the financial security. However, most tankers engaged in service along the Chinese coastline do not have insurance for liability for oil pollution, and there are disputes as to the system of direct action against insurers or financial security providers.[13] Thus, there is no actual meaning or support in the relevant legal provisions to apply the CLC to tankers engaged in transport in the Chinese coastal area.

iv. Tracking down the reasons why the international conventions exist, we should know that they are intentionally applied to deal with international civil and commercial relationships, unless a national law specially provides that conventions concluded or acceded to by a country shall also apply to domestic legal relationships (by way of Monistic or Dualistic implementation).

However, it should be noted that where oil pollution damage is caused by a collision of ships, cases with foreign-related factors refer to the oil pollution and not the collision between ships. An example of this would be a foreign bulk ship being in collision with a Chinese coastal tanker in Chinese territorial sea. Should the Chinese tanker leak oil but not the foreign bulk ship, then the CLC should not be applied in this case, because both the ship that leaked oil and the victims of that pollution were Chinese, and the polluted sea area was also in China. Thus, the conclusion is that only if the foreign ship leaked oil would this case invoke international concerns. Besides, even if the foreign factors did exist, the oil pollution relationship must be under the scope of application of the CLC. That is to say, only pollution from the 'ships' and 'oil' as specified in the CLC can justify the application of the CLC. Article2(1) of CLC 1992 provides that 'Ship' means any sea-going vessel and any seaborne craft of any type whatsoever, actually carrying oil in bulk as cargo. However, a ship, which has the ability to carry oil in bulk as cargo, is considered as a ship prescribed in CLC, only when it is actually carrying the cargo and during any voyage after the type of carriage (except there are no remains of this type of oil on the ship). Further, the CLC 1992 defines that, 'oil' means any persistent oil such as crude oil, fuel oil, heavy diesel oil, lubricating oil, whether carried on board a ship as cargo or in the bunkers of such a ship, meaning that this convention could not apply to any pollution caused by non-persistent oil or bunker oil.

Where oil pollution occurs in the Chinese territorial sea and is caused by Chinese tankers that are engaged in international transportation, strictly speaking, such an oil pollution case can be considered not to invoke international concerns. However, in practice, such tankers usually have mandatory insurance or financial security according to the provisions of the CLC, in order fully to compensate the victims of oil pollution. We suggest that in such a case the CLC, as acceded to by China, will be applied.

---

[13] Although art. 97 of Special Maritime Procedure Law of the P.R.C. provides that the parties suffering from oil pollution damage may directly bring an action against oil pollution insurers or financial security providers, this provision belongs to procedural law. There are no provisions of direct litigation in Chinese substantive law relating to oil pollution damage compensation, so that both in theory and in judicial practice, there are disputes as to whether the parties who have suffered from oil pollution damage can bring direct litigation against insurers or financial security providers according to art. 97 of this Law.

LIST OF REFERENCES

J. Collins et al (eds), *Dicey and Morris on the Conflict of Laws* (13th edn, I, Sweet & Maxwell, London, 2001).

G. Jiang, <http://www.Chinalawedu.com/news/2003_12/5/1000536033.htm>.

Y. Si et al., 'Suggestion on the Legislation of the Relationship between International Maritime Conventions and National Laws', (1999) *Annual of Chinese Maritime Law.*

G. Xu and C. Zhou, *On Legal Problems of Limitation of Liability for Compensation for Oil Pollution Damage Caused by Ships,* <http://www.ccmt.org.cn>.

APPLICATION OF LAW IN CIVIL LIABILITY FOR OIL POLLUTION DAMAGE CAUSED BY COASTAL VESSELS IN CHINA

James Hu and Yang Bo

1 INTRODUCTION

On 22 March 1999, the tanker 'Mingrangong 2' owned by China Marine Fuel Supply Fujian Co., Ltd. loaded 1,032,067 metric tonnes fuel oil at Xiamen for Dongguan. The following day, she collided with another tanker 'East Sea 209' in ballast. Consequently, the cargo of oil carried on board the 'Mingrangong 2' spilled into the waters in the region of Guangzhou and serious oil pollution occurred. Zhuhai Environmental Protection Bureau and the Ocean & Fishery Bureau of Guangdong Province filed a lawsuit as the Plaintiffs with the Guangzhou Maritime Court against the owners of the 'Mingrangong 2' for claims for oil pollution damages. In the trial of the case, the defendants filed an application for limitation of liability for oil pollution damage stipulated in the International Convention on Civil Liability for Oil Pollution Damage of 1969 (hereinafter '1969 CLC'[1]). The plaintiffs raised an objection to such an application on the grounds that the 'Mingrangong 2' was not within the scope of application of 1969 CLC, no foreign element was involved in this case and consequently only Chinese law could apply. The Guangzhou Maritime Court held that the defendants were entitled to avail themselves of the limitation of liability stipulated in 1969 CLC and granted the defendants' application.[2]

This case is a typical example that reflects the diversities of views existing both in the academic circles and in admiralty practice in China regarding the application of law to cases of civil liability for oil pollution damage caused by Chinese vessels engaged in coastal services between Chinese ports (known as 'coastal vessels'). In consideration of the large number of such cases being filed every year in China, it is of great significance, both in academics and in admiralty practice, to study and clarify the issues involved therein, mainly whether the CLC applies to such cases and, if the CLC does not apply, what law shall be applied with respect to civil liability for oil pollution damage caused by these vessels.

---

[1] China was a party to 1969 CLC with effect from 29 April 1980 in China and now is party to the International Convention on Civil Liability for Oil Pollution Damage of 1992, known as '1992 CLC', with effect from 5 January 2000 in China. China was also party to the Protocol of 1976 to 1969 CLC. China has not ratified the Protocol of 2000 of 1992 CLC. In this paper, CLC stands for 1969 CLC or 1992 CLC, as appropriate.

[2] Z. Jing, *Annual of China Maritime Trial* (People's Communications Press, Beijing, 2000) p. 515.

2 DIVERSITY OF VIEWS REGARDING APPLICATION OF LAW

In summary, there are three views in this respect, each of which has been supported by case judgements of various maritime courts in China.[3]

2.1 APPLICATION OF CLC

The 1992 CLC shall be applied to the cases of civil liability for oil pollution damage caused by coastal vessels, whether foreign elements are involved or not.[4] This view is mainly based upon the following: (a) China acknowledges the effect of an international convention that China has ratified or acceded to as domestic law and a Chinese court can apply its provision directly.[5] Furthermore, by virtue of the provisions regarding the scope of application of the CLC, the CLC does not differentiate a foreign-related case from a non-foreign-related case. Consequently, the CLC can be applied to all sea-going ships carrying oil in bulk as cargo. (b) The Marine Environment Protection Law[6] carries out the principle that an international convention that China has ratified or acceded to is generally applicable, whether the case has foreign elements or not. (c) The Notice on Implementing the International Convention on Civil Liability for Oil Pollution Damage of 1969 issued by the Ministry of Communications indicates the applicability of CLC to civil liability for oil pollution damage caused by coastal vessels. Apparently, the Court in the case of 'Mingrangong 2' adopted this view.

2.2 APPLICATION OF CIVIL LAW

This means that the General Principles of Civil Law[7] shall be applied to such cases for the following reasons:[8] (a) The General Principles of Civil Law are generally applicable to all kinds of civil liabilities including those for oil pollution to the marine environment. (b) Art.13 of the Administrative Regulations on the Prevention of Marine Pollution Caused by Ships of 1983 promulgated by the State Council does not provide that the CLC applies to coastal vessels. (c) The Maritime Code[9] shall not apply by virtue of Article 208 which provides that the limitation of liability contained in the Code shall not be applicable to

---

[3] Compare on these issues also the contribution by Han Lixin and Guan Zhengyi to this volume.
[4] S. Li, *A study on the relevant issues of application of international conventions in maritime trial, the Guide and Study on China's Foreign-related Commercial and Maritime Trials* (II, People's Court Press, Beijing, 2003) pp. 126-140.
[5] It is even maintained that since the effect of an international convention that China has ratified or acceded to as domestic law is acknowledged, the Convention has become domestic law and can be applied to the cases only in the light of its own scope of application regardless of the nature of the cases or the related provisions of domestic laws. *Ibid.* p. 138.
[6] Adopted on 23 August 1982, amended on 25 December 1999 and came into force 1 April 2000.
[7] Adopted on 12 April 1986 and entered into force as of 1 January 1987.
[8] H. Lao, *A study on the Compensation for Oil Pollution in China, Transport and Environmental Protection* (I, Transport and Environmental Protection Editorial Dept., Beijing, 1998) p. 28.
[9] Adopted on 7 November 1992 and entered into force as of 1 July 1993.

'claims for oil pollution damage under the International Convention on Civil Liability for Oil Pollution Damage to which China is a party.'

2.3 Application of the Maritime Code

This means that the Maritime Code shall be applied to such cases for the following reasons:[10] (a) In the application of international conventions in China, the principle of differentiating foreign-related cases and non-foreign-related cases is adopted, by which CLC shall not be applied to non-foreign-related cases. (b) By virtue of the provision of Article 208 of the Code as partly cited above, claims for oil pollution damage caused by coastal vessels are subject to the limitation of liability provided for in the Code and, accordingly, the Code is applicable.

In summary, the lack of uniformity of understanding as regards the application of law as stated above has caused confusion of application of law to these kinds of cases in the admiralty judicial practice in China. In view of the same, this paper attempts to discuss mainly two issues: (a) whether the CLC can be applied to the civil liability for oil pollution damage caused by coastal vessels; (b) if Chinese domestic law were not applied, then which law shall be applied.

3 APPLICATION OF THE CLC

3.1 The Application of International Conventions in China Generally

For the first instance, the question whether the CLC can be applied to civil liability for oil pollution damage caused by coastal vessels is in essence a question of application of an international convention in China in a general sense. The solution to this question should be based upon the prerequisite of a clear understanding of the general application of international conventions in China.

Traditionally in the theory of international law, there are two basic approaches as regards the relations between the international conventions and domestic law *viz.* 'monism' and 'dualism'. 'Monism' means that once an international convention is ratified or acceded to by a State, it shall automatically become a domestic law of the State and form part of the domestic legal system without the need for domestic legislation. Consequently, the convention acquires the effect of domestic law upon coming into force in that State by way of ratification, accession or otherwise. 'Dualism' means that an international convention ratified or acceded to by a State shall acquire the effect of law in that State only by way of domestic legislation by which the rules contained in the convention shall be converted into the corresponding rules of domestic law. Accordingly, the approaches of different States in enforcing an international convention can be divided into two kinds, namely, incorporation and transformation. Generally speaking, countries of the European continent like France, Switzerland and the Netherlands adopt 'monism'

---

[10] Y. Si and Z. Zhu, 'Suggestions on the relations between maritime international convention and domestic law', in *1999 Annual of China Maritime Law* (Dalian Maritime University Press, Dalian, 1999) p. 5.

or incorporation, whereas some common law countries, like the U.K., adopt 'dualism' or transformation. Moreover, some countries adopt a mixed approach by distinguishing different situations and adopting either incorporation or transformation. This is the result of interaction between the nature of the international conventions themselves and the sovereignty interests of the State concerned.

In China, the Chinese Constitution does not expressly provide the legal status of international conventions in domestic law. Some scholars of international law maintain the co-existence of transformation and direct application. Some are argue there is direct application of international conventions. Others believe that the approach of incorporation is generally adopted. By looking at the prevailing Chinese laws concerned, it can be remarked that China adopts a mixed and not uniform approach regarding the application of the international conventions to which China is a party. More specifically:

a Most of the laws which may involve foreign elements in their application, especially those in the area of civil and commercial law, stipulate the direct applicability of the international conventions with priority over domestic law in foreign-related cases. Examples are the Maritime Code, the General Principles of Civil Law, the Negotiable Instrument Law, the Law of Succession, the Maritime Special Procedure Law, the Civil Procedure Code and the Administrative Procedure Law. A provision is paragraph 1 of Article 268 in Chapter XIV 'Application of Law in Relation to Foreign-related Matters' of the Maritime Code which provides: 'If any international treaty concluded or acceded to by the People's Republic of China contains provisions differing from those contained in this Code, the provisions of the relevant international treaty shall apply, unless the provisions are those on which the People's Republic of China has announced reservation.'
b Several laws stipulate the direct applicability of international conventions with priority over domestic law, but do not indicate that such applicability is limited to cases with foreign elements. A typical example is Article 97 of the Marine Environment Protection Law which provides: 'If any international treaty concluded or acceded to by the People's Republic of China relating to Marine Environment Protection contains provisions differing from those contained in this Law, the provisions of the relevant international treaty shall apply, unless the provisions are those on which the People's Republic of China has announced reservation.' Nevertheless, the lack of such indication often leads to arguments as to whether such applicability is limited to cases with foreign elements or not, although in exceptional cases, there is no argument as to the application of specific international technical rules to cases without foreign elements. A good example of the latter is the application of the International Regulations for Preventing Collisions at Sea, known as COLREGs.
c The rules of some international conventions are transformed into domestic law without indication of their applicability. Examples are the Law on the Territorial Sea and Contiguous Zones[11] and the Law on the Exclusive Economic Zones and Continental Selves.[12] The provisions of these two Laws are basically a reproduction of the United

---

[11] Adopted on 5 February 1992 and coming into force the same date.
[12] Adopted on 26 June 1998 and coming into force the same date.

Nations Convention on the Law of the Seas, known as UNCLOS, to which China is a party. The lack of a provision on the applicability of the UNCLOS in the two Laws may lead to, at least theoretically, confusion of the application of UNCLOS in China in case of repugnancy between the laws and UNCLOS.

d   Several laws adopt both the approach of incorporation and that of transformation. So far as the Maritime Code is concerned, besides Article 268 thereof which is a reflection of incorporation, the provisions of several chapters are in fact reproductions of the relevant international conventions, whether China is a party thereto or not, and thus transformations of these conventions. Where China is not a party, a convention does not have any effect of domestic law and is thus irrelevant to the discussion in this paper.

In view of the same, the following conclusions may be reached: (a) the approach towards application of international conventions in China is, strictly speaking, neither 'monist' nor 'dualist', but is a flexible or mixed one, depending upon different statutory provisions for a specific area; (b) by virtue of the civil and commercial laws, the application of international conventions are expressly limited to cases with a foreign element only; (c) due to lack of uniformity and even ambiguity of statutory provisions, arguments and consequential diversities have arisen regarding the application of international conventions.

3.2 NON-APPLICABILITY OF CLC TO CASES WITHOUT FOREIGN ELEMENTS

As discussed above, the question where the CLC applies or whether it applies to oil pollution caused by coastal vessels shall be determined in accordance with the provisions of relevant laws which mainly include the Maritime Code, the Marine Environment Protection Law, the Environment Protection Law and the General Principles of Civil Law. *Among these laws,* the Maritime Code and the Marine Environment Protection Law are of a maritime nature, the Environment Protection Law governs all forms of pollution damage, while the General Principles of Civil Law are applicable to all kinds of tort liability including liability for oil pollution to the marine environment.

By virtue of Article 268 of the Maritime Code and Article 124 of the General Principles of Civil Law, it is clear that the CLC and other international conventions shall be applicable only to cases with foreign elements. However, as mentioned above, Article 97 of the Marine Environment Protection Law does not indicate that an international convention shall apply only to cases with foreign elements. The question then arises whether the CLC applies to cases without foreign elements by virtue of Article 97 of the Marine Environment Protection Law.

The Marine Environment Protection Law is of the nature of public law, except Articles 90 and 92 thereof which provide civil liability for oil pollution damage to the marine environment. To the contrary, the CLC, as its title stands, regulates solely civil liability for oil pollution damage and is of the nature of civil law. This would mean that the solution to this question should follow the principle of application of international conventions of the nature of civil and commercial law. In other words, their application shall be limited to cases with a foreign element. This being so, the CLC shall not apply to cases without foreign elements.

There is another reason to come to this conclusion. So far as the global limitation of liability is concerned, much lower limits of liability are provided for in the Provisions on Limit of Liability for Maritime Claims against Vessels not More Than 300 Gross Tonnage or Engaged in Coastal Carriage or Coastal Operations,[13] promulgated by the Ministry of Communications pursuant to Article 210 of the Maritime Code, which are applicable to coastal vessels in order to fit for the extent of their owners' financial capability. In the current Chinese shipping market, most of the coastal oil tankers are of small size and are owned by individuals or very small companies. Such a situation together with a comparatively low economic level now in the developing China would mean the unsuitability of the application of the much higher limit of liability contained in the CLC to cases without foreign elements, although many other provisions are suitable to apply to such cases. This also means that the restriction of the application of CLC to cases with foreign elements is of significance as a policy issue.

In view of the above, it may be concluded that the CLC is not applicable to oil pollution damage to the marine environment caused by coastal vessels in cases without foreign elements in China, although the provisions of CLC themselves do not exclude its application to cases without foreign elements in terms of its scope of application.

3.3 APPLICATION OF THE CLC TO COASTAL VESSELS INVOLVING FOREIGN ELEMENTS

Noticeably, the CLC itself stipulates its scope of application in terms of 'ship', 'oil' and place of 'pollution damage'. Thus, Article I (1) defines a 'ship' as 'any sea-going vessel and sea-borne craft of any type whatsoever constructed or adapted to for the carriage of oil in bulk as cargo' and does not exclude coastal vessels.[14] 'Oil' is defined in Article I (5) 'any persistent hydrocarbon mineral oil'.[15] By virtue of Article II of the CLC, it shall apply exclusively to pollution damage caused in the territory including the territorial sea and the exclusive economic zone of a Contracting State and to preventive measures, wherever taken, to prevent or minimize such damage.

As discussed above, the CLC applies to oil pollution damage to the marine environment within its scope of application in cases with foreign elements in China and its application does not exclude oil pollution damage caused by coastal vessels. Therefore, where oil pollution within the scope of application of the CLC is caused by a coastal vessel and with a foreign element, the application of the CLC shall not be excluded. That is to say, for the CLC to apply, the key criterion is the foreign element requirement

---

[13] Came into force as of 1 January 1994.
[14] Art. I (1): '"Ship" means any sea-going vessel and sea-borne craft of any type whatsoever constructed or adapted to for the carriage of oil in bulk as cargo, provided that a ship capable of carrying oil and other cargoes shall be regarded as a ship only when it is actually carrying oil in bulk as cargo and during any voyage following such carriage unless it is proved that it has no residues of such carriage of oil in bulk onboard.'
[15] Art. I (5): '"Oil" means any persistent hydrocarbon mineral oil such as crude oil, fuel oil, heavy diesel oil and lubricating oil, whether carried on board a ship as cargo or in the bunkers of such a ship.'

involved in the oil pollution, regardless of whether it is caused by an ocean-going ship or a coastal one.

It is well understood in China that a foreign element exists in any of the following circumstances: (a) one or more parties involved is a foreign one; (b) the cause of the case occurs in a foreign country; (c) the subject matter is located in a foreign country. Thus, where a coastal vessel causes oil pollution damage to a foreign vessel nearby, or, to the territorial sea or the exclusive economic zone of a foreign State or a preventive measure is taken there, there is a foreign element in the case. In addition, if a vessel engaged in coastal trade in China and causes oil pollution is a foreign vessel, it means that at least one foreign party is involved and thus the case has a foreign element.[16] The CLC shall apply under any of these circumstances, provided that the vessel and the oil are within the scope of application of the CLC.

3.4 ANALYSIS ON VIEWS IN FAVOUR OF THE APPLICATION OF THE CLC TO CASES WITHOUT FOREIGN ELEMENTS

3.4.1 By Virtue of *the Administrative Regulations on Preventing Marine Pollution from Ships*

Article 13 of the Administrative Regulations on Preventing Marine Pollution from Ships provides: 'Vessels engaged in international trade and carrying more than 2000 tonnes of oil in bulk as cargo shall, in addition to implementing the stipulations of these Regulations, abide by the International Convention on Civil Liability for Oil Pollution Damage of 1969 acceded to by China.' By virtue of these provisions, it is thought that the CLC shall apply to oil pollution damage to the marine environment caused by vessels engaged in international trade and carrying more than 2000 tonnes of oil in bulk as cargo, even if no foreign element is involved.[17]

The said Regulations were promulgated by the State Council on 29 December 1983 for the purpose of implementing the Marine Environment Protection Law as Article 1 of the Regulations provides. The contents of the Regulations are of administrative issues regarding the prevention of marine pollution from ships and do not stipulate civil liability for marine pollution damage at all. From the text of the 1969 CLC, what is especially relevant to ships carrying more than 2000 tonnes of oil in bulk as cargo is Article VII thereof which requires compulsory liability insurance or other financial security to be maintained for such a ship and a valid certificate attesting such insurance or other financial security to be kept onboard. Such a certificate is named a Certificate of Insurance or Other Financial Security in Respect of Civil Liability for Oil Pollution Damage in the CLC. Without such a certificate, a Contracting State shall not permit such a ship under its flag to trade. In addition, Contracting States must ensure that all such

---

[16] China has been preserving its cabotage strictly. As provided by para. 2 of Art. 4 of the Maritime Code, a ship flying a foreign flag shall be allowed to be engaged in the carriage or towage by sea between Chinese ports only after the approval by the Ministry of Communications. Such a case is quite exceptional in practice.

[17] Y. Si, 'Study on application of law in civil liability for oil pollution caused by coastal vessel', (2002) 3 *Journal of Dalian Maritime University*.

ships, wherever registered, which enter or leave their ports or offshore terminals within the territorial sea have the required insurance or other financial security.[18]

Clearly, therefore, Article 13 of the Regulations means that vessels engaged in international trade and carrying more than 2000 tonnes of oil in bulk as cargo shall be bound by Article VII of the 1969 CLC regarding compulsory liability insurance or other financial security and certificate attesting the same, but nothing more. In other words, it does not and can by no means mean that the 1969 CLC applies when determining civil liability for oil pollution from such ships regardless of foreign elements.

3.4.2 By Virtue of Governmental Notices for Implementing the 1969 CLC

For the purpose of implementing the 1969 CLC, the Ministry of Communications issued a series of Notices in 1980.[19] Among the Notices, the Notice on Implementing the International Convention on Civil Liability for Oil Pollution Damage of 1969 provides in Article 2 that if a Chinese ship engaged in domestic trade carrying more than 2000 tonnes of oil in bulk as cargo is covered by oil pollution liability insurance, the Harbour Superintendency Administration[20] may issue a Certificate of Insurance or Other Financial Security in Respect of Civil Liability for Oil Pollution Damage and if not, the shipowner should apply for an issuance of a Credit Certificate on Civil Liability for Oil Pollution Damage as a temporary measure. Otherwise, such ships should not be allowed to trade as of 1 October 1980.[21] By virtue of these provisions, it is thought that the CLC shall apply to determine the civil liability for oil pollution damage caused by coastal vessels carrying more than 2000 tonnes of oil in bulk as cargo, even if no foreign element is involved.[22]

From the contexts of all these Notices, it is clear that their contents are limited to the Certificate of Insurance or Other Financial Security in Respect of Civil Liability for Oil Pollution Damage or Credit Certificate on Civil Liability for Oil Pollution Damage. Thus, like the analysis in 3.4.1 *supra*, the Notices only mean that vessels engaged in domestic trade and carrying more than 2000 tonnes of oil in bulk as cargo should be fully or partly bound by Article VII of 1969 CLC regarding compulsory liability insurance or other financial security and certificate attesting the same, but nothing more.

3.5 INTEREST-ORIENTATED CONSIDERATION

Clearly, if the CLC is applied to a case of oil pollution caused by a coastal vessel, the limit of liability of the shipowner shall be much higher than the case in the Provisions on Limit of Liability for Maritime Claims against Vessels not More Than 300 Gross Tonnage or Engaged in Coastal Carriage or Coastal Operations. This means that the application of CLC in such a case would be in favour of the victims of oil pollution damage. Under the

---

[18] D.W. Abecassis et al., *Oil Pollution from Ships* (Stevens & Sons, London, 1985) p. 225.
[19] See Maritime Safety Administration of P.R.C., *The Collection of Maritime Laws & Regulations*, People's Communication Press, Beijing, 2000, vol. 1, 688-696.
[20] Now Maritime Safety Administration or MSA.
[21] See Maritime Safety Administration of P.R.C., *The Collection of Maritime Laws & Regulations*, People's Communication Press, Beijing, 2000, vol. 1, 695.
[22] S. Li, *supra* note 4, p. 138.

circumstances of conflicting views regarding the application of the CLC to oil pollution damage caused by coastal vessels, the maritime courts applied the CLC to cases of oil pollution caused by coastal vessels mainly or partly based upon consideration of protecting the interests of the victims, especially where the State's interests are involved.[23]

The Chinese legal system originated from the continental one. As a statutory law country, courts of law are not empowered to make law, but only to construe law in specific cases by virtue of the Chinese Institution and the Law on the Organization of the People's Court of Law. Thus, the courts of law can never overlook the statutory provisions and take any interest-orientated consideration, regardless of whose interests such consideration aims at. To the contrary, as the basic judicial rule of compliance with law, a court of law should strictly comply with law in rendering its judgement. In other words, the statutory provisions can in no case be disregarded or be substituted with abstract and arguable theory, let alone the judicial power of a court of law shall not be expanded at its will. Otherwise, the basic judicial rule may be ruined in admiralty practice.

4 APPLICABLE DOMESTIC LAWS

Currently, there is no special legal regime regulating civil liability for oil pollution damage caused by coastal vessels. The main domestic laws in this regard are the Maritime Code, the Marine Environment Protection Law, the Environment Protection Law and the General Principles of Civil Law. Even if the application of the CLC to oil pollution damage caused by the coastal vessels is set aside, the issue remains unsolved as to which domestic law shall be applied to such cases.

4.1 COMMON FEATURES OF APPLICATION

The above domestic laws share the feature of application to civil liability for oil pollution damage caused by coastal vessels.

The Maritime Code regulates relations arising from maritime transport and those pertaining to ships as provided by Article 1 thereof. However, there is no independent chapter in it regulating oil pollution damage caused by ships. The directly relevant provisions are limited to the limitation of liability for such damage, namely, those contained in the Provisions on Limit of Liability for Maritime Claims against Vessels not More Than 300 Gross Tonnage or Engaged in Coastal Carriage or Coastal Operations.

The Marine Environment Protection Law is basically of the nature of a public and administrative law. However, Articles 90 and 92 of the Marine Environment Protection Law provide civil liability for oil pollution damage to marine environment and exemptions therefrom. Paragraph 1 of Article 90 provides: 'The party responsible for pollution damage to the marine environment shall eliminate the pollution and compensate for the damage. In the case of pollution damage to the marine environment resulting

---

[23] Interest-orientated considerations are also sometimes taken into account by courts when deciding which domestic law shall be applied, viz. the Maritime Code, the Marine Environment Protection Law or the General Principles of Civil Law in cases of oil pollution caused by coastal vessels.

entirely from the intentional or wrongful act of a third party, that party shall be liable for compensation.' Article 90 provides: 'The liability may be exempted if pollution damage to the marine environment cannot be avoided, despite prompt and reasonable measures taken, when the pollution damage is caused by any of the following circumstances: (a) acts of war; (b) irresistible natural calamities; or (c) negligence or other wrongful acts in the exercise of the functions of departments responsible for maintenance of beacons or other navigational aids.' No limitation of liability is provided for in the Law.

Similar to the Marine Environment Protection Law, the Environment Protection Law is basically of the nature of a public and administrative law. However, Article 41 of the Law governs civil liability for damage to the environment and exemptions therefrom which provides: 'A unit or individual who causes an environmental pollution hazard shall be obliged to eliminate hazard and to make compensation to the units or individuals who suffer direct losses... The liability shall be exempted where pollution damage to the environment is caused by an irresistible natural calamity and cannot be avoided, despite prompt and reasonable measures taken.' Clearly, this Article, together with Article 42, which provides time-bar for litigation of claim for pollution damage to the environment, is of general application to the civil liability for pollution damage to the environment including the marine environment.

The General Principles of Civil Law as a basic law applies to all kinds of civil liability. Civil liability for oil pollution damage to marine environment is in essence a kind of tort liability. Thus, the issues of civil liability for oil pollution damage to the marine environment are within the scope of application of the provisions of Chapter VI 'Civil Liability', especially Section 3 'Civil Liability governing tort liability' of the General Principles of Civil Law. Notably, Article 124 of Section 3 provides: 'Any person who pollutes the environment and causes damage to others in violation of state provisions for environmental protection and prevention of pollution shall bear civil liability in accordance with the law.'

4.2 PRIORITY ISSUES

Where a case is within the scope of application of two or more laws, a particular rule or particular rules should be followed in determining which law shall be applied. In the Chinese jurisprudence, the rule in this respect is called the special rule of levels of effects of law. According to this rule, various laws in the same field are divided into basic laws and special ones and a special law shall have priority over a basic one.[24]

So far as oil pollution damage to the marine environment by ships without foreign elements is concerned, the General Principles of Civil Law is a basic law as compared with others, while the Marine Environment Protection Law is a basic law as compared with the Environment Protection Law. Thus, the Marine Environment Protection Law shall have priority over the General Principles of Civil Law and the Environment Protection Law with respect to civil liability for oil pollution damage caused by coastal ships without foreign elements, while the limitation of such liability shall be determined by the Provisions on Limit of Liability for Maritime Claims against Vessels not More

---

[24] W. Zhang, *Jurisprudence* (Law Press, Beijing, 1997) p. 91.

Than 300 Gross Tonnage or Engaged in Coastal Carriage or Coastal Operations. Consequently, Articles 90 and 92 of the Marine Environment Protection Law shall be applied in determining the civil liability for oil pollution damage caused by coastal ships without foreign elements and the exemptions from such liability.

5 THE NECESSITY OF A SPECIFIC LEGAL REGIME

Since CLC 1969, the international community has been emphasizing the protection of the marine environment with regard to legal regimes. Besides the up-dating of the CLC and the Fund Convention,[25] the International Convention on Liability and Compensation for Damage in Connection with the Carriage of Hazardous and Noxious Substances by Sea, 1996, known as 'the HNS Convention',[26] and the International Convention on Civil Liability for Bunker Oil Pollution Damage, 2001, known as 'the Bunker Convention'[27] have been adopted. By means of these international treaties, special regimes governing civil liability for pollution damage to the marine environment, which are different from the traditional regime governing tort liability, have been established to meet the requirements for the protection of marine environment. The features of law in this respect may be summarized as shipowners' strict liability,[28] higher limits of liability,[29] compulsory liability insurance or other financial security and a direct action against the liability insurer or persons providing other financial security[30] and so forth.

Environmental protection is State policy in China. The Chinese government has been emphasizing the protection of marine environment. So far as legislation in this respect is concerned, besides the enactment of laws such as the Marine Environment Protection Law and regulations such as the Administrative Regulations on Preventing Marine Pollution from Ships, China has ratified the CLC, UNCLOS and various international treaties shaped by IMO with respect to the prevention of marine pollution such as MARPOL 73/78. However, unlike the American Oil Pollution Act, there is no law specifically regulating civil liability for marine pollution damage. Due to the limit on the scope of application of the CLC itself and its application to cases with foreign elements and the lack of a sound legal regime in the current domestic law fitting for the requirements for regulating civil liability for marine pollution damage, there is an urgent need to establish a sound legal regime in this respect.

---

[25] International Convention for the Establishment of an International Fund for Compensation for Oil pollution Damage, 1969 and its 1984 and 1992 Protocols. China is not a party to the Fund Convention.
[26] Not coming into force yet.
[27] Not coming into force yet.
[28] Art. III (2) and (3) of CLC, art. 7(2) and (3) of the HNS Convention, art. 3(3) and (4) of the Bunker Convention.
[29] Art. V (1) of CLC, art. 9(1) of the HNS Convention. The Bunker Convention does not stipulate limits of liability and instead, art. 6 thereof provides: 'Nothing in this Convention shall affect the right of the shipowner and the person or persons providing insurance or other financial security to limit liability under any applicable national or international regime, such as the Convention on Limitation of Liability for Maritime Claims, 1976, as amended'.
[30] Art. VII of CLC, art. 12 of the HNS Convention, art. 7 of the Bunker Convention.

Such a legal regime, aiming at regulating civil liability for ship-caused marine pollution damage without foreign elements, should cover strict liability of the shipowners and other responsible parties, the scope of compensation, a higher limit of liability, compulsory liability insurance or other financial security, a direct action against the liability insurer or persons providing other financial security and the establishment of a national fund for the compensation for pollution damage to the marine environment.[31] Needless to say, in establishing such a legal regime, it is important to adopt the provisions of the CLC, the Fund Convention, the HNS Convention and the Bunker Convention, provided that they are fit for the corresponding circumstances in China. It seems appropriate that the Maritime Code, when amended in the near future, should have a separate chapter regulating civil liability for marine pollution damage caused by ships.[32]

6 CONCLUSIONS

Through the above analysis of the application of law to the civil liability for oil pollution damage caused by coastal vessel, we may come to the following conclusions:

1. The CLC is not applicable in determining civil liability for oil pollution damage caused by coastal vessel in China, if no foreign element is involved.
2. Where no foreign element is involved, Articles 90 and 92 of the Marine Environment Protection Law shall be applied in determining the civil liability for oil pollution damage caused by coastal ships and the exemptions from such liability, while the limitation of such liability shall be determined pursuant to the Provisions on Limit of Liability for Maritime Claims against Vessels not More Than 300 Gross Tonnage or Engaged in Coastal Carriage or Coastal Operations.
3. There is now an urgent need in China to establish a sound legal regime regulating the civil liability for oil pollution damage caused by ships. Such a regime shall cover strict liability of the shipowners and other responsible parties, the scope of compensation, a higher limit of liability, compulsory liability insurance or other financial security, a direct action against the liability insurer or persons providing other financial security and the establishment of a national fund for compensation for pollution damage to the marine environment.

---

[31] Art. 66 of the Marine Environment Protection Law provides: 'The State shall make perfect and put into practice the regime of civil liability for oil pollution damage caused by ships and shall establish the system of liability insurance for oil pollution damage caused by ships and the system of compensation fund for oil pollution damage caused by ships in accordance with the principle that the shipowners and the cargo owners shall jointly undertake the risks of oil pollution from ships. Specific measures for implementing the system of liability insurance for oil pollution damage caused by ships and the system of compensation fund for oil pollution damage caused by ships shall be formulated by the State Council'.

[32] Y. Si and J. Hu, *Proposed Amendments to the Maritime Code of the People's Republic of China* (Dalian Maritime University Press, Dalian, 2003) p. 521.

LIST OF REFERENCES

D.W. Abecassis et al., *Oil Pollution from Ships* (Stevens & Sons, London, 1985).

Z. Jing, *Annual of China Maritime Trial* (People's Communications Press, Beijing, 2000).

H. Lao, *A study on the Compensation for Oil Pollution in China, Transport and Environmental Protection* (I, Transport and Environmental Protection Editorial Dept., Beijing, 1998).

S. Li, *A study on the relevant issues of application of international conventions in maritime trial, the Guide and Study on China's Foreign-related Commercial and Maritime Trials* (II, People's Court Press, Beijing, 2003).

Maritime Safety Administration of P.R.C., *The Collection of Maritime Laws & Regulations*, People's Communication Press, 2000, vol. 1, 688-696.

Y. Si, 'Study on application of law in civil liability for oil pollution caused by coastal vessel', (2002) 3 *Journal of Dalian Maritime University* 1.

Y. Si and J. Hu, *Proposed Amendments to the Maritime Code of the People's Republic of China* (Dalian Maritime University Press, Dalian, 2003).

Y. Si and Z. Zhu, 'Suggestions on the relations between maritime international convention and domestic law', in *Annual of China Maritime Law* (Dalian Maritime University Press, Dalian, 1999) p. 5.

W. Zhang, *Jurisprudence* (Law Press, Beijing, 1997).

A STUDY OF THE LEGAL NATURE OF COMPULSORY CLEAN-UP COSTS UNDER CHINESE LAW

Zhao Yuelin, James Hu and Wang Hua

1 INTRODUCTION

Where an oil spill accident occurs and causes pollution damage or creates a grave and imminent threat of causing such damage, compulsory clean-up measures are often taken by a maritime administrative authority, that is, the Maritime Safety Administration (MSA) in China, to prevent or minimize such damage or threat of damage. In this paper, as in admiralty practice, the costs and/or expenses borne in such activities are called the compulsory clean-up costs. Arguments as to the legal nature of such costs have arisen in academic circles in China due to the absence of any specific stipulation in the applicable law. Consequently, confusion has been caused in the application of both substantive and procedural law in the cases of compensation for such costs. Some think that the payment of such costs is in the nature of the injuring party's responsibility, since compulsory clean-up acts are administrative measures, and therefore a maritime administrative authority should have the power to order the responsible party to pay such costs without any limitation.[1] Others think that the compulsory clean-up costs should be borne by the responsible party on account of either administrative responsibility or civil liability, since the compulsory clean up acts represent a concurrence of public and private law acts.[2] Further others think that the legal nature of the compulsory clean-up costs is different under the old and the new Marine Environment Protection Law of PRC (hereinafter referred to as 'MEPL') and that payments were as an administrative responsibility under MEPL 1982, but have become a civil liability under the MEPL as amended in 1999.[3] In practice, maritime administrative authorities often consider the compulsory clean-up costs as costs of administrative intervention. Thus, claims for these costs differ from the general maritime claim, and are not to be subject to limitation of liability.[4] With respect to the

---

[1] M. Hong, *A study on responsible party and nature of responsibility of ship's oil pollution accident*, <http://www.999abc.com>, 13 August 2003; Z. Chunchang, *A further study on the legal nature of clean-up costs in ship's oil spillage accident*, <http://www/ccmt.org.cn>, 3 December 2002.
[2] S. Zhang, *The title of the administrative authority to file a maritime pollution claim*, Proceedings of 11th Conference on Maritime Trial of PRC, Tianjin Maritime Court, 2000, 58-59.
[3] Y. Si, 'A study on applicable law in the compensation of oil pollution damages from the coastal vessels', (2002) 4 *Journal of Dalian Maritime University* 1-2.
[4] For example, in the case of Sekwang Shipping Co., Ltd. of Korea applying for limitation on fund levels for maritime claims, the Shanghai Environmental Protection Bureau, East Sea Fishery Administrative and Fishery Port Superintendence Bureau, under the Ministry of Agriculture and Shanghai Maritime Safety Administration, pointed out that they took compulsory measures as administrative functional departments of the State, evaluated the effect of environmental pollution caused by the collision of vessels, and therefore

above arguments, this paper is an attempt to clarify the legal nature of compulsory clean-up costs based on theoretical research.

2 COMPULSORY CLEAN-UP ACTS AS COMPULSORY ADMINISTRATIVE MEASURES

In accordance with the theory of administrative law in China, administrative compulsoriness is divided into compulsory enforcement and compulsory measures.[5] A compulsory administrative measure is the same as the immediate compulsory administrative measure in some continental law countries.[6] Scholars in China have used various definitions for this concept. For example, Prof. Zhan Zhongle has concluded that it is a temporary disposal of the persons, property, ordinary rights and interests of the counter-party taken by the administrative body with such powers in case of emergency or necessity, in order to facilitate and maintain the public interest, public safety, and individual safety and interests.[7] Prof. Hu Jianmiao, however, finds that a compulsory administrative measure is an act restricting the rights of persons, the acts and property of specific citizens, legal persons or organizations, taken by an administrative body pursuant to the provisions of laws and regulations for maintaining order, realizing the public administrative order, preventing the occurrence of or eliminating the existence of events endangering the society or of illegal acts.[8] In summary, compulsory administrative measures have the following characteristics: (1) being statutory, that is, the bodies to take the compulsory administrative measures are limited to the administrative authorities with such powers pursuant to the provisions of the applicable laws and regulations and the compulsory measures are taken in line with these laws; (2) specific administrative acts, that is, that the compulsory administrative measures are taken against the freedom, properties, abilities or acts of the counterpart who is either dangerous or disadvantageous to public interests, health or safety or in danger or imminent danger himself, or use of whose rights or interests is imminent required by the State or in the public interest; (3) being mandatory, that is, an administrative authority takes such measures, and the counterpart is obligated to accept them; (4) protection of public interests, that is, the purpose of compulsory administrative measures is to maintain the public administrative order, ensure public safety, maintain and promote public interests, while protecting the personal safety or proprietary interests of the counterpart himself, but eliminating their

---

the costs incurred should be taken as the costs of administrative intervention and should not be subject to limitation of liability. See, <http://ccmt.org.cn>, 11 July 2003; in another example, in the collision case between M/V 'Dainty River' and M/V 'Sinokor Tianjin', Liaoning Maritime Safety Administration also held that the payment of the clean-up costs by the responsible party should be his responsibility under administrative law. See, Z. Guan, 'On perfection of procedures for constitution of limitation fund for maritime claims from the case of the collision case of M/V 'Dainty River'', *Annual of China Maritime Law* (2003) pp. 241-246.

[5] J. Hu, *A Study on Administrative Law* (Law Press, Beijing, 2003) pp. 1-58.
[6] *Ibid.*, p. 58.
[7] Z. Zhan, 'A study on the system of administrative compulsory measures', (2002) *Journal of Gansu Politics and Law College* 2.
[8] J. Hu, *supra* note 5, 17.

hazardous act which may endanger either the society or himself; (5) being non-punitive, that is, the direct aim of compulsory administrative measures is to achieve certain administrative objectives, but not to punish the illegal actions. From the viewpoint of the differences between compulsory administrative enforcement and compulsory administrative measures, most scholars take the view that the fundamental difference is that the former is an compulsory act restricting the rights of persons, the acts or properties of the counterpart taken by a specific administration body to prevent and protect against the occurrence of events endangering society or illegal acts, and for the need to maintain the social administrative order, while the latter is a specific administrative act taken by an appropriate authority to force the counterpart to carry out his administrative obligations when he has not obeyed the sanctions decision in time.[9]

Compulsory administrative measures can also be divided into general compulsory measures and emergency compulsory measures. Emergency compulsory measures, also called emergency compulsoriness, are those measures taken by administrative bodies in cases of disaster, infectious disease and other similar situations in which it is not possible to take measures that conform with normal procedures, and as a result of which such measures may be taken without strictly obeying the general procedures for compulsory administrative measures.

Article 71 of MEPL 1999 provides: 'If vessels occur maritime incidents causing or being likely to result in major pollution damages to the marine environment, the State competent authority being in charge of maritime affairs shall have the power to take compulsory measures to avoid or decrease pollution damage. If vessels and facilities occur maritime incidents at the high see resulting in consequences of major pollution damage or threat to the sea areas under the jurisdiction of the People's Republic of China, the State competent authority being in charge of maritime affairs shall have the power to take corresponding measures necessary for pollution damages which have caused or are likely to cause'. Therefore, if a vessel is involved in a marine accident which has caused or is likely to cause serious pollution damage to the marine environment, a maritime administrative authority has the power to perform compulsory clean-up actions to prevent or minimize such pollution damage. Thus, a compulsory clean-up act fully conforms to the above characteristics of the emergency compulsory measures for the following reasons:

1. A compulsory clean-up act is a measure taken by an appropriate maritime administrative authority with powers pursuant to the provisions of MEPL. The counter-party is obliged to carry out such measures. Therefore, a compulsory clean-up act has the characteristics of a compulsory administrative measure, being statutory, specific and mandatory.
2. The purpose of such clean-up acts is to prevent or minimize pollution damage and thus to maintain State and public interests. Therefore, its aim is to maintain the social order and social condition specified by law. Accordingly, such an act is not punitive.
3. Such a clean-up act has the characteristic of an emergency situation. When a vessel is involved in a marine accident which has caused or is likely to cause serious pollution

---

[9] *Ibid.*, p. 16.

damage to the marine environment, such pollution damage endangers society and effective measures must be taken immediately to prevent or limit such damage. Therefore, based upon such an emergency, an appropriate maritime administrative authority may take the necessary clean-up actions without attesting to procedures such as approval and notification and so forth.

3 LEGAL NATURE OF THE COMPULSORY CLEAN-UP COSTS BORNE BY RESPONSIBLE PARTIES

It can be concluded from the above discussion that a compulsory clean-up act itself may be an emergency compulsory measure and is in essence a specific administrative act. However, what is the legal nature of the compulsory clean-up costs borne by the responsible party? Is it an administrative responsibility or civil liability?

As to the form of the pollution liability borne by the responsible party, the provisions in the old and new MEPL differ. Article 41 of MEPL 1982 provides that

> 'In case of a violation of this Law that has caused or is likely to cause pollution damage to the marine environment, the competent authority prescribed in Article 5 of this Law may order the violator to remedy the pollution damage within a definite time, pay a pollutant discharging fee, pay the cost for eliminating the pollution and compensate for the losses sustained by the State. The authority may also give the violator a warning or impose a fine. An involved party contesting the decision may file a suit with a people's court within 15 days after it has received the written decision. If a suit has not been filed and the decision has not been carried out upon the expiration of that period, the competent authority shall request the people's court to enforce the decision in accordance with the law.'

Article 39 of the Regulations of the People's Republic of China Concerning the Prevention of Pollution of Sea Areas by Vessels contains similar provision. In accordance with the above provisions, the responsibility of the responsible party to pay the costs for eliminating the pollution, including the compulsory clean-up costs, and to compensate for the losses sustained by the State was an administrative responsibility. However, Article 90 of MEPL 1999 provides:

> 'Those who causes pollution damage to the marine environment shall eliminate the damage and compensate the losses; in case of pollution damage to the marine environment resulting entirely from the intentional act or fault of a third party, third party shall eliminate the damage and be liable for the compensation. If the State suffers heavy losses from the damages to marine ecosystems, marine aquatic resources and marine nature reserves, the department invested by this law with the power of marine environment supervision and administration shall, on behalf of the State, put forward compensation demand to those who are responsible for the damages'.

From the provisions of this Article, it can be seen that the forms of responsibility for marine pollution are eradication of the pollution and compensation for losses. These are forms of civil liability rather than administrative responsibility. Therefore, the provisions of this Article are obviously those of civil liability for pollution damage. Although the Administrative Regulations on the Prevention of Pollution to the Sea Area by Vessels has not yet been amended in accordance with MEPL 1999, the provisions of Article 39 thereof are in conflict with the law at a higher level and shall therefore be null and void. Article 90 of MEPL 1999 does not refer to clean-up costs directly. Therefore, it cannot be deduced from this Article whether the legal nature of the compulsory clean-up costs to be borne by the responsible party are a matter of administrative responsibility or civil liability. It requires theoretical analysis.

3.1 COMPULSORY CLEAN-UP COSTS AS COSTS OF SUBSTITUTION OF PERFORMANCE UNDER ADMINISTRATIVE LAW

3.1.1 Compulsory Clean-up Acts as Indirect Compulsory Acts

Compulsory administrative measures can be divided into direct and indirect compulsoriness. Direct compulsoriness means the compulsory measures taken by an institution of administrative body are enforced directly against the body, or properties of the responsible party, so as to achieve the same status as if the responsible party had carried out his administrative obligations.[10] Indirect compulsoriness means the compulsory measures taken are to oblige the responsible party to carry out his administrative obligations or to achieve the same situation as were the responsible party to carry out his obligations using indirect means.[11] At present, Chinese law does not clearly stipulate the form of emergency administrative compulsoriness; however, scholars have the view that, except for penalties, various forms of compulsory administrative enforcement are the result of direct compulsoriness[12] and substitution of performance may naturally become the emergency administrative compulsoriness as they share the same source.[13] With respect to the compulsory clean-up acts, these kinds of compulsory measures themselves are not directly aimed at the counter party or his property (the vessel), but rather to the clean up of the pollutants spilled from the vessel. In other words, such acts relate, rather, to the pollutants. Therefore, compulsory clean-up acts are indirect compulsory means and the same as the means of substitution of performance in the compulsory administrative enforcement.

3.1.2 Compulsory Clean-up Costs as those of Substituted Performance of Carrying out Compulsory Administrative Measures

The basic procedures for general compulsory administrative measures usually include decision, desist/discontinue notification, assistance, of counter-parties'/party's

---

[10] H. Hu, *supra* note 5, 136.
[11] *Ibid.*
[12] Direct compulsoriness is not the same as emergency administrative compulsoriness. Direct compulsoriness is a method within the category of administrative compulsoriness and is applicable to both compulsory administrative enforcement and emergency administrative compulsoriness. Emergency administrative compulsoriness is a kind of administrative system and is an exception or special form of the administrative compulsory enforcement. Therefore, direct compulsoriness and administrative emergency compulsoriness are not concepts at the same level. See J. Hu, *supra* note 5, 291.
[13] J. Hu, *supra* note 5, 291.

performance, enforcement and collection of costs.[14] As regards emergency compulsoriness, as it originates from emergency situations, if an administrative body has to carry out the procedures of requesting suspension, assistance and/or expecting a counterpart's performance, time may be delayed and the purpose of preventing damage may not be achieved efficiently. Therefore, only the procedures of decision and enforcement may be considered compulsory clean-up acts. The collection of costs is not a pre-requisite for all compulsory administrative measures. Normally, it only exists in the enforcement of a decision of compulsory administrative measures by means of substitution of performance. By contrast, direct compulsoriness is an act within the duty of the administrative body, the costs of which have already been paid by the counter-party through payment of taxes. Therefore, if the costs were collected for compulsory administrative enforcement, it would constitute double payment by the counterpart.[15]

Based on the above analysis, the authors take the view that the compulsory clean-up costs are the costs of an appropriate authority by means of substitution of performance for enforcing compulsory administrative measures under the administrative law. The reason for adding the word 'means' here is only to differentiate such costs from those of substitution of performance in compulsory administrative enforcement. However, there is no difference in nature between such costs and those of substitution of performance. Article 7 of the Regulations of the People's Republic of China Concerning the Prevention of Pollution of Sea Areas by Vessels also stipulates:

> 'Where a marine accident has occurred to a vessel which has caused or is likely to cause a major environmental pollution damage, the Harbour Superintendency Administration may decide to take whatever compulsory steps necessary to avoid or mitigate any pollution damage, such as compulsory cleaning-up or compulsory towage. All the expenses incurred therefrom shaal be borne by the owners of the vessels concerned.'

Combining the above analysis and these provisions, the authors have the view that, under administrative law, compulsory clean-up costs should be regarded as costs of substitution of performance. As the costs of substitution of performance, paying such costs should be the responsibility of the counter-party under administrative law.

There are not many provisions directing how to collect the costs of substituted performance in national laws worldwide, with the exception of the laws of Germany and Japan.[16] The Administrative Execution Law of 1998 of Taiwan contains a special chapter on the enforcement means for monetary payment obligations under public law, including the payment of the costs of substituted performance.[17] However, the current Chinese law does not have a clear stipulation in this respect. Therefore, if a counter-party fails to carry out his payment obligation under administrative law, the appropriate authority is only competent to file an application with a court for enforcement. However, before filing such

---

[14] J. Hu, *supra* note 5, 240-246.
[15] J. Hu, *supra* note 5, 245-246.
[16] For example, art. 19 of the Administrative Compulsory Execution Law of 1953 of Germany and art. 5 of the Administrative Substitution Performance Law of 1948 of Japan. See, J. Zhang, *Study on materials of domestic and foreign administrative compulsory law* (Law Press, Beijing, 2003) pp. 96-97.
[17] W. Jin, *Administrative Compulsory Legal System of China (sic)* (Law Press, Beijing, 2003) pp. 494-495.

an application, a maritime administrative authority has the power to detain the vessel in order to provide security for the payment of such compulsory clean-up costs, by virtue of Article 19 of the Maritime Traffic Safety Law.[18]

3.2 TORTIOUS LIABILITY UNDER THE CIVIL LAW

3.2.1 Compulsory Clean-up Costs within Tortious Liability

The 'essentials of tortious liability' mean the essential conditions for the imposition of tortious liability. These essentials are different in the law and judicial practice of different countries. National laws as represented by the French Law hold that the essential elements are the fact of damage, fault and causation, while national laws as represented by German Law hold that, in addition to the three French elements, there is the further essential element of the illegality of the act.[19] Scholars in China have different opinions on this issue with some holding that there is a three-part doctrine while others argue for a four-part doctrine. With regard to liability for environmental pollution damage, as a strict liability regime is generally acknowledged, it means that fault is not an essential element. Therefore, the three- or four-element doctrine becomes a two- or three-element doctrine respectively.[20] However, insofar as the liability for compulsory clean-up costs is to be borne by the responsible party, it is if the three-part doctrine is applied that the essential elements of tortious liability for environmental pollution damage are fully met, that is, the illegality of environmental pollution act, the fact of damage and a causal link between the illegal act and the fact of damage.

The illegality of the environmental pollution act means that such an act is a violation of the law or of regulations related to environmental protection. As stated above, the prerequisite for a maritime administrative authority to take clean-up acts is that a vessel is involved in a marine accident which has caused or is likely to cause serious pollution damage to the marine environment. An act of spillage or potential spillage itself is prohibited by law and regulations for marine environmental protection and is therefore illegal.

National laws seldom define the nature of damage itself. An exception is the Civil Code of Austria, which defines it as any harm to an individual's property, right or body.[21] Chinese law only provides that parties who through their fault encroach upon the property or body of another person shall bear civil liability without defining the term 'damage'.[22] Scholars generally express the opinion that damage means any disadvantageous effect,

---

[18] Art. 19 of the Maritime Traffic Safety Law of PRC provides that 'The competent authority shall have the right to forbid a vessel or an installation from leaving the harbour or order it to suspend its voyage, change its route or cease its operation under any one of the following circumstances: ...(4) if it has not paid the fees that are due or furnished appropriate security to the competent authority or the department concerned, ...'.
[19] L. Wang, *Study on Tort Liability Principles* (China Politics and Law University Press, Beijing, 2002) p. 370.
[20] For different doctrines on the essentials of liability for tortious liability and liability for environmental pollution infringements, see D. Cao, *Environmental Tort Law* (Law Press, Beijing, 2000) pp. 158-165.
[21] Art. 1293 of the Civil Code of Austria.
[22] §2 of art. 106 of the General Principles of Civil Law of PRC.

though a specified act or occurrence, to the rights or profits protected by the tort law.[23] Normally, preventive measures are taken to prevent or minimize the damage or loss when the infringement has occurred. Under normal circumstances, preventive measures are taken by victims. However, there are certain circumstances under which such measures are taken by others. Whoever takes preventive measures, it is commonly accepted in laws worldwide that there should be compensation for the costs arising from such measures. In other words, the costs of preventive measures are constituted in the fact of damage. However, the particular items of damage to be compensated depend on the result or likely result flowing from the act of infringement. The costs of a clean-up act for the prevention or minimization of marine pollution damage, prepaid by a maritime administrative authority, are actually the costs of preventive measures prepaid by the authority representing the State. Thus, it is a kind of detriment to the interests of the State and constitutes 'damage' in fact.

As to causation in tortious liability, some scholars take the view that it is the causation between the illegal act and the fact of damage which is critical, while others take the view that it is the causation between the act or things carried out by the actor and the fact of damage. In determining causation, some scholars hold the equivalent causation doctrine while others hold a certain causation doctrine. However, whatever kind of causation doctrine is applied and however the causation is determined, the requirement of causation between the infringing act and the damage caused has been popularly accepted worldwide. In Anglo-American tort law, the wrongdoer's duty of care toward and liability to compensate the salvor has become a general principle of tort law following the decision in the case of *Wager* v. *International Railway Company in New York, USA*.[24] The German prejudication doctrine also recognizes equivalent causation in emergency rescue cases. The dominant ideology in Germany is, in addition to the right to claim based on the management of affairs without mandate, that the salvor also has the right to claim compensation in tort law.[25] The Draft Civil Code of China also respects the above opinions.[26] Therefore, compulsory clean-up costs are, by nature, losses or damage sustained by a maritime administrative authority when it carries out emergency rescue procedures and such losses or damage have a causal link with the infringing act of environmental pollution.

In sum, compulsory clean-up costs are a part of the damage or losses caused by an act of oil pollution infringement inasmuch as causation exists between them. Therefore, they meet the essentials for tortious liability such that the responsible party shall bear the

---

[23] L. Wang and L. Yang, *Tort Law* (Law Press, Beijing, 1996) p. 55.
[24] Z. Wang, *Tort Law* (I, China Politics and Law University Press, Beijing, 2001) pp. 210-212.
[25] *Ibid.*, p. 212.
[26] Art. 7 of the draft Civil Law, for discussion at the 31st Meeting of NPC Standing Committee dated 12 December 2002 provides that 'If a person suffers damage from preventing or stopping encroachment on the property or person of a third party, the infringer shall bear liability for compensation and the beneficiary may also give appropriate compensation'. Art. 6 of the draft Tort Law of Civil Code prepared by Law Science Research Centre of China Renmin University provides: 'If a person suffers damage from preventing or stopping encroachment on state or collective property, or the property or person of a third party, the infringer shall bear responsibility for compensation. In the absence of infringer or full compensation from the infringer, the beneficiary shall give appropriate compensation (*sic*)'. See, L. Wang, *Study on Tort Law of Civil Code* (People's Court Press, Beijing, 2003) p. 3 and p. 13.

compulsory clean-up costs. Consequently, there is a legal basis for the responsible party to bear civil liability for such compulsory clean-up costs.

3.2.2 Compulsory Clean-up Costs within the Scope of Civil Compensation Liability

Article 134 of the General Principles of the Civil Law of 1986 provides 12 forms of civil liability: cessation of infringement, removal of obstacles, elimination of dangers, return of property, restoration of the original condition, repair, reworking, replacement, compensation for losses, elimination of ill effects, rehabilitation of reputation and extension of an apology. However, not all of the above forms of civil liability apply to marine pollution damage. Article 90 of the MEPL 1999 provides that 'The responsible party who caused the marine pollution damage shall remove the danger and obstacles and compensate for the losses'. Therefore, it is removal of the danger and any obstacles, as well as compensation for losses that are the forms of civil liability for marine pollution damages. The form of removal of the dangers and obstacles implies the definition of cessation of infringements, removal of obstacles and elimination of dangers stipulated in Article 134 of the General Principles of the Civil Law.[27] In ordinary situations, civil liability for the removal of obstacles and the costs thereof shall be borne by the infringer. This includes requiring the party who is causing or has caused pollution to immediately stop causing such damage or to prevent, remove or minimize such damage, or any damage which is likely to occur. Therefore, this form of civil liability is more positive and effective in prevention of marine pollution damage. A compulsory clean-up act itself is a compulsory administrative act and its aim is to prevent and minimize damage to marine environment. Therefore, compulsory clean-up acts are, in essence, measures taken by a competent authority for and on behalf of the responsible party to mitigate oil pollution damages as necessary in the prevailing circumstances. Subsequently, the clean-up costs should be borne by the responsible party. In other words, the responsible party shall compensate for such costs as are advanced by the competent authority. Therefore, such civil liability is, substantively, a form of compensation. On the other hand, as full compensation is a general principle in compensation for damages. the infringer must also compensate any victims for all actual property damage or losses caused by the act. Accordingly, the scope of compensation covers all the necessary and reasonable costs paid by the victims for the recovery of their rights and for the minimization of damages or losses.[28] Therefore, clean-up costs, as the costs of preventive measures, are within the scope of compensation.

Another related question is whether an appropriate competent authority is entitled to represent the State as a body with civil rights. The answer to this question is positive. Article 90 of the MEPL 1999 clearly provides that 'The department exercising superintendence and administration power pursuant to this Law is entitled to represent the State to claim against the responsible party.' The law of the former Soviet Union also

---

[27] D. Cao, *supra* note 20, 216.
[28] L. Wang and L. Yang, *supra* note 23, 327.

provided that the Ministry of Water Conservancy was entitled to represent the State to claim compensation against an infringer in the area of marine environmental protection.[29]

Furthermore, from the provisions of Article 21 of the Special Maritime Procedure Law of 1999,[30] it can be assumed that procedural law in China already prescribes that the costs of preventive measures as including clean-up costs as part of maritime claims. This also means that compulsory clean-up costs are within the scope of civil liability.

In summary, payment of compulsory clean-up costs by the responsible party can be treated as a matter of civil liability under Chinese civil law.

3.3 CONTRIBUTION TO THE REALISATION OF UNIFORMITY OF FAIRNESS AND EFFICIENCY

Every legal system has its own values of existence and the differences of values among them are only the attitudes of legislators towards the various values of legal systems. English legal scholars Stan and Chandler stated in their book *Values of Law in Western Society* that the basic values of law in western society should be 'order, fairness and self-freedom'.[31] In China, Prof. Zhang Wenxian and Prof. Sun Guohua, et al., consider that 'order, freedom, efficiency and justice' should be the basic values of law.[32] Prof. Shen Zongling considers that the values of law 'mainly rest on two categories, that is, justice and interests' and points out the general principles which should be complied with, while law regulates the conflicts between various interests, namely, concurrently caring for the interests of the State, collectively and individually; protecting the benefits of both major and minor interests, the long and short-term, integral and partial interests; efficiency being preferred while concurrently ensuring fairness; and successfully selecting the best approach.[33] As a result, although discrepancies exist between the scholars' expressions of the basic values of the law, efficiency and fairness are always among them. Accordingly, these two basic values of law are to be pursued in the construction or choice of legal system in relation to clean-up costs.

Efficiency and fairness are social values which are not only mutually suitable, but which can also contradict each other. From the viewpoint of suitability, on the one hand, if social resources are distributed in accordance with the standard of efficiency, then social economic development will be promoted and the total social welfare will be increased, based upon which, it is likely that a higher level of fairness will be realized, namely, common wealth. On the other hand, if efficiency is preferred absolutely and fairness is overlooked, a wide income gap may be caused among members of society, which would

---

[29] D. Cao, *supra* note 20, 170.
[30] Art. 21 of Special Maritime Procedure Law of 1999 provides that 'With respect to the following maritime claims, an application may be filed for the detention of a ship: ... (4) Damage or threat of damage caused by the ship to the environment, coast or relevant interested persons; measures adopted to prevent, diminish or eliminate such damage; compensation paid for such damage; expenses for reasonable measures actually adopted or to be adopted to restore the environment; losses caused or likely caused by such damage to a third party; and damage, expenses or losses of a similar nature to those specified in this paragraph. ... (*sic*)'.
[31] Z. Shen, *Jurisprudence* (Beijing University Press, Beijing, 2001) p. 49.
[32] W. Zhang, *Jurisprudence* (Law Press, Beijing, 1999) p. 281; S. Guohua and Z. Jingwen, *Jurisprudence* (Chinese People's University Press, Beijing, 1999) p. 59.
[33] Z. Shen, *supra* note 31, 51-62.

lead to social instability and have an adverse influence and even substantially harmful effect on efficiency.[34] Considering such contradiction, the principle of fairness is inclined to the equality of the interests of social members. However, high efficiency of social economic development may easily be ignored. Efficiency addresses the rapid development of the social economy, whereas the expansion of differences of various interests may be ignored.[35] Therefore, insofar as efficiency and fairness are concerned, efforts should be made to enlarge their suitability and to lessen the contradiction for the purpose of achieving a unification of efficiency and fairness.

For the sake of efficiency, if the payment of compulsory clean-up costs by the responsible party is treated as a responsibility under administrative law, it would be beneficial to increase the efficiency of maritime administrative authorities in the recovery of the clean-up costs. When a maritime administrative authority collects the costs from substitution of performance, the general procedure of administrative levy should be complied with, such that the maritime administrative authority makes a written administrative decision levying the compulsory clean-up costs, indicating the amount and period for payment and served on the counterpart. If the counterpart does not fulfil his responsibility to pay within the fixed time limit, both fully and voluntarily, then the administrative levy may be imposed in the form of a compulsory charge by means of making a mandatory order detaining the vessel so as to compel the counterpart to pay, or alternatively to provide security for payment, or, otherwise an application may be filed with the people's court for a compulsory enforcement order. Therefore, if the payment of compulsory clean-up costs by the responsible party is treated as an administrative responsibility under administrative law, then a maritime administrative authority may levy such costs using its administrative powers and civil proceedings, like litigation and arbitration, can be avoided.

However, in the interest of fairness, only by means of civil liability can the interests of the State, as well as the justifiable interests of the polluting vessel and the third parties be best protected. The reasons can be described as follows:

1 Insofar as the protection of the interests of the State are concerned, *prima facie*, it seems more favourable for recovery of the clean-up costs to treat the payment of the compulsory clean-up costs as an administrative responsibility of the responsible party. This is because, on the one hand, a maritime administrative authority may compel the responsible party to pay the costs within a specified time by exercising its administrative powers and, on the other hand, by virtue of the general principle that public law shall have precedence over private law, such clean-up costs may be recovered prior to civil claims. Furthermore, with regard to such costs, the responsible party cannot avail itself of a limitation of liability, as stipulated in the Chinese Maritime Code and in the related international conventions, and, consequently, the clean-up costs may be compensated both sufficiently and fully. However, this is often not the case in reality. Where the responsible party is financially capable, even if the payment of the clean-up costs is treated as a part of the regime of civil liability, the

---

[34] W. Zhang, *supra* note 32, 316.
[35] Z. Shen, *supra* note 31, 62.

maritime administrative authority may file an application with a maritime court for detention of the vessel on the basis of the maritime claim for such costs in accordance with Article 21 of the Special Maritime Procedure Law of PRC, in order to secure the recovery of such costs. In other words, the results of recovery would be the same, whether the payment of such costs by the responsible party is treated as part of the regime of either administrative responsibility or civil liability. However, where the amount of the clean-up costs is considerable and the responsible party becomes financially incapable or insolvent, and especially where the vessel sank in the accident, there may be no vessel to be compulsorily detained by the maritime administrative authority. In such a case, the maritime administrative authority would be unable to recover such costs by judicial means due to a lack of enforceable property of the responsible party. However, treating the payment of the compulsory clean-up costs as part of the regime of civil liability may be more favourable to the recovery of such costs. This is because where the vessel which actually carried more than 2000 tonnes of oil in bulk as cargo, it is covered by the 1969/1992 International Convention on the Civil Liability for Oil Pollution Damages ('CLC'), and the clean-up costs are covered by compulsory oil pollution liability insurance or other financial securities, by which the liability insurer, or the guarantor of other financial security, may be held liable for such costs with only very limited exceptions. In addition, the maritime administrative authority may bring a claim against the insurer or the guarantor directly. Meanwhile, a double-level compensation scheme for such oil pollution damages is soon to be adopted in China, that is, those damages not compensated for under the CLC may soon be compensated for or indemnified by the national oil pollution compensation fund which is to be established in China, and despite that China is not a party to the 1969/1992 Fund Convention. Consequently the international oil pollution compensation fund established thereunder is unavailable. As a result, even if the responsible party becomes financially incapable or insolvent, sufficient or appropriate compensation for clean-up costs is still possible. By contrast, if the payment of the clean-up costs is treated as part of a regime of administrative responsibility, such costs are not covered by compulsory liability insurance or other financial security or by the oil pollution fund. Civil liability for the clean-up costs arising from the polluting vessel, even if not covered by CLC, are normally covered by the shipowners' protection and indemnity insurance (P&I policy).[36] However, if the payment for the clean-up costs is treated as part of a regime of administrative responsibility, it is arguable whether such costs are covered by P&I policy. Meanwhile, a maritime administrative authority is not entitled to bring a claim against the liability insurer, or

---

[36] §11 of Rule 5 of the 'Protection and Indemnity Risks Covered' of China Shipowners Mutual Assurance Association Rules provides that 'The liability, losses, damages, costs and expenses set out below when they are caused by or incurred in consequence of the discharge or escape from an entered ship transporting oil or any other hazardous substance, or the threat of such discharge or escape: ... 3. The costs of any measures reasonably taken for the purpose of avoiding or minimizing pollution or any resulting loss or damage together with any liability for loss of damage to property caused by measures so taken. 4. The costs of any measures reasonably taken to prevent an imminent danger of discharge or escape from the entered ship transporting oil or any hazardous substance which may cause pollution. 5. The costs or liabilities incurred as a result of compliance with any order or direction given by any government or authority for the purpose of preventing or reducing pollution or the risks of pollution'.

guarantor of other financial security directly due to a lack of an administrative legal relationship between them. Furthermore, if the payment of the clean-up costs is treated as part of a regime of administrative responsibility, and where there is no enforceable property of the responsible party, in China it can be argued successfully that a judgement of Chinese court rendered under administrative law by which the responsible party is obliged to pay, such costs shall be recognized and enforced by a court in a foreign country where the responsible party has property. Nevertheless, if payment of the clean-up costs is to be treated under a regime of civil liability, a civil judgement or arbitration award can also easily be recognized and enforced by a foreign court under the CLC or the 1958 New York Convention, to which China is a party.

2 Insofar as the protection of the interests of the polluting vessel is concerned, if the payment of the clean-up costs is treated as part of a regime of administrative responsibility, the responsible party is not entitled to limit its liability, and shall pay in full. Where the amount of such costs is very high, such an administrative responsibility may cause the shipowners to go bankrupt. However, if it is treated under a civil liability regime, the responsible party is entitled to limit its liability under the CLC or Chapter XI 'Limitation of liability for maritime claims' of the Chinese Maritime Code, as the case may be.

3 Insofar as protection of the interests of third parties is concerned, if the payment of the clean-up costs is treated as part of a regime of administrative responsibility, by virtue of the general principle that public law has precedence over private law, such costs are compensated for before the compensation for other oil pollution damages arising out of the same accident. Thus, where the responsible party is financially incapable, it means that the rights of third parties who bring claims for such oil pollution damage are prejudiced.

In summary, treating the payment of the clean-up costs as within a regime of administrative responsibility of the counter or injuring party, may, in some cases, increase the efficiency of recovery of such costs by a maritime administrative authority. However, such a treatment may also prejudice the justifiable interests of the responsible party and third parties, and even perhaps the interests of the State. Furthermore, in many cases such efficiency is very difficult or even impossible to attain, or is attained at the cost of sacrificing the justifiable interests of the responsible party or the third parties. By contrast, if the payment of compulsory clean-up costs is treated as part of a regime of civil liability, the interests of the State, the responsible party and third parties would be better protected as a whole and the unification of efficiency and fairness would be better attained.

3.4 THE INTERNATIONAL TENDENCY TO TREAT PAYMENT OF COMPULSORY CLEAN-UP COSTS AS CIVIL LIABILITY

Oil pollution damage, as stipulated by the CLC, includes the costs of preventative measures, which means that any reasonable measures taken by anyone to prevent or minimize pollution damage is covered. Obviously, the word 'anyone' includes a maritime administrative authority that has taken such preventative measures.[37] Paragraphs 7 and 9 of Article 1 of the Convention on the Civil Liability for Bunker Pollution Damage of 2001 have the same provisions as the CLC. Insofar as the legislation of other nations is concerned, the Oil Pollution Act of 1990 (OPA) of the United States expressly stipulates that all clean-up costs paid by federal authorities, the states or Indian tribes, in accordance with Section 311 of the Federal Water Pollution Control Act as revised by OPA or the High Sea Intervention Act or state law, are within the scope of liability for damages.[38] In the United Kingdom, the Merchant Shipping Act of 1995 stipulates that the port authority and Secretary of the State have the power to take compulsory measures to prevent or minimize oil pollution damages when necessary. The costs of such preventative measures come within oil pollution damages.[39]

4 CONCLUSIONS

As discussed above, the following conclusions can be drawn:

1 From the viewpoint of administrative law, compulsory clean-up costs in the case of oil pollution from vessels are in the nature of the costs of substitution of performance by way of compulsory administrative measures, and a maritime administrative authority has the power to require that the counterpart pay such costs as part of his administrative responsibility.
2 From the viewpoint of tort law, compulsory clean-up costs meet the essential elements of tortious liability and the responsible party may bear civil liability for such costs.
3 Treating the payment of the compulsory clean-up costs as part of a civil liability regime rather than as part of a regime of administrative responsibility may better protect State interests and the justifiable interests of the responsible party and third parties, and the unification of efficiency and fairness may be attained much more easily. Such treatment is the trend in international legislation.
4 To solve current disputes regarding the nature of the compulsory clean-up costs in the case of pollution damage to the marine environment, it is advisable to classify the payment of such costs as part of the civil liability of the responsible party as regards pollution damage in Chinese law.

---

[37] §6 and 7 of art. 1 of the 1992 Protocol of International Convention on Civil Liability for Oil Pollution Damage, 1969.
[38] Subsecs. (a) and (b) of S.1002 'Factor of liability' of Oil Pollution Act of 1990 of the United States.
[39] Ss. 137, 153 and 154 of the Merchant Shipping Act of 1995 of the United Kingdom.

LIST OF REFERENCES

D. Cao, *Environmental Tort Law* (Law Press, Beijing, 2000).

Z. Guan, 'On perfection of procedures for constitution of limitation fund for maritime claims from the case of the collision case of M/V 'Dainty River'', *Annual of China Maritime Law* (2003) pp. 241-246.

M. Hong, *A study on responsible party and nature of responsibility of ship's oil pollution accident*, <http://www.999abc.com>, 13 August 2003.

J. Hu, *A Study on Administrative Law* (Law Press, Beijing, 2003).

W. Jin, *Administrative Compulsory Legal System of China (sic)* (Law Press, Beijing, 2003).

Z. Shen, *Jurisprudence* (Beijing University Press, Beijing, 2001).

Y. Si, 'A study on applicable law in the compensation of oil pollution damages from the coastal vessels', (2002) 4 *Journal of Dalian Maritime University* 1-2.

G. Sun and J. Zhu, *Jurisprudence* (Chinese People's University Press, Beijing, 1999).

L. Wang, *Study on Tort Liability Principles* (China Politics and Law University Press, Beijing, 2002).

L. Wang, *Study on Tort Law of Civil Code* (People's Court Press, Beijing, 2003).

L. Wang and L. Yang, *Tort Law* (Law Press, Beijing, 1996).

Z. Zhan, 'A study on the system of administrative compulsory measures', (2002) *Journal of Gansu Politics and Law College* 2.

C. Zhang, *A further study on the legal nature of clean-up costs in ship's oil spillage accident*, <http://www/ccmt.org.cn>, 3 December 2002.

J. Zhang, *Study on materials of domestic and foreign administrative compulsory law* (Law Press, Beijing, 2003).

S. Zhang, *The title of the administrative authority to file a maritime pollution claim*, Proceedings of 11th Conference on Maritime Trial of PRC, Tianjin Maritime Court, 2000, 58-59.

W. Zhang, *Jurisprudence* (Law Press, Beijing, 1999).

Z. Wang, *Tort Law* (I, China Politics and Law University Press, Beijing, 2001).

DISCUSSION ON THE NATIONAL CLAIM SYSTEM FOR OIL POLLUTION DAMAGE FROM SHIPS

Ma Jing-jing and Du Jiang

1 INTRODUCTION

The oceans are mankind's most valuable resource, and the economy in the 21st century depends on the marine economy. However, with mankind's increasing development and utilization of the oceans and their resources, the marine environment is deteriorating, and so protecting the oceans against all pollution has already aroused the attention of many. In marine environmental pollution, oil pollution from ships is already one of the most threatening contaminations, and China is a country both with wide sea areas and extensive maritime interests. As such, it has a realistic need to establish a proper national claim system for oil pollution damage from ships. However, there is little theoretical research, and few legislative and judicial practices in the system. In the light of this, and on the basis of affirming the need to set up a national claim system for oil pollution damage from ships, this system is considered from the aspects of the establishment of the subject and practices of the national claim, as well as the scope of the national claim and so forth, with the aim of being helpful in the development of practices in China in compensation for oil pollution damage.

2 THE NEED TO SET UP THE NATIONAL CLAIM SYSTEM FOR OIL POLLUTION DAMAGE CAUSED BY SHIPS

With the continuously rapid development of China's national economy, there is a great demand for oil, as a kind of strategic material. Since 1993, oil production has been unable to meet the demand, with the result that China has become a pure importer of oil. Now, in China, the amount of oil shipped by ocean has leapt to third place in the world, second only to the US and Japan.[1] While promoting the economic growth of our country, the swift development of the seaborne oil trade also attracts the potential threat of oil spills, especially at present as more and more Very Large Crude Carriers (VLCC) come into operation, bringing a greater risk of oil spills. According to the statistics, from 1976 to 1999, there were 51 serious oil spill accidents on the Chinese coast, increasing by over two per year on average. Since 1994, serious oil spill accidents have increased by five-seven per year.[2] To make matters worse, 38 such accidents took place in 2000 alone.[3] All

---

[1] 'The super killer of marine pollution with the petroleum revealed, Our country deeply hurt (*sic*)', Xinhua Website, 14 March 2000.
[2] 'Lingding Ocean 6.0 actual combat', (2000) *China Maritime* 3.

of these cannot but be a dangerous sign. Because of the characteristics of oil pollution, such as spreading widely, staying longer and being difficult to deal with once the oil has spilt from the vessel, it causes a calamitous influence on the marine environment. China has many maritime interests, including aquaculture, the fishing industry, maritime communications and transportation businesses, the tourist industry, and so on, all of which rely on the marine environment. If that environment is seriously polluted, it affects many parties in the development and utilization of the marine resources, and, at the same time, it will certainly cause the owner of the resources, the country, inestimable economic losses. For example, in the incident of the ship the 'Feoso Ambassador' that occured in Qingdao in 1983, economic losses of RMB 17 million resulted;, the accident of the ship the 'Patriot', which hit a submerged reef in Weihai in 1996, caused losses of more than RMB 50 million;[4] the collision of the ship the 'Tasman Sea' in the Bohai Gulf in 2003, leading to a large amount of crude oil overflow, brought China environmental economic losses of over RMB 100 million.[5]

The reason for the frequent occurrences of ship oil pollution accidents is China lacks a national claim mechanism for oil pollution. Data show that, among the 29 serious tanker oil spill accidents between 1973 and 2000, there were seven between foreign tankers, all of which were compensated, to an average of about RMB 8.28 million per accident, and with the largest amount of compensation being RMB 17.75 million. Among the 22 accidents between Chinese tankers, however, only nine were compensated, accounting for only 30 percent of the losses, the accident compensation averaging only RMB 1.53 million on average, and not more than RMB 5.5 million.[6] Compared with developed countries that have already set up the national mechanisms for compensation of oil pollution damage, in China, not only is the proportion of accidents that receive compensation disproportionate, but the amounts of compensation are also low. Furthermore, since the country, as one of the oil pollution victims, cannot get due compensation, there are insufficient funds for polluted marine environment to be reconstructed, which has a negative impact on the sustainable development of the national marine resources.

The international community has, in legislation on oil pollution, regularly issued relevant international conventions from the 1960s onwards. Many developed countries, such as the UK, the US, Japan, Korea and Canada, have enacted special pollution laws. Reviewing China, up to now there is still no single special law related to compensation for oil pollution damage from ships, and the regulations involved in this respect appear only sporadically in different laws and administrative rules, all of which are of general principle. Indeed, legislation for damage claims is only at a starting stage. For example, according to China's National Oil Spill Contingency Plan, carrying out claims for and compensation of oil pollution damage should comply with the 'International Convention on Civil Liability for Oil Pollution Damage 1992', and the 'concrete' method stipulated

---

[3] K. Li, *China Ship Survey* (2001) 6.
[4] Y. Si et al., *Special Subject Study on Maritime Law* (Dalian Maritime University Press, Dalian, 2002), p. 436.
[5] 'The super killer of marine pollution with the petroleum revealed, Our country deeply hurt', Xinhua Website, 14 March 2000.
[6] 'Perfect the mechanism and defend the oil pollution (sic)', *People Daily*, Aug 5th 2002.

by the State Council about insurance of shipping oil and the system of a Fund for Compensation of Oil Pollution Damage. Obviously, these abstract and principle stipulations have weakened the judicial power to enforce the law in practice, and the results of disputes over oil pollution damage are not satisfactory. In addition, in respect of the international conventions, China has only ratified the 'International Convention on Civil Liability for Oil Pollution Damage, 1969' (hereafter referred to as the 1969 CLC) and its 1992 protocol, and has not participated in the 'International Convention on the Establishment of an International Fund for Compensation for Oil Pollution Damage, 1971' (hereafter referred to as the 1971 Fund), which is closely connected with the 1969 CLC in international mechanisms for compensation for oil pollution, nor has it established such a domestic fund.

Therefore, in order to protect the marine environment and its resources, to maintain an ecological balance, and to promote sustainable development of the economy and society, China must set up a national claim mechanism for oil pollution, reducing the possibility of marine pollution to a minimum level.

3 ESTABLISHMENT OF THE SUBJECT OF THE NATIONAL CLAIM FOR OIL POLLUTION DAMAGE FROM SHIPS

3.1 QUALIFICATION OF THE COUNTRY TO PROPOSE THE OIL POLLUTION CLAIM

In claims after oil pollution damage from ships, firstly, the subject of the civil rights following the oil spill accident should be clarified, that is, the proper plaintiff in the civil litigation. Pursuant to the traditional theory of qualified litigant participation of the continental legal system, the 'litigant' in civil litigation means a directly interested party who takes legal action because, due to the dispute, he is involved in the interplay between civil rights and obligations, as determined by the court arbiter. This is termed the 'Theory of Directly Interested Party'. This theory adopts the concept and criterion of the substantial party, and emphasizes their identity as between the civil litigation participants and civil subjects, that is to say, such that only the person who has a substantial right to his case is entitled to bring an action in court. The civil procedure law of China simply adopts this traditional theory. For instance, Article 108 of the *Law of Civil Procedure of the PRC* provides: 'the plaintiff is a citizen, a legal person or other organisation who has direct stakes with the case'. It is obvious that only when his rights and interests are infringed, or he enters civil disputes with others, a citizen, legal person or other organization may go to the people's court as a plaintiff, and request the court to exercise its jurisdiction to protect his civil rights and interests. Those people or organizations with no direct stakes, therefore, have no right to bring litigation to court, and at most they have subsidiary rights in litigation.

Based on this traditional theory, the country never suffers any direct losses, and so as concerns the question of whether the country can become a 'subject' in a civil claim or not, there are many differences in judicial practice. For instance, in the oil pollution case of the ship 'Hai Cheng', in 1997, the defendant shipowner of 'Hai Cheng' claimed that 'The Fishery Supervision Department of Guangdong province, Zhanjiang detachment' is an administrative authority, no law and regulation stipulates that fishery administration is the operator, user or beneficiary of state-run fishery resources, so he has no title to go to court due to his failure to any losses [sic]'.[7] In the case of the ship 'Min Ran Gong 2', in 1999, the shipowner, with China Shipping Fuel Supplies Co. Ltd of Fujian, stated in defence that 'This case is a general civil litigation, as stipulated in Law of Civil Procedure of PRC, the civil litigant participants are citizens, legal persons and other organizations with equal status, the country is not the participant of civil litigation [sic]'.[8] If based on the traditional theory, both shipowners' defences are reasonable. Just because of this, during litigation for compensation for oil pollution damage brought by the administrative authority, the defendants would always put forward such objections to the plaintiffs. That said, in the two above-mentioned cases, in courts' opinions, the qualification of the plaintiff was legitimate, on behalf of the country, and the administrative authority gained entitlement to claim compensation, and so the objections of the defendants were overruled. Thus it can be seen, there is a conflict between the existing theory and judicial practice in the definition of the country's qualification as a subject. How, then, should this contradiction be resolved?

There is an argument that it is necessary to introduce the 'Theory of Action for the Civil Public Good'. The public good generally refers to the aggregation of benefits enjoyed by non-specific social subjects; and thus an action for the civil public good means use of judicial activity by 'the society', in the way of litigation, to safeguard such public benefits.[9] The modern theory of action for civil public benefit expands the scope of qualified litigant participants such that it includes not only persons who have direct stakes in a civil dispute, but also persons who, while without direct stakes, bring actions for another person's civil rights and interests. The latter mainly refers to a person who has the right to manage and dispose of such civil rights and interests, namely, as long as the citizen, public representative organizations or government offices consider that the purported tortfeasor has infringed the civil rights of the public, they may bring an action. Therefore, government offices and organizations can become subjects in civil litigation. Shipping-related oil pollution mostly harms the interests of non-specific coastal subjects, and is a behaviour which violates the public good, consequently allowing the country to assist through a claim.

Considering legislative developments of countries all over the world, it is a tendency in recent judicial practices that litigant participants have been extended to include the officials, government offices, consumer protection groups, trade federations and associations acting on behalf of the public good. In order to set up a judicial redress system which will be effective in protecting the public good efficiently, above all, the scope of the litigant participants must be so expanded.

Is it possible, then, to find the legal basis for the country to be a subject in claims from either existing domestic laws and/or relevant international conventions? The Constitution of the PRC, Article 9, Paragraph 1 states 'Mineral resources, waters, forests, mountains,

---

[7] D. Qi, 'On Plaintiff in the Disputes arising from Compensation for Oil Pollution Damages (sic)', 1 *Journal of Ocean University of China*, 2004, 10-12.
[8] 'Case of Oil pollution compensation for damages of the mouth of Zhu Jiang Through adjudicating in the first trial (sic)', *China Ocean News*, 4 April 2004.
[9] D. Qi, *supra* note 7, 10-12.

grassland, unreclaimed land, beaches and other natural resources are owned by the state' And, further, the Law of the People's Republic of China on the Administration of Sea Areas, Article 3 reads 'the sea area is owned by the State, the State Council exercises the ownership of the sea area in the name of country. Any unit or individual is forbidden to seize, purchase or in any other ways to illegally transfer the sea area'. These provisions show that the country is the exclusive legal owner of the natural resources, and has complete ownership of the marine resources. An oil pollution accident, which infringes upon the ownership of, rights to use and benefits from, the marine resources of the country, falls into tort law. As to such a tort, the country is thus entitled to participate in litigation.

Although the 1969 CLC has no concrete provisions about the subject of claims, according to the aim of the convention, as long as the person has suffered damages from oil spilt or discharged from the ship, he has the option to claim. 'Person' in the context of the 1969 CLC means any individual or partnership or any public or private body, whether corporate or not, including a State or any of its constituent subdivisions.[10] The CMI Guidelines on Oil Pollution Damage, though not an international convention with legal effect, explains the range of compensation for oil pollution damage and, on the whole, it can be concluded that a subject in a claim of oil pollution damage may include the owners or manager of the natural resources.

It is the United States' Oil Pollution Act 1990 (hereafter referred to as the OPA 1990) that most clearly classifies the country as a claimant subject. The section 'definition' prescribes that 'claimant' refers to any person or government who presents a claim for compensation under the Act; and 'person' means any individual, corporation, partnership, association, state, municipality, commission, or political subdivision of a state, or any interstate body. According to the distribution of rights, the natural resources can be divided into four kinds: (1) to the United States Government for natural resources belonging to, managed by, controlled by, or appertaining to the United States; (2) to any State for natural resources belonging to, managed by, controlled by, or appertaining to such State or political subdivision thereof; (3) to any Indian tribe for natural resources belonging to, managed by, controlled by, or appertaining to such Indian tribe; and (4) to the government of a foreign country for natural resources belonging to, managed by, controlled by, or appertaining to such country. Namely, then, the President, or the authorized representative of any State, Indian tribe, or foreign government, may act on behalf of the public, Indian tribe, or foreign country as trustee of natural resources to present a claim for and to recover damages in regard of the natural resources.

Considering this, in the opinion of the authors, aside from the new theory of action for the civil public good, and majority views on the qualified litigants in civil litigation held by the international community, and from the relevant Chinese law, a country can certainly become a claimant for compensation after ships oil pollution.

---

[10] G. Xv and C. Zhou, 'Subject Position of Administration Authority in lodging claim for oil pollution damages (*sic*)', (2000) *Annual of China Admiralty Judgments* 274.

3.2 EXERCISE BY THE ADMINISTRATIVE AUTHORITY OF A RIGHT TO FILE CLAIMS ON BEHALF OF THE COUNTRY

3.2.1 Administrative Authority Entitlement to bring an Action as Trustee of National Property

Marine environmental pollution caused by oil spillages directly infringes on national property, such as marine ecological resources, marine aquatic resources and so on and damages national interests, so that, whether based on theory or in law, the country is entitled to submit a civil claim. However, considering the technological nature of the law, it is not feasible for the country to participate in the litigation in its own name, and thus the country must choose a representative to act in the proceedings in its name. The country comprises a number of organizations, such as agencies of State power, military bodies, the administrative authority, the judicial authority, and so on. The administrative authority means the State body, set up by the will of the ruling class of the State, according to the constitution and regulations concerning the rules of organization authorized by the State, which enjoys and exercises the national executive power in accordance with the law, and is responsible for the organization, management, supervision and command of national affairs.[11] It is the trustee of national property, and has thereby been authorized and awarded the rights of supervision and management of the natural resources through legislating in the country. According to Article 5 of the Marine Environment Protection Law of the People's Republic Of China (hereafter referred to as Marine Environment Protection Law), the responsibility for the supervision and administration of marine pollution falls to the competent administrative department in charge of environment protection under the State Council, the competent State Oceanic administrative department in charge of marine affairs, the competent State administrative department in charge of maritime affairs, the competent administrative department in charge of fisheries, the environmental protection department of the Armed Forces and the coastal local Governing bodies above the county level. That is, that all the above-mentioned administrative authorities are entitled to present claims on behalf of the country in the event of damage arising from marine environmental pollution. The so-put authority 'on behalf of' means that, in personality and in administrative authority, they coincide with the country. Therefore, litigant actions of the administrative authority are those of the country. The administrative authority exists as the executor of the litigation right of the country, following the will of the country in safeguarding national interests.[12] The participation of the administrative authority, in its own name, in the legal procedure, has the same validity as the participation of the country itself.

---

[11] H. Luo, *Administrative Law,* (Beijing Press, Beijing, 1996), p. 63.
[12] Z. Li and H. Wang, 'Analysis of Subject Position of Administration Authority in lodging claim for coastal oil pollution damages on behalf of country (*sic*)', (2003) 4 *Journal of Dalian Maritime University* 1.

3.2.2 Equality of the Administrative Authority as a Civil Subject in Claims for Damages from Oil Pollution

The administrative authority has a dual identity in law, that is, both administrative subject and civil subject. When carrying out administration, the administrative authority exercises its binding administrative powers, conferred by the country, displaying the subordinate relationship of dependence and management. At this moment, it is an administrative subject acting to, for example, make administrative penalties. When the natural resources in the country are encroached upon, the administrative authority may file claims by virtue of its identity as the trustee of national property, and the right it claims is national ownership. In such an action, the administrative authority and any defendants are of equal status, and enjoy the rights and fulfil the obligations of litigation equally. At that moment, it is a civil subject. Consequently, when oil pollution damages national marine interests, the administrative authority should have an identity equal to that of a civil subject in the action. In the case of the ship the 'Min Ran Gong 2', the defendant submitted 'the plaintiff is the administrative department which has not an equal civil relationship with the respondent, so he is not entitled to file a claim for the loss of national resources towards the civil litigation participant.' His defence was rejected, pursuant to the Law of Civil Procedure of PRC Article 49, Paragraph 1, which states 'Citizens, legal persons and other organisations may act as litigants in civil proceedings'. Since the administrative authority belongs to the body of 'legal persons', and thus complies with the stipulation of the *Law of Civil Procedure of PRC*, therefore the administrative authority can make known the civil claim to the person liable for the oil pollution, on behalf of the country, in accordance with its functions and powers.

Meanwhile, the Marine Environment Protection Law, in force from 4 April 2000, stipulates, in Article 90, Paragraph 2, that 'For damages to marine ecosystems, marine fishery resources and marine protected areas which cause heavy losses to the State, the department invested with power by the provisions of this law to conduct marine environment supervision and administration shall, on behalf of the State, put forward compensation demand to those held responsible for the damages [sic]'. This represents the first time that legislation has expressly prescribed that the administrative authority can represent the country as subject in claims of oil pollution in the environment and for compensation for damages.

4 PRACTICE IN NATIONAL CLAIMS OF OIL POLLUTION DAMAGE FROM SHIPS

Through the above analysis, it can be seen that there is sufficient evidence in theory to grant the competent administrative authority the right to claim for oil pollution damage on behalf of the country, but that to exercise such a claim efficiently in judicial practice the most important condition is to determine the specific competent administrative authority.

As stipulated in Article 5 of the Marine Environment Protection Law, different administrative authorities are granted different administrative powers according to the different functions they oversee. The State Bureau of Environmental Protection exercises unified supervision and administration over nationwide marine environmental protection work, provides guidance, coordination and supervision and is responsible for nationwide work to prevent and control marine pollution damage caused by land-based pollutants and coastal construction projects. The Oceanic Administrative Department is responsible for nationwide environmental protection work to prevent and control marine pollution damage caused by marine construction projects and the dumping of wastes in the sea. The Maritime Safety Administration (hereafter referred to as MSA) is responsible for the supervision and administration of marine environmental pollution caused by non-military vessels inside port waters and non-fishery and non-military vessels outside port waters under their jurisdiction. The Fishery Bureau is responsible for the supervision and administration of marine environmental pollution caused by non-military vessels inside fishing port waters and fishing vessels outside fishing port waters. The Environmental Protection Department of the Armed Forces is responsible for the supervision and administration of marine pollution caused by all military vessels. As can be seen, China adopts the management system called the 'Five competent administrative authorities supervising and administering the sea area'. To an extent, this results in overlap between some departments' functions and management that, while not duplicated, can go unmonitored.[13] In exercising the right to a national claim, this situation embodies the phenomenon that at times the departments fall over each other to seek compensation for oil pollution damage, while at others mutual prevarication in claiming results in no department showing any interest in litigation, which causes disorder in the civil proceedings.

At the same time, Article 5 of the Marine Environment Protection Law only provides general principles without further clarification or specification, leaving the door open to uncertainty, with different administrative authorities seeking compensation for the same oil pollution damage, as can now be seen in some judgements. In the case of the ship the 'Feoso Ambassador' in 1983, the qualified administrative authorities competent to represent the country and bring claims were the Environmental Protection Department and the Aquatic Products Department.[14] In the case of the ship the 'Hai Cheng' in 1997, the

---

[13] Ibid.
[14] S. Zhang, 'Subject Qualification of Administration Authority in lodging claim for marine pollution damages, Tianjing Maritime Court', (the 11th National Conference on Admiralty Judgments, 2002), vol. 9, p. 52.

competent authority was the Fishery Supervision Department.[15] In the case of the ship the 'Min Ran Gong 2' in 1999, the competent departments were the Environmental Protection Department and the Aquatic Products Department.[16] Finally, in the case of the ship the 'Tasman Sea' in 2003, the competent offices were the Oceanic Administrative Department and the Fishery Supervision Department.[17] Therefore, only when solving the problem of which competent administrative authority enjoys the right to claim compensation for oil pollution damage on behalf of the country, will the disorder in litigation to be eliminated, the impairment of the national interest and public good avoided, and the severe damage to the State's image caused by abuse of power by some administrative authorities be curtailed.

In confirming which competent administrative authority may represent the whole nation in bringing claims, different opinions exist. Firstly, there is the view that nowadays China implements a multi-centric management system in marine environmental protection, pursuant to the Marine Environment Protection Law of the PRC and the 'three-definitions scheme' of the State council. The MSA is entitled to file a claim for the fees of pursuing trouble-making ships and the costs involved in cleaning up the oil, the Marine Office claims for damage to the marine ecological resources, the Aquatic Products Bureau claims related to aquatic resources, the Management Institution of Nature Reserves claims for nature reserve damages, and the Tourism Administration and Forestry Bureau claims concerning tourist or mangrove resources. Further, competent provincial administrative authorities can file claims for pollution damage that extend over regions, and neighbouring and corresponding competent provincial administrative authorities can claim for pollution damages that extend over provinces.[18]

The second viewpoint contends that shipping pollution accidents cause losses and damage as such to the industrial and agricultural, the tourist and the fishery departments, each of whom has its respective compensation rights. Hence, the preferred approach is for the country to establish a unified law enforcement agency to represent the whole country, and to be exclusively responsible for lodging claims for such damages as the country has suffered.[19]

Third, it is argued that the approach should be 'in accordance' with Article 5, Paragraph 3 of the Marine Environment Protection Law, which states 'The competent State administrative department in charge of maritime affairs shall be responsible for the supervision and administration of marine environment pollution caused by non-military vessels inside the port waters and non-fishery vessels and non-military vessels outside the port waters under their jurisdiction, and be responsible for the investigation and handling of the pollution accidents. In the event of a pollution caused by a foreign vessel navigating, berthing and anchoring and operating in the sea area under the jurisdiction of the People's Republic of China, officers in charge shall board the vessel in question to

---

[15] Y. Si et al., *Study Course of Case in Maritime Law* (Knowledge Press, 2003), p. 275.
[16] S. Zhang, *supra* note 14.
[17] *See* relevant reports at <http://www.ccmt.org.cn>, <http://www.enorth.com.cn>.
[18] F. Xv, 'On Civil Subject of Claim for Marine Environmental Damages (*sic*)', <http://www.ccmt.org.cn>, 2002, 11.
[19] C. Zhang, 'Discussion on National Indemnity mechanism of pollution accidents caused by ships in Our Country (*sic*)', <http://www.ccmt.org.cn>, 2002, 12.

examine and handle the case. Should a pollution accident caused by vessels result in fishery damages, the competent administrative department in charge of fisheries shall be invited to take part in the investigation and handling of the accident'. Thus, it is argued that clearly the MSA and fishery bureau are competent to file claims on behalf of the country for marine environmental pollution damage from the oil spills.[20]

These three different views form a valuable basis for discussion. The first expounds the notion of a multi-centric management system, which strictly defines the authorities of each department to file claims, and favours claims made separately. This suggestion seems, on the surface, to lay down an exhaustive and orderly system, but in fact, a number of drawbacks exists. Firstly, oil pollution damages from ships often affect many different aspects, such as marine ecological resources, the aquatic resources and accident handling costs, and so on. Because there is no precise standard of classification, finding concrete loss in every respect is difficult and there is some overlap; thus, the method of filing separate claims lacks flexibility in practice. Second, if the bodies claim respectively, every department is bound to attach more importance to their own departmental interests, which will certainly not benefit the full safeguarding of the interests of the country and society. Third, an oil pollution accident will thus generally cause a succession of departmental claims, which will in turn necessarily raise the costs of litigation, increase the burden of the suit, and reduce the efficiency of the court's justice. Meanwhile, it leads to the responsible parties, like foreign owners of polluting ships or the P&I club and so forth, in confusion as they will not know how to adapt to China's national law and regulations. Consequently, the judicial image of China around the world may be damaged.

Analyzing the second viewpoint, the starting point for establishing a unified law enforcement agency is attractive, for the reasons that by a single administrative authority representing the country in claims, oil spill accident matters may be settled in time, the country receives sufficient compensation as well as the social interests being wholly safeguarded. At the same time, it shows a healthy trend of allowing only one administrative agency to protect China's interests.[21] The result of 'economy' should ideally be cost-effective legislative and administrative reforms. For instance, it must amend the existing laws and regulations and redistribute the power of every department; in addition, it should grant such a unified administrative agency a legal status and authority, such that it obtains approval in law. In addition, it is necessary to change the setting of the current relevant administrative authorities. These suggestions involve overhauling the system, and it would not be possible to fulfil in a short time. It would, therefore, be difficult to implement.

In interpreting the provision of the Marine Environment Protection Law, the third viewpoint explains the concrete supervision and management roles of each department, and considers the qualified administrative authorities for representing the country in claims to be the MSA and fishery bureau. Compared to the above-mentioned views, it may be richer in feasibility and rationality under the current law and administrative

---

[20] Z. Li and H. Wang, *supra* note 12, 1.
[21] For instance, the provision of the United States 'OPA' 1990 is that 'The President, or the authorised representative of any State, Indian tribe, or foreign government, shall act on behalf of the public, Indian tribe, or foreign country as trustee of natural resources to present a claim for and to recover damages to the natural resources'.

structure. However, in the authors' opinion there are still flawed points in this position. From the provisions of the Marine Environment Protection Law, it is known that the nature of the marine resources (such as the ecological resources, marine aquatic resources and ocean nature reserves) divide, without good grounds, the administration authorities. There are, however, crucial threads concerning pollution sources (such as from land, coastal construction projects or ships), and consideration of the designation of the object (such as a merchant vessel, fishing boat or military ship) in dividing the administration authorities.[22] Thus, pursuant to Article 5, Paragraph 3 of the Marine Environment Protection Law, management of oil pollution damage from ships should belong to the MSA. Conversely, there is a lack of evidence to infer the fishery bureau holds a right of management from the provision 'should a pollution accident caused by vessels result in fishery damages, the competent administrative department in charge of fisheries shall be invited to take part in the investigation and handling of the accident'. Considering the provision of Article 5, Paragraph 4 that 'the competent State administrative department in charge of fisheries shall be responsible for the supervision and administration of marine environment pollution used by non-military vessels inside the fishing port waters and the fishing vessels outside the fishing port waters, and be responsible for the protection of ecological environment in the fishing zones and examine and handle fishery pollution cases beyond the pollution accidents mentioned in the previous clause [sic]', the fishery bureau has no right to settle claims for oil pollution accidents arising from merchant vessels. In sum, the authors would argue that the qualification of a competent administrative authority is exclusive, and is under the MSA's sole jurisdiction.

5 THE SCOPE OF A NATIONAL CLAIM SYSTEM FOR OIL POLLUTION DAMAGE FROM SHIPS

The principal existing international legislation concerning compensation for oil pollution damages from ships is the 1969 CLC, the 1971 Fund and their 1992 Protocols. The agreed compensation items, according to those international conventions, the OPA 1990, the UK's Merchant Shipping (Oil Pollution) Regulations 1995, and the 2000 Claims Manual of the International Oil Pollution Compensation Funds, include: loss or damage caused outside the ship by contamination resulting from the escape or discharge of oil from the ship; the costs of preventive measures and any further loss or damage caused by such preventive measures. Disputes remain, in practice, in the areas of compensation for clean-up charges, pure economic losses and punitive damages, which are ambiguous in the legislation. We will try, then, to further specify the scope of national claims for shipping-related oil pollution damage by defining the nature of these three claim items.

5.1 COMPENSATION FOR THE CLEAN-UP CHARGES

In China, there are two opposing attitudes, which differ over the issue of whether the clean-up charges of the administrative authority may be satisfied in terms of civil

---

[22] C. Zhang, *supra* note 19.

procedure. Some point out that the cleaning actions, which should be considered specifically executive in nature, derive from executive authority, and the charges and other relevant expenditures should be collected from those who are liable in an executive procedure.[23] Others hold that, clean-up charges being an indispensable part of claiming for oil pollution damages, the cleaning operations, which are carried out by the administrative authority, are actually for the purpose of avoiding expanding the tortious liability of the accused. Accordingly, they draw the conclusion that it is right to impose clean-up charges on the polluter, and that the administrative authority should be entitled to institute civil proceedings for compensation.[24]

The authors prefer the latter approach, in that, firstly (and most saliently) the nature of the clean-up charges involve a great deal of energy, material resources, various equipment, specialized technology and vessels. Individuals cannot manage this, which usually results in the administrative authority taking over such charges. The cleaning actions, performed by the authority, have the status of administrative functions, which, from another point of view, are measures to prevent or reduce extension of the losses. As far as the principles of tort law are concerned, when infringement occurs, the charges and expenditures caused the innocent victims in preventing extension of the losses should be the infringer's responsibility. Therefore, the administrative competent authority has the right to make a claim for the clean-up charges, which should be paid by the infringer, in the form of civil litigation, if the cleaning operations have successfully prevented or reduced extension of losses, and the charges incurred thereby are reasonable. Second, taking into consideration international legislation on oil pollution, the OPA 1990 divides the losses which are claimable into two parts, the first of which is clean-up charges; the 2000 Claims Manuals of International Oil Pollution Compensation Funds provide criteria that have been accepted by the member States, to claim compensation for oil pollution. Part Three of the Claims Manuals of the IOPC Fund 2000 on 'acceptable claim[s]' lists 'clean-up operation & property losses', provides that the clean-up measures should be sensible, and stands as an important reference for all countries dealing with compensation for oil pollution damages. Accordingly, the clean-up charges and expenditure, when disbursed reasonably by the responsible administrative authority, should be classified as claimable items.

5.2 COMPENSATION FOR PURE ECONOMIC LOSS

Pure Economic Loss means the impairment of earning capacity resulting from the damage of property or natural resources. This concept originates with the classification of loss in common law countries and there is no such concept in civil law countries. A focus of discussion in recent years has been whether the person liable should have to pay compensation for pure economic losses.

Focusing on the relevant legal precedents of the UK and the US, it is not difficult to see that these countries have very different attitudes as regards this question. British courts fear causing too many endless claims, and insist on the principle of not compensating for

---

[23] *Ibid.*
[24] Z. Li and H. Wang, *supra* note 12, 1.

pure economic loss under the principles of lack of physical damage or remoteness of damages.[25] America traditionally formed the 'Robins Dry Dock Rule',[26] which contended that in maritime tort, only when the property suffers physical damages could compensation for the damage be awarded and economic losses such as profit and so forth be covered, equally denying compensation for pure economic loss. Subsequently, however, with the courts in the US often breaking this traditional rule, the prevalent attitude is to accept claims for pure economic loss. For example, the OPA 1990 shows that, notwithstanding any other provision or rules of law, the tortfeasor is liable for damages equal to the loss of profits or impairment of earning capacity incurred by any claimant due to the injury, destruction, or loss of real property, personal property, or natural resources. That is, compensation for pure economic loss is acknowledged at law. Meanwhile, in the case *Ballard Shipping Co.* v. *Beach Shellfish*, and a case related to Ballard Shipping Co., Judge Duggan felt that the OPA 1990 undoubtedly stipulates the principle of compensation for pure economic loss in oil pollution cases, and most legal reviewers also go beyond the Robins Dry Dock Rule in the course of interpreting OPA 1990.[27] However, besides this the CMI Guidelines on Oil Pollution Damage state that 'Pure economic loss will be treated as caused by contamination only when a reasonable degree of proximity exists between the contamination and the loss. In ascertaining whether such proximity exists, account is to be taken of all the circumstances, including (but not limited to) the following general criteria: i. the geographic proximity between the claimant's activities and the contamination; ii. the degree to which the claimant is economically dependent on an affected natural resource; iii. the extent to which the claimant's business forms an integral part of economic activities in the areas which are directly affected by the contamination; iv. the scope available for the claimant to mitigate his loss; v. the foreseeability of the loss; and vi. the effect of any concurrent causes contributing to the claimant's loss.' The IOPC Fund holds a similar opinion.[28] Though the Guidelines and the Fund regard the existence of a reasonable degree of proximity between the pollution and the damage as a precondition for the compensation for pure economic loss, they admit, at least, that compensation for pure economic loss may be awarded.

---

[25] Y. Si et al., *supra* note 4, 423.

[26] In that case, the TC/P charterer and the shipowner agreed that the ship chartered should enter the dry dock for maintenance once every half year, during which time the hire cost ceased to be due. However, when the ship was put in the dry dock for repair, the ship's screw propeller was damaged due to the negligence of maintainance company. The charterer brought an action for the loss of ship operation during the period of maintenance. Judge Holmes found that the charterer had no interest in the maintenance contract concluded by the shipowner and the dry dock company, and a third party, who wanted to sue for breach of contract must provide evidence as to the direct interest between himself and the contract. Therefore, the Court denied the charterer compensation.

[27] 32 F. 3d 630, 1994 AMC 2717 (1st Cir.1994), Y. Si, *supra* note 4, 423.

[28] Fund considers that a proper degree of connection must exist between the contamination and the loss. In ascertaining whether such connection exists, all the circumstances must be considered, including: i. the geographic proximity between the claimant's activities and the contamination; ii. the degree to which the claimant is economically dependent on an affected natural resource; iii. the degree to which the claimant has the supply sources or business opportunity rather than the destroyed business and iv. the extent to which the claimant's business forms an integral part of economic activities in the areas which are affected by the contamination.

As in the theories of many civil law countries, it is very difficult to find the concept of pure economic loss in Chinese theories on the compensation of civil torts. This conforms to China's endorsement of the principle of full and complete indemnity for losses resulting from tortious acts. This is in order to protect the victim's rights and interests wholly and penalize the illegal act of the tortfeasors. Moreover, in 1991, in the 'Answer to the Problem the Operating Deficit Arising from the Marine Casualty by Domestic Ships Should Be Placed into the Scope of Compensation of Accidents of Damages at Sea [sic]' the Supreme People's Court pointed out that 'According to Article 117 Paragraph 2,3 of the General Principles of the Civil law of the PRC anyone who damages the property of another person shall compensate the victim the actual loss resulting from action of tort, including the acquirable future interests of victim in normal cases [sic].' Although the loss of the victim's foreseeable interests is no more extensive than pure economic loss in meaning, it reflects the value that pure economic loss should be at least partially indemnified. Nevertheless, with the unification of maritime legislation, it may be argued that ultimately China will join the international oil pollution fund in the field of compensation for oil pollution damage. Since the Fund has already admitted compensation for pure economic loss to a certain extent, China should also take an active approach to this kind of compensation. When marine resources are destroyed, the definite competent administrative authority is able to claim pure economic loss on behalf of the whole country, as the owner of the natural resources.

5.3 PUNITIVE DAMAGE

Punitive damage is one of the most disputed features in contemporary tort law. The US 'Model Act on Punitive Damages' gives it the definition of 'nothing more than be used for punitive or deterring money in favour of applicants [sic]'.[29] Forasmuch as, as to punitive damages, the courts sometimes adjudge the compensation quantum beyond the factual damages to victims to penalize the tortfeasors.

Compensation for damages is usually restricted in the civil litigation brought in continental law jurisdictions, and compensation for damages takes a 'resume [the] original state' principle. Therefore, a punitive damages principle is difficult to implement among continental law jurisdictions, with the exception of criminal litigation procedures.[30] Chinese scholars and courts essentially agree with the approach of these countries. Namely, they consider that civil liability differs from administrative and criminal liabilities and thus that civil compensation should scrupulously abide by the principle of compensating the actual damages that occur, to ensure maximally restoring of the property and the physical and moral states to their original condition using monetary compensation or use of other property of tortfeasors as compensation.[31]

However, some authors argue that in order to prevent and limit intentional tortious behaviour or that of gross fault, as well as to effectively implement liability in tort and

---

[29] L. Wang, *Civil Law-Tort Act* (Publishing House of People's Court, Beijing, 2003) p. 274.
[30] U. Si, *supra* note 4, 426.
[31] W. Peng, *Civil Law* (Publishing House of China University of Political Science and Law, Beijing, 1998) p. 634.

thus the duty to educate and manage seamen of shipowners, punitive damages should be implemented and strengthened by way of a common principle of confirmation of extension of damages for compensation concerning maritime torts.[32] The authors do not share this opinion. Firstly, although maritime law displays a trend towards aggravating the shipowners' liability and a development towards equity and rationalization,[33] shipping holds the balance for economic development in the world and in individual countries. Currently, irrespective of whether international or Chinese shipping is concerned, there is still the view to preserve shipowners' interest as an important standard. If China affirms the punitive damages principle, then the shipowner's liability will be aggravated without measure and work against building a great shipping country. Furthermore, in America, where in the early nineteenth century a system of punitive damages compensation was established, such punitive damages compensation has not been included among the compensation items enumerated in the OPA 1990. This reflects the negative attitude to the above principles which the US government now holds in the area of oil pollution accidents. Also, as to claims for compensation in the international oil pollution fund, in practice, there has not been any fully recognized punitive damages compensation system. Finally, due to the limitation of the liability of the shipowner, in most instances where there are numerous parties claiming compensation and large sums of compensation are due, punitive damages compensation becomes meaningless.[34] In conclusion, it is still too early to implement a system of punitive damages, and the principle should be excluded from extensions of national claims.

Several important aspects covering national claims systems for shipping-related oil pollution damage were discussed above. Considering that China has entered into the 1969 CLC but is not a member of the 1971 Fund, which made it impossible to operate according to the international oil pollution compensation mechanism, or in the manner like the dual mechanisms of national and international oil pollution in nations like Canada, it is recommended establishing a respective national claim mechanism as soon as possible.

The US is one of the most dominant countries that rely on a national mechanism for oil pollution compensation,[35] and China can use that beneficial experience as a reference. Relevant law and regulations should be established and improved for oil pollution compensation, and a national claims institution should be created whose representative is the MSA. When damage from oil pollution of vessels come at great cost to the nation, the MSA shall actively unite all other concerned departments that bear the responsibility for investigation and collection of evidence. For instance, the Ministry of the Ocean provides oceanic environment inspection files and brings civil claims to prevent loss to national interests. At the same time, a national compensation fund should be set up, using as an example the Oil Spill Liability Trust Fund managed by the US national fund centre against pollution, which can pay clean-up expenses occurring after a shipping oil spill and

---

[32] R. Deng, *Research on Basic Theoretical Matters Concerning Tort Act of Ships* (Publishing House of Law, Beijing, 1999).
[33] Y. Si et al., *supra* note 15, 29.
[34] Y. Si et al., *supra* note 4, 430.
[35] Y. Si et al., *Study on Tendency and countermeasures of International maritime legislation* (Publishing House of Law, Beijing, 2002) p. 241.

during the evaluation and investigation of accidents and compensation for damage.[36] In this way, the interests of national marine resources can be safeguarded and the marine environment maintained efficiently.

6 CONCLUSION

To conclude, with the prerequisite of explaining the need to establish a national claim system for oil pollution damage from ships, this paper has discussed some vital issues about such a national claim system. The following conclusions may be drawn:

i With the theory of litigation for the civil public good, and the provisions of International Conventions and domestic law, as well as judicial practice, the country is qualified to become a subject in claims for oil pollution damage from ships.
ii In the practical operation of claims for oil pollution, the MSA is entitled to lodge a claim on behalf of the country as the exclusive competent administrative authority.
iii As regards concrete claim items, clean-up charges of the competent administrative authority and pure economic losses can be included in the extension of claim, but punitive damages cannot be and are an exception.
iv Under current circumstances, China ought to establish actively a practical and efficient mechanism for national claims to safeguard the interests of national marine resources.

---

[36] Y. Si et al., *supra* note 35.

LIST OF REFERENCES

R. Deng, *Research on Basic Theoretical Matters Concerning Tort Act of Ships* (Publishing House of Law, Beijing, 1999).

Z. Li and H. Wang, 'Analysis of Subject Position of Administration Authority in lodging claim for coastal oil pollution damages on behalf of country', (2003) 4 *Journal of Dalian Maritime University* 1-4.

H. Luo, *Administrative Law* (Beijing Press, Beijing, 1996).

N. Ma, 'Compensation of the Pure Economic Loss in Oil Pollution Damage', <www.ccmt.org.cn>, 2002, 11.

W. Peng, *Civil Law* (Publishing House of China University of Political Science and Law, Beijing, 1998).

Y. Si et al., *Detailed Study on Maritime Law* (Dalian Maritime University Press, Dalian, 1999).

Y. Si et al., *Study on Tendency and countermeasures of International maritime legislation* (Publishing House of Law, Beijing, 2002).

Y. Si et al., *Special Subject Study on maritime Law* (Dalian Maritime University Press, Dalian, 2002).

L. Wang, *Civil Law-Tort Act* (Publishing House of People's Court, Beijing, 2003).

M. Xie and M. Lan, 'Study on Pure Economic Loss in Ships Oil Pollution Damage', (2002) *Annual of China Admiralty Judgments*, 229.

F. Xv, 'On Civil Subject of Claim for marine Environmental Damages', <www.ccmt.org.cn>, 2002, 11.

G. Xv and C. Zhou, 'Subject Position of Administration Authority in lodging claim for oil pollution damages', (2000) *Annual of China Admiralty Judgments*, 274-279.

D. Qi, 'On Plaintiff in the Disputes arising fromCompensation for Oil Pollution Damages', (2004) 1 *Journal of Ocean University of China* 10-12.

C. Zhang, 'Discussion on National Indemnity mechanism of pollution accidents caused by ships in Our Country', <www.ccmt.org.cn>, 2002, 12.

S. Zhang, 'Subject Qualification of Administration Authority in lodging claim for marine pollution damages, Tianjing Maritime Court', The 11th National Conference on Admiralty Judgments, 2002, 9, 315.

A STUDY ON THE TYPES OF LIABILITY OF THE INSURER FOR OIL POLLUTION AND THAT OF THE PARTY LIABLE

Chen Pingping

1 INTRODUCTION

There is no detailed and comprehensive provision in respect of issues concerning compulsory liability insurance for oil pollution in the General Principles of the Civil Law of China (hereinafter as 'GPCL'), in the Insurance Law of China[1] (hereinafter as 'IL'), in the Maritime Code of China (hereinafter as 'MC') or in other laws of an administrative nature such as the Marine Environment Protection Law (hereinafter as 'MEPL'), or the Fisheries Law (hereinafter as 'FL') in China. Therefore, in legal practice, it is quite difficult for a court to apply a sound legal basis at trials when cases arise out of oil pollution damage. It is hoped that the legislative authority concerned will pay attention to perfecting the IL and the MC in amendments thereto and to the Oil Pollution Law (hereinafter as 'OPL') which will be enacted in the future.

2 PRESENTATION OF THE PROBLEM

Motor tanker 'T' (MT T), owned by company 'O' for which an insurance company 'C' undertook to provide liability insurance for oil pollution, sank in the sea and a great deal of diesel oil leaked out from MT T and caused oil pollution in the China sea area. As a result, some farm animals in the area died. Investigation revealed the cause of the accident to be that the MT T was severely undermanned when the accident occurred. In another words, the misconduct of the shipowner constituted the gross fault which had caused the accident. The farmers who suffered loss from the incident later filed a joint lawsuit in tort against O and C for compensation for damages. Although the facts of the case were clear, controversy arose regarding the kind of liability C and O should respectively bear and how to differentiate the liability between C and O toward the victims. Generally, there were three opposing opinions regarding this issue.

The first opinion held that the relationship of compensation for damage arising from the oil pollution (a tortious act) between O and the victims who had suffered damage differed wholly from the contractual relationship of liability insurance between O and C. Therefore, it was inappropriate to try two kinds of legal matters within one case. The

---

[1] The Insurance Law of China mentioned in this article refers to the new Insurance Law as amended in October 2002.

precondition that an insurance company should indemnify the victim is that it should undertake the liability for compensation for the insured. Because O was at fault for the shortage of seaman on the MT T, O had violated the warranty clause of the policy, so C's liability to O should be exempted. As a result, C is not liable for the victim's damage claims.

The second opinion held that Article 97 of Special Maritime Procedure Law of China (hereinafter as 'SMPL') provides that 'a claim may be made by a victim in respect of oil pollution directly against the insurer who underwrites the oil pollution liability'. It, however, is a rule of procedural law. So far, there is no provision regarding what kind of liability should be undertaken by the insurer who underwrites the liability for oil pollution among the current substantive laws in China.[2] Therefore, the lawsuit filed by the farmers directly against the insurance company should be dismissed.

The third opinion held that Article 97 of the SMPL provides that a claim may be made by a victim directly against the insurer who underwrites the oil pollution liability. However, there is no corresponding provision in the substantive laws on the same issue, so a legislative authority should interpret the law. It was further proposed to follow the trend of international environmental protection and the development of liability insurance for oil pollution and thus to apply a joint and several liability between the insurer and the insured regarding oil pollution, in order to ensure adequate protection of the victims' legal rights and interests.

3 RELATIVE CONCEPTS OF LIABILITY INSURANCE

3.1 ORIGIN AND DEVELOPMENT OF LIABILITY INSURANCE

It is widely held that liability insurance is compensation provided by the insurer to the third party to whom the insured is legally liable. Article 50(2) of IL provides a similar rule. According to a liability insurance contract, the applicant for insurance (the insured) must pay the agreed premium to the insurer. When the insured is involved in an accident that results in a third party suffering injury or damage, the insurer shall be obliged to undertake the insured's liability by paying insurance indemnity pursuant to the stipulations of the insurance contract. Because the subject matter of liability insurance is the liability of the insured towards a third party with the purpose of paying for the loss experienced by the insured that arises from compensating the third party, liability insurance is also known as third-party liability insurance.[3] Liability insurance is treated by the industry as the third or final stage in the development of insurance industry. It is also an express demonstration of the insurance industry's direct participation in the

---

[2] At present, there are four opinions in the field of theory and the field of justice in respect of the types of liability of the insurer of oil pollution, i.e., direct compensation liability, substantive compensation liability, compensation liability, joint and several liability. This is the author's conclusion following various discussions with the experts and scholars of Dalian Maritime University, Shanghai Maritime University and many judges of the Xiamen Maritime Court, the Guangzhou Maritime Court, the Fujian Higher People's Court and the Supreme People's Court in China.

[3] H. Zou, *Liability Insurance* (Law Press, Beijing, 1999) p. 30.

development of social activities.[4]

The principal difference between liability insurance and life insurance or general property insurance is the final object of protection. Liability insurance protects the insured's liability toward the third party who has suffered a loss caused by the insured instead of the loss of the insured itself. Because the insured of liability insurance is legally obligated to pay compensation to the aggrieved third party, the liability insurance indemnity is also specific, which means the liability insurance indemnity only exists when the third party claims damages.

In the second half of the 19th century, industrial damage commonly existed in industrialized countries and had caused unexpected harm to the society. Whilst people were harvesting the fruits of the industrialization, human beings had also suffered crisis of a much greater scale and with a much more severe impact than ever and resulting in a series of social problems. The problem was complicated in respect of environmental pollution. If the insurance industry rigidly withheld the traditional doctrine of fault-based liability insurance, the channel to provide the victims with legal remedies would inevitably be blocked. Article 25 of Prussian Railway Law promulgated in Germany in 1838 is considered the first legislative mechanism of the doctrine of no-fault liability.[5] Later, many countries followed the trend and eventually made the doctrine of non-fault liability one of the standards of allocation of tortious liability.

The appearance of liability insurance and social insurance is related to the comprehensive exercise of the doctrine of non-fault liability in the field of tort laws. The insurance system had provided the basis of compensation for the fulfilment of the doctrine of no-fault liability, which resulted in the trend to socialize undertaking damage. The essence and function of the insurance system lies in transferring and dispersing the risks and losses arising therefrom, which corresponds with the basic thought of a reasonable allocation of unfortunate harms as advocated by the doctrine of no-fault liability.[6] Therefore, liability insurance has come into existence by adapting to the need to distribute risks (compensation risks) after the industrial revolution. Since the 19th century when liability insurance was first set up in France, liability insurance has developed into an insurance business with a relatively independent theoretical system and a high value of application. This kind of insurance has been used to eliminate the risks of economic loss undertaken by the insured and protect the insured from stress, inconvenience and aggravation that it had to bear to formulate defences against the claims raised by victims.[7]

Essentially, the traditional liability insurance aimed at making up the loss suffered by the insured arising from its liability for compensation to the victim. In the modern world, liability insurance has paid increasing attention to the protection of victims by developing from pure compensation for losses to indemnifying the insured for its liability to the victim as its object of indemnification. Modern liability insurance has extended its coverage by giving the victim the status of third party beneficiary so as to establish its position to protect the victim. The loss indemnified by the liability insurance is the

---

[4] H. Zhang and C. Zheng, *Insurance* (Beijing University Press, Beijing, 2000) p. 312.
[5] M. Cao, *Environmental Tort Law* (Law Press, Beijing, 2000) p. 143.
[6] *Ibid.*, pp. 155-156.
[7] J.F. Christ, *Fundamental Business Law* (American Technology Society, Chicago, 1944) p. 276, quoted from *Ibid.* by H. Zou, *supra* note 3, 33.

insured's liability to the third party for compensation borne rather than any loss arising from undertaking compensation liability. It has become the trend in liability insurance to place the rights and interests of the victim on a position of priority.[8] From the of liability insurance in the US where the liability insurance system was very well developed, in the first stage of the development of liability insurance, liability insurance was only deemed to indemnify the insured for losses. Even though the third party may have won the lawsuit against the insured, it could not demand compensation directly from the insurer. In the second stage, in the judgement or after an agreement was reached between the third party, the insured and the insurer determining the insured's liability, the third party might garnish the debt or subrogate the creditor to request payment from the underwriter. In the third stage, at the same time as the third party's right of direct action against the insurer was recognized, the validity of defences made by the insurer against the insured in respect of the third party should be excluded. If the said defence did exist and was tenable, after the insurer made insurance indemnity to the third party, it might pursue the insured for recovery in accordance with the insurance policy reached between the insurer and the insured.[9]

3.2 SUBSTITUTE LIABILITY FOR COMPENSATION OF THE LIABILITY INSURER

As a method to distribute and transfer the compensation liability undertaken by the insured, certain restrictions have been set up in the liability insurance as conditions for the victim to obtain actual indemnity. Aside from the compensation liability that, pursuant to the law, cannot be transferred via liability insurance (for instance, the insured caused damage to the victim intentionally), or the compensation liability not covered according to the policy, the insurer should undertake the liability for compensation when the insured is liable to compensate the victim or the victim demands indemnity from the insured. From this point of view, the liability insurer undertakes the liability for compensation borne by the insured and is in a position to pay compensation to the victim on behalf of the insured.[10] This is called substitute liability for compensation instead of direct compensation liability for the victim.

The victim suffers loss because of the insured's act and the insured himself should take up civil liability for the victim. The insurer is not the undertaker of civil liability for compensation claimed by the victim. Instead, the insurer is only obligated to indemnify the insured for the loss that had already occurred or is likely to occur because of the undertaking of compensation liability according to the terms of the insurance policy. In essence, the compensation liability borne by the insurer is the liability for payment as stipulated in the insurance policy. The development of liability insurance filled a function of protecting the victim's interests as the public interests. The victim may enjoy and acquire the insured's interest pursuant to the terms of the insurance policy or laws that require the insurer to undertake liability for indemnity. The insurer may also pay the indemnity directly to the victim. Article 50(1) of the IL provides that 'with respect to

---

[8] H. Zou, *supra* note 3, 47.
[9] Z. Liu, *Insurance Law* (Taiwan Sanming Press, Taiwan, 1995), pp. 173-176.
[10] H. Zou, *supra* note 3, 61.

damages caused by the insured to a third party in case of liability insurance, the insurer may, in accordance with the law or stipulations of the insurance policy, make compensation directly to the third party.'

3.3 VICTIM'S RIGHT OF DIRECT ACTION

In conformity with the trend in liability insurance to 'protect the victim's interests', third party victim's right of direct action have also been established. The direct action is recognized for the following two reasons: (1) to avoid the tortfeasor's non-performance of compensation to the victim; (2) to avoid the tortfeasor's delayed compensation to the victim. The victim's right to claim indemnity against the insurance company directly is to achieve the actual and quick protection and remedy for the victim.[11]

Although there are many different opinions regarding the theoretical basis for establishing the victim's right of direct action, these are two of the mainstream theories.

3.3.1 Doctrine of Original Acquisition (Doctrine of Statutory Rights)

According to this doctrine, the victim should acquire in accordance with laws and at the time of the loss the independent rights equal to those of the insured. This doctrine is cited from the interpretation of direct action made by French scholars. Article 53 of the Insurance Law of France, enacted in 1930 provided that 'with respect to pecuniary damage suffered by the victim on account of the insured, as long as this damage amount has been within the insured amount and has not been compensated, the insurer shall not pay the total or partial compensation payable to any party other than the victim.'[12] According to this theory, some scholars are of the view that the direct action is a direct right recognized by the direct right of action. Establishing the creditor's right should be on a legal basis. The victim acquires its right as creditor through substantive law.[13]

The doctrine of original acquisition of statutory rights provides an interpretation to the legality of the right to claim directly. Unfortunately, there is no further explanation for this reasoning.

3.3.2 The Doctrine of Payment without Evasion of Liability

'Without evasion of liability' means the avoidance of exemption of liability. According to this doctrine, the nature of insurance liability payment rests with the avoidance of evasion of liability by the tortfeasor so as to guarantee the tortfeasor's payment. The victim's right to direct action is established exactly with the purpose of preventing a tortfeasor from exempting its liability, the existence of which means that the insurance company is in a statutory position of joint and several surety regarding the debt of compensation for

---

[11] W. Li, *Study on Japanese Automobile Accident Damages Legal System* (Law Press, Beijing, 1997), pp. 238-239.
[12] U. Toshio, *Direct Right of Claim of Victim, Present Situation of Compensation of Traffic Accident* (Administration Press, 1979); quoted from W. Li, *supra* note 11, 244.
[13] K. Kouichiro, *Direct Right of Claim of Victim of Liability Insurer* (Contemporary Damages Lecture (8), 1973) edition of Japan Comment Commission, quoted from W. Li, *supra* note 11, 245.

damage on account of the tortfeasor.[14] In the light of provisions of joint and several guarantees in the GPCL, the principal debtor undertakes the debt jointly and severally with the guarantor. As a creditor, the victim may claim compensation against either the tortfeasor as a debtor or the insurance company as a joint and several guarantor.

The doctrine of payment without exclusion has not elucidated the nature of the debt enjoyed by the victim, but it makes it clear that the essence of the direct action is the claim for compensation for damage against the insurance company within the limit of the insured amount instead of a claim of insurance indemnity. Under the principle of strict liability, both the right of direct action and the victim's claim for compensation for damage against the tortfeasor are absolute rights. The insurer should thus not refuse payment arising from the victim's direct action for reason of the insured's illegal intention nor can the tortfeasor refuse the payment to the victim for reason of the victim's lack of fault or intention.

Each of the two doctrines has its advantages and disadvantages, but the doctrine of original acquisition lays particular emphasis on the procedural basis of the victim's right to claim directly, while the doctrine of payment without evasion of liability emphasizes more the substantial content of the right. Therefore, in comparison, the doctrine of payment without evasion of liability appears much stronger than the doctrine of original acquisition in respect of the protection of the victim's rights and interests.

3.4 RESTRICTION ON THE INSURER'S RIGHT TO CALL ON DEFENSES

The restriction on the insurer's right to call on defences is also an important element in the developing trend of liability insurance concerning the protection of the victim. It particularly reflects the aspect that the insurer may not always assume no liability for insurance indemnity if the liability arises from the insured's intentional act.

In China, because the market of liability insurance is not yet fully developed, there are no comprehensive provisions regarding the distinctiveness of liability insurance within the current IL. In fact, it is combined with general property insurance. There are only simple stipulations in Articles 50 and 51 of the IL. From Article 28(2) in the same law, it follows that 'if the applicant, the insured or the beneficiary intentionally gives rise to the occurrence of an insured risk, the insurer is entitled to terminate the insurance contract, and to bear no obligation for indemnity or payment of insurance compensations. Except as otherwise provided in Article 65(1) of this law, the insurer will not refund the premiums either.' The conclusion that may thus be drawn is that the insurer should be exempted from the liability of indemnity under this circumstance. If reliance is placed on this clause to handle the issue of the pollution of the marine environment caused a shipowner's deliberate acts under the oil pollution liability insurance, it will be found that the protection to the victims is both unfavourable and insufficient.

In accordance with judicial practice in respect of insurance in some countries, such as the US, the doctrine of strict interpretation should apply to the question of whether the insured's intentional act falls into the category of excluded liabilities. Compensation

---

[14] K. Kouichiro, *Status of Victim (A Third Party) of Liability Insurance and Direct Right of Claim* (Automobile Premium Calculating Commission Printing Room no. 22), p. 21; quoted from W. Li, *supra* note 11, 246.

liability caused by the intentional acts of the insured should be limited to the compensation liability caused by the insured's specific intention to do harm to a third party. This is called the excepted liability.[15] The liability insurer should not refuse to assume insurance liability because of the insured's intentional acts under the following circumstances: (1) the compensation liability for damage caused by a person of legal incapacity; (2) damage caused unexpectedly by the insured's intentional acts; (3) damage caused by the insured's act in self-defence; (4) the intentional act of one of the joint and several tortfeasors; or (5) the right to direct claim enjoyed by the third party against the insurer without defence.[16] When the insured's intentional acts result in environmental pollution as provided in the liability insurance contract, US courts have held that when the insured's intentional acts are the cause of the environmental pollution at issue but the consequence of environmental pollution is not the intended result of the insured's intentional acts, the insurer should not refuse to assume insurance liability in respect of the insured's civil liability for environmental pollution because the insured acted with intent.[17] That is, the insured's intentional or even illegal acts or criminal conduct cannot serve as an excuse for the insurer to claim an exemption from liability.

Comparing the IL in China with the judicial application of insurance law in the US, it can be found that the latter restricts the excepted liability on account of intentional acts within the range of compensation liability that the insured should assume due to its specific intention to cause damage to the third party. Obviously, judicial practice in the US appears more powerful in respect of the protection of third-party interests than the relevant provisions in China.

4 ISSUES RELATED TO COMPENSATION LIABILITY FOR OIL POLLUTION DAMAGE

4.1 THE CURRENT STATUS OF LEGISLATION IN RESPECT OF OIL POLLUTION DAMAGE IN CHINA

The ocean is the most precious resource of human beings. However, at the same time when the shipping industry was developing rapidly, it brought with it pollution to the sea. As a result, it heavily affected human exploitation and utilization of the sea, and even caused certain damage to our living environment. Among the many sources of marine oil pollution, the vessel is a major one. Oil cargo, bunkers and other oil materials discharged intentionally by vessels during the daily operation constitute 70 percent of the total amount of the discharged oil. In case of an accident, a great deal of oil material leaks from the vessel causing marine pollution. Though it is less than the intentionally discharged oil, the results are usually more severe and concentrated. In March 1967, approximately 90,000 tones of crude oil leaked from the Liberian MT Torrey Canyon off the coast of

---

[15] H. Zou, *supra* note 3, 209.
[16] H. Zou, *supra* note 3, 209-211.
[17] N. Lockett, *Environmental Insurance Liability* (Cameron May, London, 1996), p. 20; quoted from H. Zou, *supra* note 3, 79-80.

Britain and severely polluted British and French coastal waters. This aroused much international attention. It also triggered the establishment of the 1969 International Convention on Civil Liability for Oil Pollution Damage (hereinafter as 'CLC 69'). From 1967 to 1996, 54 major oil pollution accidents occurred and in each case the amount of oil that leaked exceeded 10,000 tonnes. Among the accidents, was one that occurred in America in 1989 and caused total losses as high as USD 8,000 million.[18]

From the 1970s, a series of international conventions regarding prevention of pollution by vessels came into force in quick succession. Domestic laws and regional regulations were also promulgated endlessly. Prevention of oil pollution and determination of liability for oil pollution are two topics with the quickest development and became more popular than any other topic in international maritime legislation. China is a large oceanic country. The production value of its marine industry is increasing at an average annual rate of approximately 20 percent. Unfortunately, China's legislation regarding the prevention of oil pollution and the determination of liability for oil pollution is quite deficient, so that it is difficult to find a sound legal basis to decide an individual case involving oil pollution damage. To present, the legal system of marine environmental protection and prevention of marine pollution established in China comprises the following four parts: (1) various provisions of substantive laws, such as Article 9 of the Constitution,[19] Article 124 of the GPCL, and Article 90 of the MEPL (as amended in 1999);[20] (2) international conventions to which China has acceded, for example, the CLC 69, the 1973 International Convention for the Prevention of Pollution from Ships, the 1978 Protocol Relating to 1973 International Convention for the Prevention of Pollution from Ships (73/78 MARPOL); (3) international non-governmental treaties entered into by some shipping companies, such as the 1969 Tanker Owner's Voluntary Agreement Concerning Liability for Oil Pollution (TOVALOP), and the 1971 Contract Regarding an Interim Supplement to Tanker Liability for Oil Pollution (CRISTAL); and (4) administrative regulations promulgated by the authorities concerned for the implementation of above laws, conventions and agreements, such as the Rules of Administrative Regulation for Prevention of Marine Pollution from Ships, and the Notice of Strict Implementation of 73/78 MARPOL, Annex I.[21] In the aforesaid legal system, most of the stipulations in domestic laws are principled outlines or administrative in nature without providing a sufficient guidance in trials of civil cases in respect of oil pollution damage. While the application of international conventions is restricted in many aspects, the handling of cases related to the compensation for oil pollution damage is still lacking a sound legal basis.

---

[18] Y. Si, *International Maritime Legislation Trend and Policy Study* (Law Press, Beijing, 2002), p. 220.
[19] Art. 9 of the Constitution provides 'Waters are natural resources owned by the state, appropriation or damaging natural resources by any organization or individual by whatever means is prohibited'.
[20] Art. 90 of the Marine Environment Protection Law provides 'any party that is directly responsible for a pollution damage to the marine environment shall relieve the damage and compensate for the losses'.
[21] Y. Si, *supra* note 18, 226.

4.2 THE PRINCIPLE OF IMPUTATION OF COMPENSATION LIABILITY FOR OIL POLLUTION DAMAGE

The imputation of liability refers to the standard on which the tortfeasor's civil liability shall be determined and prosecuted after the occurrence of the tortious act results in damage to the third party. Imputation means that the tortfeasor assumes the consequence of its acts. Insofar as a victim is concerned, it also means indemnifying the victim for this loss.[22]

The CLC 69 provides that the owners of vessels laden with persistent oil cargo shall assume civil liability for oil pollution damage resulting from the oil leaked or discharged from vessels, irrespective of whether the owners, masters, crewmembers or other employees are at fault or not. The liability for oil pollution as provided in the CLC 69 is strict, regardless of the shipowners' negligence or not, but it is not absolute liability. The shipowners may be exempted from liability under three circumstances: (1) oil pollution results from war, hostility, civil wars or unavoidable and special natural disasters; (2) oil pollution is caused wholly by the third party's (especially the third party victim itself) intentional acts; (3) oil pollution is caused by the governmental authorities' negligence or fault not to maintain navigation lights or other navigational equipment.[23] In the International Convention on Liability and Compensation for Damage in Connection with the Carrying of Hazardous and Noxious Substances by Sea of 1996, it is provided that owners of vessels carrying poisonous and toxic materials causing accidents shall assume compensation liability for damage arising therefrom, but if the accidents are a series of accidents with the same cause, then the shipowner who initially caused the accident shall assume the liability. The principle of strict liability for oil pollution is also adopted in the 1990 Oil Pollution Act of the US (hereinafter as 'OPA 90'), which also provides several circumstances of exemption of liability: (1) natural disasters; (2) warlike acts; (3) the third party's act or omission (if there is a contractual relationship between the liable party and the third party, while the act or omission of the liable party's employee or agent is related to this contractual relation, the liable party shall not be exempted from liability); or (4) oil pollution caused by the claimant's gross negligence or malicious conduct.[24]

Article 42 of the 1982 MEPL is the first law in China to establish a no-fault liability for environmental torts.[25] Article 41 of Law on Prevention and Control of Water Pollution of 1984 has a similar provision. The subsequent Law of Environmental Protection and other rules on prevention and control of environment pollution also follow the same system as above. Article 106(3) of the GPCL clearly provides for no-fault liability by stipulating 'civil liability shall still be borne even in the absence of fault, if the law so stipulates.' Article 124 of this law also provides that 'any person who pollutes the environment and causes damage to others in violation of state provisions for

---

[22] Z. Wang, *Civil Law Theory and Cases Study* (China University of Political Science and Law Press, Beijing, 1998), vol. 5, p. 259.
[23] L. Yang, *Admiralty Law* (Dalian Maritime University, Dalian, 1999), p. 275.
[24] D. Zhao, *International Maritime Law* (Beijing University Press, Beijing, 1999), pp. 500-501.
[25] Art. 42 of the Marine Environment Protection Law (1982) provides that 'units or individuals who have suffered damage caused by marine environmental pollution shall be entitled to claim compensation from the party which caused the pollution damage…'.

environmental protection and the prevention of pollution shall bear civil liability in accordance with the law.' This is the legal basis of the no-fault liability applicable to environmental torts. The GPCL, the basic law with respect to environmental protection, and special regulations on environmental protection are the principal sources for no-fault liability. This indicates the internal coherence, integrity and logic of our legislation and the latest trend in tort law. It conforms to the international trend of the no-fault liability principle that applies to environmental torts and the general rules of legislation in respect of environmental protection in many countries. This principle is adopted in judicial practice and becomes widely acceptable to scholars, that is, it is a mainstream principle.[26]

In short, the principle of no-fault liability, that is, strict liability, has been adopted in the standard of imputation of liability for marine oil pollution damage. Once marine oil pollution has occurred, aside from these explicit legal exemptions, the party concerned should assume civil liability even if it is not at fault. The establishment of a retrospective strict liability principle aims at enhancing the victim's chances of being indemnified and urging the shipowners positively to prevent oil pollution from occurring. This principle is of great significance for the protection of the marine environment.[27]

4.3 THE PROMOTION OF COMPULSORY LIABILITY INSURANCE FOR OIL POLLUTION

Generally, a liability insurance contract is concluded voluntarily between the applicant for insurance (the insured) and the insurer, unless when it is stipulated in laws that liability insurance is compulsory. Without a compulsory stipulation, parties can decide whether to buy, agree on the amount of the coverage, whether to sell, what is covered by the insurance policy and what is exempt from coverage. However, in some industries with an obvious public welfare character, the liability insurance system established based on free will is deficient in that it does not provide full protection of the victim's interests. Examples are the insurance for compensation for damages caused by motorized vehicles and oil pollution. These industries concern personal and environmental safety, and act as the guarantor of man's existence and quality of life. If the applicant and the insurer are to decide whether to buy and agree to the insurance policy, the effect will be a drastic reduction of the protection of the interests of victims. Liability insurance, social insurance and loss apportionment are three important aspects that must considered for the development of the civil liability system. If these three aspects of protection are still insufficient to protect the interests of victims, it is necessary to promote a compulsory liability insurance scheme as the additional method of protection of the victims in a certain area.[28] Promoting compulsory liability insurance for marine oil pollution conforms to the strict liability for oil pollution damage. In order to ensure the implementation of a compensation system for oil pollution and to secure adequate compensation for the victims following an incidence of oil pollution, legislation for compulsory liability insurance against shipowners must be enacted. Through the application of compulsory liability insurance, the insured's liability for oil pollution at sea could be transferred to

---

[26] M. Cao, *supra* note 5, 155.
[27] J. Wei, *Maritime Law* (Jilin People's Press, 1996), p. 376.
[28] H. Zou, *supra* note 3, 49.

various insurers or other mutual insurance associations. Thus, the party liable for oil pollution will be able to survive the excessive liability for compensation. The shipowners must procure insurance for the liabilities for oil pollution damage, or alternatively other financial securities. In order to ensure the implementation of a compulsory liability insurance system, each ship should possess a Certificate of Insurance or Other Financial Security in Respect of Civil Liability for Oil Pollution Damage, otherwise the ship will be prohibited from operating.[29] Article 97 of the MEPL of 1999 provides that the relevant stipulations of the international conventions concluded or acceded to by China shall apply to the issues of compulsory liability insurance.[30] According to the OPA 90, all vessels exceeding 300 gross tonnage or any other vessels transshipping or lightering to the water areas under US jurisdiction through the US exclusive economic zones must possess the certificate of financial responsibility which should be sufficient to indemnify the maximum financial liability. This amount excludes the civil, administrative or criminal penalty that that may be imposed on the party liable. The party's financial capability may be certified by means of insurance certificate, surety bonds, letter of guarantee, letter of credit, self-guarantee qualification certificate or other financial status certificates.

According to the compulsory liability insurance for oil pollution, a ship (the applicant for insurance or the insured) cannot operate without liability insurance for oil pollution or other financial security. However, the liability insurer's burden for the payment of the indemnity is nearly absolute if the covered risk occurs. This is the only way the compulsory liability insurance will conform to the principle of protecting the innocent victim to the greatest extent.

4.4 THE SYSTEM OF DIRECT ACTION

The direct action is a special legal system which allows the injured party to file a lawsuit directly against the liability insurer or the financial guarantor. It is to protect the third party who suffers damage from the tortfeasor's act. In civil actions in case of the tortfeasor's death, bankruptcy, dissolution, liquidation or if the subject held liable legally disappears, the creditor will probably lose its legal protection. In order to protect the injured party, based on the theory that liability insurance is primarily established in the injured party's interest, the legal system which allows the injured party to file a lawsuit directly against the liability insurer came into existence around the world. For example, in the US, federal law and the majority of state laws allow the injured party to file a lawsuit directly against the liability insurer.[31] Pursuant to the OPA 90, the claimant may claim compensation against the guarantor. The insurer may enjoy the right of defence according to laws. But if the oil pollution accident is caused by the liable party's vicious conduct, the

---

[29] Y. Si, *supra* note 18, 227-228.
[30] Art. 28.2 of the Marine Environment Protection Law 1982 clearly provides that 'any vessel carrying more than 2,000 tones of oil in bulk as cargo shall hold a valid 'insurance or other financial security certificate for civil liability against oil pollution damage', or a 'credit certificate for civil liability against oil pollution damage', or furnish other financial credit guarantees'. However, the Marine Environment Protection Law 1999 as recently amended gives a general stipulation in respect of this in art. 66.
[31] W. Wang, 'Several Points of Study in Respect of Perfecting Maritime Insurance Legislation of Our Country', (1998) *Chinese Maritime Law Annual*.

guarantor may refuse to indemnify.

Article 97(1) of the SMPL in China provides that 'in respect of oil pollution damage caused by a ship, a claim may be made by an aggrieved part either against the owner of the ship causing oil pollution or directly against the insurer who underwrites the shipowner oil pollution liability, or against the person who has provided financial security therefore.' While Article 97(2) of the same law further provides that 'where the insurer who underwrites the oil pollution liability of the shipowner or the person who has provided financial security therefore is sued in an action, such insurer or person is entitled to demand the owner of the ship causing oil pollution to join the proceedings.' This provision establishes by means of law that the victim in respect of oil pollution may file a lawsuit directly against the insurer who underwrites the oil pollution liability or the person who has provided financial security. This provision is enacted with reference to the 1992 Protocol Relating to the CLC 69 (hereinafter as 'CLC 92'). According to the CLC 92, in addition to having a claim against the shipowners, the victim of oil pollution is also entitled to claim compensation against the insurer who underwrites the oil pollution liability or against other financial guarantors. However, neither the insurer nor the financial guarantor is actually the true subject held liable for oil pollution. Hence, the CLC 92 gave them more possibilities to raise defences.[32]

After a vessel has caused oil pollution damage, a victim may file a lawsuit to claim compensation directly against the shipowners (the insured), as well as against the insurer who underwrites the oil pollution liability or against any other person who has provided financial security. This is called the direct action. Based on this, even though the shipowner might be insolvent due to the oil pollution accident, the victim may still be adequately indemnified on time to fulfil the compulsory liability insurance requirement. This is of great significance to ensure the victim's right to compensation. In case of a direct action, the insurer who underwrites the oil pollution liability or the person who has provided financial security has the right to request the shipowner be joined to the proceedings, to raise defences that the shipowner can use, except those arising following the shipowner's bankruptcy or liquidation. The insurer is also entitled to claim its right of limitation of liability for compensation. All these rights will not be affected by the shipowner's actual fault. If the damage is proved to have been caused by the shipowner's intentional wrongful act (for instance, the shipowner was fully aware of the unseaworthiness of the ship at the time of sailing, but set sail anyway), the insurer who underwrote the oil pollution liability or the person who has provided financial security may bear no liability for compensation. However, the insurer of the oil pollution liability or other financial guarantor may not raise any defence against the victim which otherwise the shipowner might be entitled to raise because the said defence occurs only between the insurer and the insured, or between the guarantor and the insured. It is not valid to extend it to the third party victim.[33]

Chinese laws of procedure have established the direct action system to allow the victim to bring a claim against the liability insurer. However, it is indeed difficult to find any substantive law to determine how the liability insurer should undertake liability and

---

[32] X. Gao, *Ship Oil Pollution Damages System* (Maritime Law Study, 2001), vol. 2, p. 174.
[33] J. Wei, *supra* note 27, 381-382.

what types of liability should be assumed between the insurer and the liable party. The only exception is Article 206 of the MC, which provides that the liability insurer may be entitled to the limitation of liability in respect of maritime claims. According to this provision, victims may be able to file a lawsuit directly against the liability insurer or to have the liability insurer be joined in the proceeding as co-defendant or as a third party during the litigation.

5 NON-TYPICAL JOINT AND SEVERAL LIABILITY BETWEEN THE INSURER FOR OIL POLLUTION AND THE PARTY LIABLE FOR OIL POLLUTION

5.1 NON-TYPICAL JOINT AND SEVERAL LIABILITY IN THE FRAMEWORK OF THE CURRENT LAW

Under the present framework of substantive laws, the insurer for oil pollution and the party liable for oil pollution do not share joint and several liability, though they may share non-typical joint and several liability toward the victim.

Because Article 97(1) of the SMPL indicates a legislative intent to make both the insurer and the liable party bear joint and several liability toward the victim, it became, in practice, the legal ground for those who believe the insurer and the liable person should bear joint and several liability. However, there is no official legal opinion regarding the legislative intent of the SMPL, although some persons who participated in the legislation of the SMPL expressed on various occasions their personal opinions on the legislative intent of 'Joint and Several Liability' of the said article.[34] However, the SMPL is only procedural law. Even though it is not impossible to infer a substantive intent from procedural law, it is unjustifiable to infer and apply substantive law via the legislative intent of a procedural law without an explicit provision in substantive law. Furthermore, it is inconceivable to render a judgement on based on procedural law rather than substantive law. Apparently, much improvement is necessary in this regard.

In the legal relation of creditors' rights and debts, the enjoyment of rights and the assumption of obligations are subject to the capacity of the individual party. However, in practice it is very common for several parties to assume the same debt towards one party. If this is the case, the law particularly empowered the creditor with the right to simultaneously or later demand all or partial debtors, or part performance of payment of the debt to ensure the fulfilment of the interests of the creditors. It conforms with the civil law in respect of joint and several liability, that is, to have one debtor to bear other debtors' portion of the debt. The legislative intent of joint and several liability is to protect the interests of the creditor. As the debtor's liability is aggravated, the legislation in most countries provides that the joint and several debt does not apply unless it is expressly

---

[34] Lecture of Guangzhou Training Class on The Special Maritime Procedure Law in March 2000, Lecture of The State Judge Institute on April 2002 as well as Forum on Foreign Maritime Trial Held in Wuyi Mountain, October 2002.

stipulated in the law or clearly agreed upon by the parties.[35] That is the statutory nature of the joint and several debt that aims at preventing any abuse of the law. Both the GPCL and the MC in China have made express provisions on the application of the joint and several liability.[36] In some circumstances, the debtors shall be bound by joint and several liability, but the provisions of the law on joint and several liability do not cover all of these circumstances. Joint and several liability between the insurer for oil pollution and the party liable for oil pollution seems to fall in this grey area.

A non-typical joint and several debt is a debt of the same contents but with a different cause of action, though each of the debtors can be requested to make a full performance. Once a debtor has fulfilled the debt as a whole, the rest of debtors will be discharged by the creditor from further payment. The legislative intent[37] of the non-typical joint and several debt is the same as that of the joint and several debt. They both aim at better protection of the creditors' interests. Both debts are similar in some respects, for instance, more than one debtor is liable for the same debt, each debtor is liable for the full amount of the debt, and the whole debt will be extinguished upon the full payment by any single debtor. However, the further analysis reveals some obvious differences between the non-typical joint and several debt and the joint and several debt.

Firstly, the causes of the two debts are different. Usually the joint and several debt arises from the contractual agreement or the common tortious acts. While the non-typical joint and several debt arises from different legal facts, the debt owes the same creditor the same debt. It is a merging of debts in a broad meaning. The creditor is entitled to claim against each debtor for the fulfilment of the whole debt and any debtor's fulfilment of the whole debt shall exempt the liability of the remaining debtors'.

The non-typical joint and several debt can be classified into four types: (1) Several independent acts of tort give rise to non-typical joint and several debt due to coincidence; (2) One's non-performance of a debt concurs with the other's act of tort and thus gives rise to a non-typical joint and several debt; (3) Concurrence of one's non-performance of a debt with other's non-performance of a debt gives rise to non-typical joint and several debt; and (4) non-typical joint and several debt can also be established through the agreement by contract, non-performance of other debts or tortious acts.

The nature of the debt to the victim owed by the insurer for oil pollution and the party liable for oil pollution perfectly conforms to the fourth of the four above-mentioned forms of non-typical joint and several debt. The debt owed by the insurer for oil pollution to the victim, that is, payment of insurance indemnity, is based on the contract for liability insurance agreed between the insurer for oil pollution and the party liable for oil pollution (i.e., the insured); but the debt owed by the party liable for oil pollution to the victim is

---

[35] S. Nian, 'Analysis on Non-typical Joint and Several Liability', (2000) *Judicial Study* 164.

[36] Application of the General Principles of the Civil Law mainly includes: (1) Joint and several liability of the partners in the individual partnership (art. 35); (2) Joint and several liability of the economic association of enterprise and institution (art. 52); (3) Joint and several liability of agency (art. 65.3, art. 66.3-4 and art. 67); (4) Joint and several liability of debtor and guarantor in case of non-performance of debt (art. 89); and (5) Joint Tortious liability in case of joint torts (art. 103). Application of joint debt in the Maritime Law mainly includes: (i) Joint liability of carrier and actual carrier (art. 63); (ii) Joint liability of the shipowners of both vessels of collision to a third party suffering the loss of life or personal injury (art. 169.3).

[37] X. Kong, *Civil and Commercial Law New Problem and Case Study* (People's Court Press, 1996), p. 134.

based on the tort of oil pollution. The causes that give rise to the said two debts are different. If the insurer has fully compensated the victim for oil pollution, the victim cannot claim further compensation from the shipowner.

Second, the purposes are different. There are usually communications or a common purpose between various debtors of the joint and several debt. However, the non-typical joint and several debt does not require a common purpose. It arises from the occasional concurrence of different legal facts, which results in the same contents of the debt.

There is no common purpose between the insurer's liability and the insured's liability to the same victim in respect of the oil pollution damage, that is, no substantive connection exists between the two liabilities. The performance of any one of the debtors will extinguish the whole debt, and the creditor is not allowed to claim against the other debtors after the debt is fully extinguished.

Third, the legal basis is different. Joint and several debt applies strictly the principle of legality. Joint and several debt does not apply unless expressly agreed between the parties or clearly stipulated by law. The non-typical joint and several debt is based on different legal facts and respectively independent and the concurrence of its legal relationship is occasional which cannot be anticipated by the legislators. In practice, its application is subject to the judge's right of discretion. The law is unable or it is unnecessary to provide for it expressly and furthermore no agreement between the parties is available.

At present, no substantive law concerning the liability for oil pollution provides that the insurer of oil pollution and the party liable for oil pollution shall assume joint and several liability. Therefore, courts should not render judgements that apply the principle of joint and several liability to the insurer and the insured in respect of oil pollution damage.

Fourth, the proportion of debt among debtors is different. There usually exists a liability-sharing relationship among the joint and several debtors. Each of the debtors shall assume an obligation in proportion to its limit of liability or the agreement of the contract. Therefore, the debtor who has performed the entire debt is entitled to claim recovery from other joint debtors for the proportion beyond his share. There is no liability-sharing proportion among the debtors of a non-typical joint and several debt.

It can thus easily be established that the liability between the insurer of oil pollution and the party liable for oil pollution is a non-typical joint and several debt. The insurer's claim against the party liable after paying the indemnity to the third party is not based on the debt-sharing relationship between them, but on the assumption of the final liability by the party liable under certain circumstances (such as the intentional act of or gross negligence of the party liable).

Fifth, the external effects are different. With respect to joint and several debt, such matters as the claim for performance, exemption, commingling, statute of limitations, and delay in acceptance of payment due to one of debtors generally have an absolute effect on other debtors. However, the non-typical joint and several debt has independent debts that arise from different legal facts. The act of one of the debtors generally has no effect on the other debtors. However, if the creditor exempts the final person liable from liability, it applies to the other debtors as well.

As a third party, the victim can either bring a claim for compensation for oil pollution against the liability insurer or against the party liable for oil pollution based on tort. Any act of the victim to the insurer has no effect on the party liable for oil pollution. However, if the victim exempts the party liable for oil pollution from compensation, the exemption also applies to the insurer of oil pollution.

In summary, the liabilities assumed by the insurer for oil pollution and the party liable for oil pollution conforms fully to the standards of the non-typical joint and several debt. If there is no legal provision providing that the insurer for oil pollution and the party liable for oil pollution bear joint and several liability to the victim in the substantive law nor is there a special agreement between the parties, it undoubtedly lacks ground to assert that the said insurer and liable party are jointly and severally liable. However, under the circumstances that there are obvious gaps in the present laws and regulations, methods to properly supplement such laws and regulations can be applied. As for the above-mentioned case, it would conform with the spirit of law to interpret broadly the legislative intent of Article 97 of the SMPL and provide a legal remedy to the victim as much as possible. It is necessary to emphasize that both systems, the non-typical joint and several debt and the joint and several debt, are in essence identical in respect of the protective strength for the creditor.

5.2 THE LEGISLATIVE TREND OF THE JOINT AND SEVERAL LIABILITY

As mentioned above, the liability of the ordinary liability insurer is not direct compensation liability but substitute compensation liability. However, since the compulsory liability insurance for oil pollution has a strong public welfare character, both the obligation of paying indemnity to the victim by the insurer and the liability of compensation of the party liable (the shipowner) should be strengthened so as to promote transforming the substituting compensation liability of the insurer to joint and several liability.

Firstly, from the statements made at various occasions by some of the Members of the Supreme People's Court who participated in the legislation of SMPL, to begin it follows that Article 97 of the SMPL contains the legislative intent to make the insurer of oil pollution and the party liable jointly and severally liable. That is greatly helpful to the protection of the victim's interests. The only obstacle existing at present is the non-existence of a corresponding rule in substantive law. In the subject case, the opinion that a lawsuit should be dismissed for lack of legal grounds seems reasonable; however, under the current circumstances, this would undoubtedly be considered mechanical and dogmatic.[38] Furthermore, to dismiss a lawsuit merely due to lack of legal provisions obviously violates the principle that the judge may not refuse trial.

Second, if the law does not require that the insurer and the party liable assume joint and several liability, but confirms that there is substituting compensation liability or compensation liability, (i.e. the insurer only compensates the difference between the full amount and the part the insured paid) it would completely contradict the fundamental purpose of liability insurance and the system of compulsory liability insurance. Since the liability insurance has been established to benefit the third party and to take the liability borne by the insured, the liability insurer should not refuse to compensate on the ground

---

[38] J. Yang, *Administrative Litigation and Judicial Activity-Study on the Suit Brought by Liu Yanwen against Beijing University, Degree Assess Commission, Law Study* (Law Press, 2001), vol. 4, p. 14.

that there is no contractual relationship between the liability insurer and the third party after the occurrence of the insured accident. Since the oil pollution is covered by the compulsory insurance, it is reasonable and acceptable for the insurer of oil pollution to assume a liability that is heavier than the liability in case of ordinary liability insurance. In addition, it would be most favourable to protect the interests of the third party if the insurer for oil pollution and the party liable for oil pollution were to assume joint and several liability within the scope of the terms of policy.

6 PERFECTION OF LEGISLATION REGARDING COMPENSATION FOR DAMAGES OF SHIP OIL POLLUTION

The forgoing analysis regarding non-typical joint and several liability between the insurer and the party liable is theoretical. Although a theoretical study is helpful, it requires express provisions in the form of a substantive law through legislation in order to resolve the problem. Because the non-typical joint and several liability system is still at the stage of theoretical study, it is premature to launch legislative procedure now. It might be feasible to secure joint and several liability between the insurer and the insured by means of legislation. Article 63 of the MC has provided a good example in respect of the assumption of the existence of a joint and several liability between the contractual carrier and the actual carrier.[39] Because the IL was amended in October 2002, there is unlikely to be another amendment after such a short period. As a result, the State Council is inclined to incorporate the compensation liability system for ship oil pollution in Marine Environment Protection Law. However, the making of the Provisional Rule on the Oil Pollution Insurance has also entered into a stage of analysis and assessment. Therefore, either amending the MC or making the Provisional Rule on the Oil Pollution Insurance will undoubtedly be a good occasion to make supplementary provisions on the joint and several liability of the insurer and the party liable.

6.1 THE SUBSTANTIVE CONTENT OF THE DIRECT ACTION

A model for reference in respect of the provisions of the liability insurance is available and it is the eighth amendment of the Insurance Law in Taiwan, dated 9 July 2001. The amendment contains a new second paragraph to Article 94, which reads 'After ascertainment of liability of compensation for loss to the third party, the third party may, within the scope of the amount insured and in accordance with his recoverable proportion,

---

[39] As regards carriage of goods by sea, what the carrier bears is contractual liability but what the actual carrier bears is tortious liability. Both the basic reasons and the characters of the two liabilities are different. The liabilities seem to belong to the non-typical joint and several liability. However Maritime Law provides that the carrier and the actual carrier shall be jointly and severally liable. The said provision is certified as effective through practice and is most favorable to the protection of the legal interests of the holder of a bill of lading. See W. Li, 'Study on Application of art. 63 of the Maritime Law of China', (1999) 1 *Chinese Maritime Law Commission Message* 9-12.

directly claim against the insurer for compensation.'[40]

The procedural provisions concerning the victim's right of direct action in the event of an oil pollution accident in China have been established in Article 97 of the SMPL. Nevertheless, Paragraph 1 of Article 50 of the IL also provides that the insurer may, in accordance with the provisions of law or the terms of an insurance contract, directly indemnify a third party for loss or damage caused to him by the insured of a liability insurance contract. However, this provision only relates to the relationship between the insurer and the insured. It ignores the interests of the third party victim. From this viewpoint, the liability insurance is merely a pure indemnity policy.

6.2 OIL TANKER TO BE COVERED BY COMPULSORY INSURANCE

The CLC 92 only requests oil tankers carrying above 2,000 tonnes of persistent oil (approximately equal to 1,000 gross tonnes) to be covered by compulsory oil pollution insurance. However, the reality in China is that there are plenty of small tankers and these ships are involved in a large number of accidents. Additionally, most small shipowners are of poor ability to indemnify losses. The statistics show that from 1973 to 2001, among the top 22 serious accidents whose oil leakage exceeded 50 tonnes involving Chinese oil tankers off China's shoreline, over 60 percent (or 13 out of 22 cases) occurred involving oil tankers of less than 1,000 gross tonnes. According to the same records, there are more than 2,500 oil tankers throughout the country; about 80 percent of these are small ships below 1,000 gross tonnes. Among these small oil tankers, only 51 ships have oil pollution liability insurance coverage, which means that most small oil tankers have no financial ability to indemnify in the event of a large accident.[41] In addition, the CLC 92 has limited its application to the area of persistent oil. Obviously, damage caused by persistent oil to the marine environment is greater than that caused by non-persistent oil. For example, persistent oil adheres more easily to marine plants since it does not dissolve in water. Moreover, the adhesive duration is greater. It can easily cause large-scale pollution in the process of floating and spreading, and it is not easily cleaned up. By contrast, the volatility of non-persistent oil is stronger than that of persistent oil and its adhesive duration is shorter. However, it has strong and acute poisonous effects and it may cause a greater scale of instant death to marine plants than that caused by persistent oil. Accordingly, it is proposed that, in China, both persistent oil and non-persistent oil carriers be covered by compulsory insurance.

One of the differences between voluntary liability insurance and compulsory liability insurance is that the former has rights to defend against claims of the third party's direct action, while the latter has in essence no right to defend against the third party. In voluntary liability insurance, a third party's direct action shall be confirmed in the form of an insurance contract while in compulsory liability insurance the direct action is based on specific legal provisions. As for iiability insurance, the differentiation of the basis of

---

[40] Art. 94 of the original Insurance Law of Taiwan provides that before the insured compensates a third party for the loss caused by the accident for which the insured shall be liable, the insurer cannot affect part or all of the compensation to the insured.

[41] H. Liu, 'Establish and Exercise the Ship Oil Pollution Damages System of Our Country as Soon as Possible', (2003) 1 *Chinese Maritime Law Commission Message* 4.

effect inevitably brings about the differentiation of the scope and the forms of liability. In the wake of the statutory direct right of claim is the limitation on the insurer's right of defence. In this regard, the Japanese Automobiles Damage Compensation Law has provided a good reference. Article 14 of this law provides that the insurance company can be exempted from liability for the damage caused by the insured's vicious intent or based on the insurance contract with the exception of the occasions as specified in Article 82(2). It can be found that even if one of the occasions as stipulated in Article 14 of the above-mentioned law for exemption of the insurer's liability due to vicious intent of the insured applies, the victim still has the right to claim against the insurance company. The insurance company may formulate a defence against the claims of the insured on grounds of exemptions, but it cannot defend against the victim's direct action, that is, exemptions can only apply to the insured and not against the victim.[42]

As the oil pollution liability insurance is closely related to public policy, the terms of the policy shall be interpreted in a manner that is unfavourable to the insurer so that the third party may enjoy more favourable rights against the insurer than the insured. The subject matter of the oil pollution liability insurance is the liability of the insured shipowner toward a third party. As long as the shipowner has shown no causes that may exempt him from liability, his liability to compensate will stand. Even if the shipowner has committed faults, general negligence, gross negligence or even vicious intent which gave rise to the occurrence of the event, the liability insurer shall bear the duty to indemnify the third party and is only entitled to claim against the insured for recovery after indemnifying the victim. That is,, the right of defence enjoyed by the oil pollution liability insurer is extraordinarily limited not to exceed the limit of the rights of defence enjoyed by the shipowner. Obviously, the oil pollution insurer's right to enjoy a limit on compensation liability will not be lost because of privity between parties or actual fault of the party liable for the oil pollution.

6.3 ESTABLISHMENT OF A SCHEME OF PAYMENT ON ACCOUNT

The Japanese Automobile Damage Compensation Law also provides that, before a mediating authority or a judicial authority determines the scope of liability and indemnity, the victim is entitled to request the insurance company to effect payment on account of a certain amount. This is called a compensation payment on account. The First Aid System in the American automobile damage compensation scheme is similar to the Japanese scheme and the execution in advance scheme of the Civil Procedure Law of China is also similar.

As the loss caused by oil pollution is large in scope and great in amount, and the victim must usually pay a heavy legal fees to secure a judicial remedy while suffering from great loss, it places an extra burden and hardship on the victim when seeking its legal right of protection. Therefore, the payment on account system should be established as part of the oil pollution damage compensation system. When the facts and liability for the oil pollution are clear, except the amount of compensation may still be determined, the

---

[42] W. Li, *supra* note 11, 239.

oil pollution liability insurer can be requested to pay a certain amount of indemnity to the victim in advance so as to provide convenience to the victim to seek a judicial remedy.

7 CONCLUDING REMARKS

Judging from the attention paid by the international community to marine oil pollution, the catastrophe of MT Torrey Canyon in 1967 triggered the first climax and was a direct cause of the conclusion of the CLC 69. The catastrophe of the MT Amoco Cadiz in 1978, which leaked 220,000 tonnes of crude oil off the French coast, triggered a second climax of international attention. People realized the insufficiency of the compensation amount as provided by the CLC 69. As a result, the CLC 84 and the CLC 92 Protocols came into existence.[43] In 1989, about 30,000 tonnes of crude oil leaked from the MT Exxon Valdez at Alaska, US, which had spurred the enactment of the comprehensive, stringent and highly controversial OPA 90 by the US Congress. That is the third tide of attention. In short, the development of the oil pollution damage compensation system has indicated a higher level of intent to protect public interests. The birth and development of new environment protectionism has strengthened the principle of strict liability and extended the application of compulsory insurance, which has heightened the possibility of protection of the interests of the victim of oil pollution. China should exercise a series of administrative legislation, criminal legislation, and in particular civil legislation[44] to enrich the contents of the oil pollution damage compensation system, to secure the exercise of compensation that has arisen from oil pollution. Considering that the existing legal framework for environmental protection and liability insurance law in China is rather simple and underdeveloped, the current laws should be amended or supplemented to protect the legitimate interests of the innocent victim to the maximum and protect the source of marine environment as well, so as to maintain a continuous social development.

---

[43] The CLC 84 Protocol did not enter into force due to the absence of an American accession, but the amount of the limitation of liability determined by this Protocol is reiterated in the CLC 92 Protocol.

[44] It is not enough to protect the environment only by Criminal and Administrative Law. Both Criminal and Administrative Law play an important role in the protection of the environment, but in the meantime they have their own limitations. The Criminal Law is just the supplementary measure to protect the environment, which cannot rid the harm to the environment nor compensate the loss of the victim. The respect of the Environmental and Administrative Law is that the environmental and administrative authority shall administer according to law and in the event the private right is harmed, such authority shall try to recover and fill the harmed legal interests of the victim. The Civil Law plays a special role in the protection of environment, the purpose of which is to fill the loss suffered by the victim, remedy the loss already caused and simultaneously make the person that caused pollution to assume some economic burden. See M. Cao, *supra* note 5, 5-6.

LIST OF REFERENCES

M. Cao, *Environmental Tort Law* (Law Press, Beijing, 2000).

J.F. Christ, *Fundamental Business Law* (American Technical Society, Chicago, 1944).

X. Gao, *Ship Oil Pollution Damages System* (Maritime Law Study, 2001), vol. 2.

X. Kong, *Civil and Commercial Law New Problem and Case Study* (People's Court Press, Beijing, 1996).

K. Kouichiro, *Direct Right of Claim of Victim of Liability Insurer* (Contemporary Damages Lecture (8), 1973) edition of Japan Comment Commission.

K. Kouichiro, *Status of Victim (A Third Party) of Liability Insurance and Direct Right of Claim* (Automobile Premium Calculating Commission Printing Room no. 22).

W. Li, *Study on Japanese Automobile Accident Damages Legal System* (Law Press, Beijing, 1997).

W. Li, 'Study on Application of art. 63 of the Maritime Law of China', (1999) 1 *Chinese Maritime Law Commission Message* 9-12.Z. Liang, *Civil Complementary Liability Relevant Problems Study* (People Justice, 4th edn, 2003).

H. Liu, 'Establish and Exercise the Ship Oil Pollution Damages System of Our Country as Soon as Possible', (2003) 1 *Chinese Maritime Law Commission Message* 4.

Z. Liu, *Insurance Law* (Taiwan Sanming Press, Taiwan, 1995).

N. Lockett, *Environmental Insurance Liability* (Cameron May, London, 1996).

S. Nian, 'Analysis on Non-typical Joint and Several Liability', (2000) *Judicial Study* 164.

Y. Si, *International Maritime Legislation Trend and Policy Study* (Law Press, Beijing, 2002).

U. Toshio, Direct Right of Claim of Victim, Present Situation of Compensation of Traffic Accident (Administration Press, 1979).

W. Wang, 'Several Points of Study in Respect of Perfecting Maritime Insurance Legislation of Our Country', (1998) *Chinese Maritime Law Annual*.

Z. Wang, *Civil Law Theory and Cases Study* (China University of Political Science and Law Press, Beijing, 1998), vol. 5.

J. Wei, *Maritime Law* (Jilin People's Press, Changchun, 1996).

J. Yang, *Administrative Litigation and Judicial Activity-Study on the Suit Brought by Liu Yanwen against Beijing University, Degree Assess Commission, Law Study* (Law Press, 2001), vol. 4.

L. Yang, *Admiralty Law* (Dalian Maritime University, Dalian, 1999).

H. Zao and C. Zheng, *Insurance* (Beijing University Press, Beijing, 2000).

D. Zhao, *International Maritime Law* (Beijing University Press, Beijing, 1999).

H. Zou, *Liability Insurance* (Law Press, Beijing, 1999).

JUDICIAL AUTHENTICATION OF MARINE POLLUTION DAMAGES

Wu Lijing and Zhang Min

1 INTRODUCTION

Marine Pollution Damages refers to the direct or indirect introduction of materials or energies into the ocean causing harm to marine organisms, human health, the fishing industry and other legal marine activities, as well as the quality of sea water, thereby contaminating the water and threatening the environment.[1] With the rapid development of coastal industries, agriculture and marine transportation, large quantities of pollutants have been emitted to the ocean. For example, statistics show that in marine transportation oil spill accidents internationally between 1965 and 1997, there were 79 of extraordinary seriousness, each of which discharged over ten thousand tonnes of oil, the total amount of spilled oil reaching 4,146 million tonnes. In China, a total of 53 oil spill accidents of over 50 tonnes of spillage occurred, and the total amount of oil spilt was 29 thousand tonnes. Consequently, through both local legislation and international cooperation, coastal countries are taking various measures to prevent, reduce and control marine pollution. They are also developing corresponding compensation mechanisms to ensure prompt and adequate compensation for the damage caused in marine environmental pollution accidents. However, the premise of such compensation is a proper judgement of the cause-effect relationship and the aftermath of the damage, like the pecuniary losses for fishing, marine aquaculture and tourism, and so on, and expenditure for the restoration of the damaged environment.

In the process of litigation for marine pollution damages compensation, the purported polluter and the aggrieved party will often argue over the existence of pollution damages, as well as the extent and degree thereof, which may increase the difficulty in settling the case. The analysis and assessment of the impact of pollutants on marine animals, the economy and environment always involve professional expertise in marine physics, marine chemistry, aquaculture and marine creatures. Such knowledge is beyond the realm of law, and so it entails professional testing, discrimination and judgement by experts. This is the so-called 'Judicial Authentication' of Marine Pollution Damages.

Judicial Authentication helps to determine scientifically the degree of pollution damages and reasonably recognize the aggrieved party's amount of loss so that both parties' legal rights are protected, and the process of adjudication is benefited. This kind of authentication is still novel in the Chinese judicial system, and as such there is much to be desired in the study of it. Based on a summary of the characteristics and features of such judicial authentication, this article discusses the ways to avoid distortion of the

---

[1] Marine Environment Protection Law of the People's Republic of China, art. 95.

authentication report and analyzes its importance in relevant lawsuits with the purpose of improving the use of judicial authentication in marine pollution compensation cases.

2 THE CURRENT SYSTEM OF JUDICIAL AUTHENTICATION OF MARINE POLLUTION DAMAGES IN CHINA

2.1 THE CURRENT JUDICIAL AUTHENTICATION SYSTEM IN CHINA

In a narrow sense, the Judicial Authentication System refers to a procedural system for testing; more broadly, it also includes the systems of examining and applying the judicial authentication conclusions. As the following chapters will concentrate on the examination and application of authentication conclusions, the Judicial Authentication System mentioned here is confined to its narrow sense, including right of authentication determination, authentication subjects and the selection and appointment of expert witness. Due to the simplicity in traditional regulations for Judicial Authentication System and the defects in the operative system, the objectivity of the authentication report cannot be adequately guaranteed.[2] To counter this problem, the Chinese Judicial Department issued The Regulations for Judicial Authentication System Process (Trial) on 31 August 2001, and on 22 February 2002, the Chinese Supreme Court adopted the People's Court Commission Judicial Authentication Regulations. These measures started the process of reforming and perfecting the traditional Judicial Authentication System with the purpose of realizing justice and efficiency in modern lawsuit legislation.

2.1.1 Authentication Determination System

The Authentication Determination System mainly deals with the issue of who determines whether to conduct authentication or not, and under what circumstances the authentication should be conducted, or, in other words, how to start the authentication process. The two different legal systems have different regulations regarding this question. In those countries where the continental legal system is adopted, it is usually the judge who decides to conduct the authentication and commission expert witnesses, while in countries adopting the Anglo-American legal system, it is the parties themselves who take the decision and commission the expert witness.[3] China used to implement the authoritative authentication system,[4] but is now beginning to learn from other countries. It now grants

---

[2] Z. Liu and J. Tang, 'Assumptions on Judicial authentication System Reform and Legislation Improvement of China', (1995) 5 *Chinese Law* 22-27; H. Huang, 'Attempt to probe the Judicial authentication System Reform in China', (2001) 10 *People's Procurator* 20-22.
[3] F. Zhang, 'Looking at the Ownership of Expert Authentication Commission from the Comparison between the two types of Testimonies', (2000) 7 *People's Procurator* 19-20.
[4] Codes of Criminal Procedure of People's Republic of China, art. 119; Civil Procedure Act of People's Republic of China, art. 72; Administrative Procedure Act of People's Republic of China, art. 35.

two parties the right to apply for judicial authentication,[5] and allows the parties to commission the expert witness.[6] These regulations are undoubtedly more reasonable.

2.1.2 Authentication Subject System

'Authentication subject' refers to the subject that accepts the authentication commission on his own behalf, carries out authentication actions and is responsible for the authentication report. It is usually stipulated in other countries that the subjects of judicial authentication are either natural persons or corporations. In these countries, natural persons are essential and independent subjects for authentication and the corporations can only become authentication subjects under special circumstances.[7] For a long time, the Judicial Authentication Subject System in China had its own characteristics. The subject who accepted the commission and answered for the authentication report, and it was an authentication organization, rather than a natural person, that would carry out the authentication. That is, the subject being commissioned was always different from the subject who actually carried out the authentication. Consequently, the reasonableness of this system has long been questioned.[8] The People's Court Commission Judicial Authentication Regulation brought about a breakthrough in this system, making it clear for the first time that a natural person can accept, on his own behalf, a commission and carry out judicial authentication.[9]

2.1.3 Expert Witness Selecting and Appointing System

The 'Expert Witness Selecting and Appointing System' is the system that stipulates how to select the authentication subject, that is, an expert witness for a particular case. Different countries have different regulations for the possession and distribution of the rights to select and appoint an expert witness.[10] In China's past practices, it was the authentication organization that exerted the right, and the legal staff had no substantial right.[11] In order to change this unreasonable situation, the People's Court Commission Judicial Authentication Regulations draws on the successful experiences of France, Italy, Japan and other countries in establishing the Judicial Authentication Name-List System, returning the right of expert witness selection and appointment to the joint force of the judge and the parties,[12] thus complying with practices accepted worldwide. It is certain

---

[5] Supreme Court's Regulations about Civil Procedure Testimony, art. 25, 26; Temporary Regulations about People's Court Judicial authentication, art. 2.
[6] Supreme Court's Regulations about Civil Procedure Testimony, art. 28.
[7] F. Jian, 'Researches on Some Issues in Judicial authentication', (2001) 4 *Law and Medical Science* 235.
[8] J. Wang and J. Wu, 'Reviews on Basic Theories of Judicial authentication (2)', (2002) 5 *Forum for Law* 84-92.
[9] People's Court Commission Judicial authentication Regulations, art. 5.
[10] It varies from country to country. Refer to J. Wu, 'Researches on Some Issues in Judicial authentication', (2001) 4 *Law and Medical Science* 234-235.
[11] This situation results from the expert subject system of China. As the subject of testimony, the testimony organization accepts the commission and selects and appoints expert witness. Please refer to Judicial authentication Procedure Regulations (Trial), art. 14 and art. 20.
[12] People's Court Commission Judicial authentication Regulations, art. 3 and art. 5.

that the implementation of this system will have a far-reaching effect on the Chinese Judicial Authentication System.

2.2 SYSTEM OF JUDICIAL AUTHENTICATION OF MARINE POLLUTION DAMAGES

The system of judicial authentication of marine pollution damages is a kind of civil authentication and is subject to the judicial authentication system mentioned above. For those special issues that need authentication, the Maritime Court has the right to commission authentication; the party bearing the burden of proof may apply to the Maritime Court for authentication; each party can also commission expert witnesses himself.

Qualified authentication organizations and professionals can, with documentation, apply to the authentication department of the Maritime Court, and, after being examined by Courts, such organizations and professionals will be enlisted into the Maritime Court Expert Witness Name-List and granted the expert qualification. According to the principle of combining the choice of the parties and the appointment of the court, the Maritime Court will organize for both parties selection of the expert witness.

Compared with those relatively well-developed kinds of authentications, such as legal medical authentication, legal psychosis authentication, dactylogram authentication and handwriting authentication and so on, marine pollution damages authentication is still a new concept, having, for the time being, no uniform regulations for the methods and standards used in China. Thus, in each case it is up to the expert witness to choose proper materials, design experiments, build models and refer to the relevant criteria having been issued by the country.

For instance, in the case of judicial authentication of marine creature resource damages, factors taken into consideration include the type, feature and quantity of the pollutants; the drift, transportation route and effect of marine pollutants; locale and water situation; water function and the type of environmentally sensitive resources; the clearing of the pollutants, and so on.

When selecting evaluating factors, experts often choose those pollutants with high levels of influx and volume and strong toxicity (thus the greatest dangers), and usually adopt the national 'Standards for Seawater Quality' and 'Standards for Fishery Water Quality' when selecting the evaluating standards. Having surveyed the marine animal resources, the structure of fishery resources and the composition of seawater aquaculture, experts can design, according to the facts specific to each case, simulated oil film pollution experiments and experiments showing the toxicity of pollutants on marine animal resources. On the basis of analyzing and processing the data, in accordance with the calculation method for pollution damages disseminated by the State, and combining his own professional knowledge and experience, the expert witness can evaluate the impact on the environment from an economic point of view and make both quantitative and qualitative analyses of the pollution damages to decide the final cost of pollution damages.

3 ANALYSIS OF OBJECTIVITY AND AUTHENTICITY OF MARINE POLLUTION DAMAGES AUTHENTICATION REPORTS

As the technological basis of a lawsuit, judicial authentication plays a vital role in some complicated litigations for marine pollution damages compensation. Since the authentication report has a special authority, it should be scientific, objective and accurate. However, there is a possibility of distortion in authentication reports. In order to ensure the justice of law, it is necessary to further study the causes of authentication report distortion and ways in which distortion can be prevented and eliminated. Authentication report distortion is closely related to the features of judicial authentication and authentication reporting.

3.1 FEATURES OF JUDICIAL AUTHENTICATION

Judicial authentication is the activity by which an expert witness applies certain techniques and professional knowledge to test, authenticate and judge a specialized problem.[13] The process of authentication involves four phases: experiment-observe-explain-judge.[14] The subject matters of such experimenting and observation are authentication materials like the type and amount of pollutants, the body of marine animals affected by pollutants and the polluted seawater, and so on. The aim is to expose fully and to discover the features of these subject matters of authentication (materials or samples). At the stages of explaining and judging, the expert witness uses scientific theories to analyze the appraisable features, including what those features reveal and how they come into being. On this basis, the expert witness judges whether there are enough reasons to draw either positive or negative conclusions according to fixed standards and finally draws up an authentication report on whether the pollutants damage the seawater and animal resources, and, if so, to what extent they do. Through this authentication process, it can be seen that judicial authentication is a form of recognition, a subjective activity, during which the expert witness uses science, special experience or professional knowledge to make a factual judgement on professional problems.

Judicial authentication has scientific features, in that design of the experiment, modes of observation, selection of appraisable features, the grounds for explanation and standards for judgement are all determined with regard to scientific theories or special experience. At the same time, though, the authentication can be subjective, because the expert witness' intrinsic ideas or emotions may influence his observation, explanation and judgement. Thus, it is not possible to ensure objectivity. This can be the case in particular for judicial authentication of marine pollution damages, for there are no uniform authentication methods or judging standards, and therefore the authentication is more likely to be affected by individual subjective factors.

---

[13] Decisions of Standing Committee of National People's Congress About Judicial authentication Management (Draft), art. 1.
[14] Specifics of judicial authentication procedures. J. Wu, 'Researches on Some Issues in Judicial authentication (1)', (2001) 8 *Law and Medical Science* 168-169.

3.2 ESSENCE AND FEATURES OF THE AUTHENTICATION REPORT

3.2.1 Essence of the Authentication Report

The essence of judicial authentication determines the essence of the authentication report. Since judicial authentication is the activity in which scientific knowledge and special experience are employed to explain objective phenomena in order to draw a deductive conclusion, the authentication report is not a direct reflection of objective fact, but a deductive result of scientific explanation and judgement of the authentication materials. 'It is still an imitative fact even if it is taken as a evidential fact'.[15] The special regulations made for judicial authentication in some countries further demonstrate the particularity of its quality. 'British Law stipulates that the witness is supposed to tell the facts he observes without personal opinions which refers to conclusions drew from those facts [Sic]'.[16] In America, authentication reports are termed 'expert testimony'. According to American testimony rules, the expert testimony is different from general testimony in that it includes the experts' own opinions and is therefore called opinion testimony. Here the difference between 'personal opinion' and 'the facts he observes' lies in the fact that the former contains deduction as well as observation.

3.2.2 Features of the Authentication Report

Determined by its essence, the authentication report has the following features:

3.2.2.1 Dualism of Subjectivity and Objectivity

Since the authentication report is the conclusion of judicial authentication, subjectivity in judicial authentication results in subjectivity of the authentication report. Its objectivity is not negated, however, as the authentication materials and the scientific theories employed in authentication are objective, or, in other words, the subjectivity is restricted by objective facts and rules.

3.2.2.2 The Dual Tendency of Authenticity and Distortion

The scientific feature of judicial authentication determines that its conclusions are more reliable compared to other testimonies. For example, it is more reliable to determine seawater quality through data on oil content, dissolved oxygen (DO) and chemical oxygen demand (COD), than through observation with the naked eye. However, there is still a possibility of distortion in the authentication report because more elements are involved in the authentication process than general evidence-obtaining measures. Besides, in the course of experimenting, observing, explaining and judging, the witness' own subjective factors can also lead to distortion.

---

[15] J. Wu, *supra* note 14, 171.
[16] D. Shen, *British and American Acts of Testimony* (Zhong Xin Press, Beijing, 1996).

3.3 POSSIBILITIES FOR THE DISTORTION OF THE AUTHENTICATION REPORT AND THEIR REMOVAL

As an important kind of testimony in litigation, the authentication report should be objective and authentic. However, as is mentioned above, there is still a possibility of distortion. In order to ensure an equitable trial and judgement, it is necessary to eliminate such distortion in the authentication report.

3.3.1 Analysis of the Causes of Distortion

At different stages, the possible causes of distortion are as follows:

3.3.1.1 Inauthentic Materials

If the materials lack suitability, the authentication based on such inadequate and unreliable materials will certainly lead to a distorted conclusion. For instance, if the the aquaculture sample is sent in for testing after having been processed by the aqua-culturist rather than being taken directly from the polluted seawater, the authentication report will not be reliable.

3.3.1.2 Improper Process

Any improper treatment in different stages will lead to a distorted conclusion. This can be further divided as follows:

1 Experimental Distortion: caused by the witness' improper design or operation of authentication experiments.
2 Observational Distortion: caused by the witness' mistakes in observing materials or experiments.
3 Explanatory Distortion: caused by lack of scientific bases in the witness' explanation of authentication materials and/or experiment phenomena.
4 Judgemental Distortion: caused by improper standards that the expert witness uses.

3.3.1.3 Incorrect Theories used in Authentication

As there is no 'absolute truth' in the world and science itself may have falsehoods, the scientific theories applied in authentication may not be absolutely accurate. Moreover, scientific knowledge should be 'applicable' in the particular case, that is, it should be relevant to the case and thus apply to the facts in dispute.

3.3.1.4 Subjectivity of the Expert Witness

Different expert witnesses may hold different academic views and/or emotional tendencies in the authentication of the same case, which may affect the authentication activity and lead to distortions in its conclusions.

For the above-mentioned reasons, there is a possibility of authentication distortion. However, the authentication report is different from other forms of testimonies and its distortion is covert. The problems that the authentication report deal with and the scientific theories employed in authentication are too specialized to be understood by laypeople, including the judges. Meanwhile, people's belief in the reliability of science makes them overlook the difference between science and its application, and so they take the authentication report, which applies scientific techniques, as science itself. Such an attitude tends to result in blindness to its flaws.

3.3.2 Prevention of Authentication Report Distortion

Prevention of authentication report distortion requires the removal of the distortion-inducing factors and avoidance of distortion in the course of authentication. With the analysis of the aforesaid reasons which lead to authentication report distortion, the following measures can be taken to prevent it:

3.3.2.1 Selection and Appointment of Proper Expert Witness

Firstly, the expert witness should have the relevant professional qualities and knowledge, which can be ensured by use of the Expert Witness Name-List System.

Second, the expert witness should be objective and disinterested, without a strong subjective tendency. To achieve this, both parties and the judge should enjoy a joint right of expert witness selection and appointment.

Finally, the expert witness should have no conflicts of interests in the case. At present, a system is used to avoid a conflict of interests of the expert witness to ensure the neutrality of the expert witness.

3.3.2.2 Strict Conformance to the Regulations and Procedural Rules

Owing to subjectivity, in the course of authentication it is inevitable for the expert witness to make errors, and even mistakes or faults. A strict conformance to the regulations and procedural rules of authentication could to some extent avoid mistakes and realize the justice of the procedure.[17]

Firstly, before the authentication, the expert witness should examine the materials and refuse those either not suitable for authentication or not conforming to requirements, in order to eliminate the distortion of authentication materials.

Second, there should be two or more expert witnesses in the process of the authentication of one item. When samples need to be taken on the spot, the client and other people involved should be informed and be present at that time. In addition, the experiment results obtained, and the standards used, should be contained in the judicial authentication report.

---

[17] At present, there are no specific regulations for the procedures of Marine Judicial authentication and relevant authentication activities can refer to Judicial authentication Procedure Regulations (Trial), Legal Department, 2001, art. 92.

3.3.2.3 Judicial Authentication Review

Before being formally presented, the authentication report should be checked by an senior expert witness, with an advanced technical post, for the purpose of avoiding distorted conclusions resulting from the subjectivity of the expert witness.

4 THE ROLE OF THE AUTHENTICATION REPORT IN LITIGATION FOR MARINE POLLUTION DAMAGES COMPENSATION

4.1 EVIDENTIAL EFFECT OF THE AUTHENTICATION REPORT

In litigation for Marine Pollution Damages Compensation, one of the most difficult parts is to determine the extent and degree of the pollution damage. On the one hand, the interests of the aggrieved party should be protected and prompt and full compensation should be made. On the other hand, it should also be noted that some fishermen and aquaculturists see pollution accidents as money-making opportunities and deliberately make false statements of their losses to obtain illegal profits. As has been mentioned, determination of damages requires professional knowledge and special experience in marine physics, chemistry and biology, and so it is wise to commission professionals to perform judicial authentication, which is vital to clarify the facts of the case and to settle it properly.

For instance, in a case of an oil spill from a foreign ship, the claimant sought compensation of RMB 10 million, taking into consideration pollution damages to seawaters and marine animals, as well as the clearance fees. The expert witness, though, using formulae to calculate the amount of oil spill, made an evaluation on the effect of the petroleum hydrocarbon on the marine environment, marine animals and marine ecology, and concluded that the cost for clearance and compensation was only approximately RMB 2 million.[18] In another case, the defending party insisted that there were no damages. The expert witness made biological determinations and oil film pollution simulated experiments in two fishery bases and applied the joint method of marine biological determination and the oil spill extension mathematical model to evaluate the influences of pollutants on marine animals and the fishery environment. The demonstration of the existence, extent and degree of pollution damages in the authentication report resulted in negotiations between the two parties based on the facts and finally they came to an agreement that stipulated a total compensation of USD 600 thousand.[19]

It has been shown that judicial authentication in litigation for marine pollution damages compensation is an effective measure in solving technical problems and that clarifying facts with the help of science, technology and professional knowledge, plays an important role in lawsuits. Owing to its high value as proof, judicial authentication is

---

[18] Please refer to the *Authentication Report of Shandong Marine Judicial authentication Center*, Shandong Marine Department, 2002, no. 16.

[19] 'Hai Li' Oil Spill Pollution Damages Compensation Case, Chapter II of the *Attachment to China Ships Oil Spill Emergency Plan*, Marine Bureau of People's Republic of China, March 2003.

termed the 'king of testimony' and as a 'scientific guard' for the justice of law.[20] However, as the scientific features and authority of the authentication report were emphasized, it should also be recognized that there are some limits to the evidential value thereof.

The legal effect of the authentication report varies from country to country. In countries of the Anglo-American legal system, the parties appoint their own expert witnesses and there is no such a concept as an 'authentication report'. The expert advice known as expert testimony is a kind of testimony and does not bind the judge. In the continental legal system, the expert witness is often appointed by the judge and carries out authentication activities according to legal authorization. For example, in France the old Roman proverb 'an expert witness is the judge of the facts' is still widely followed, while in Japan, it is stipulated that the authentication report restricts the judge and that the judge should give explanations in his verdict if he does not adopt the authentication report.[21] According to Chinese law, the authentication report is parallel to witness testimony, and only after they have been examined in court can they be taken as proof. If the parties dissent from the opinion, and if the legal conditions are satisfied, the special issue should be re-authenticated.[22] Once taken as proof by the court, however, the authentication report has greater power than ordinary written documents, audio evidence and witness' oral testimonies.[23]

4.2 STATUS OF THE EXPERT WITNESS IN LITIGATION

The status of the expert witness varies in countries of the two different legal systems, depending on the respective procedural system. In countries of the Anglo-American legal system, the expert witness has the same status as a witness in a lawsuit. The only difference is that expert witness should have relevant knowledge or experience. Unlike ordinary witness, though, an expert witness is given the right to state personal opinions, that is, an expert witness may state his own views and give conclusive suggestions, setting forth the grounds for the conclusions, to the judge and the jury.[24] Continental law distinguishes between ordinary witnesses and expert witnesses, and grants a higher position to the latter.[25] For instance, in France, the expert witness is called the 'scientific judge', while in Germany, the expert witness is deemed an assistant of the judge.[26]

The Chinese procedural system also distinguishes between ordinary witnesses and expert witnesses, and grants an expert witness the rights related to authentication activities, such as checking records, interrogating the parties, taking part in surveys and

---

[20] Judicial authentication – King of Testimonies. <http://www.southcn.com>.
[21] P. Tian, 'On the Expert Authentication Conclusions', (2000) 6 *Modern Law* 26-27. L. Yang, 'On the Effect of Expert Authentication Conclusions in Procedures', (2001) 4 *University of Electronics and Technology Journal (Social Science)* 66-69.
[22] Supreme Court's Regulations about Civil Procedure Testimony, 2001, 33, art. 27.
[23] *Ibid.*, art. 77.
[24] P. Tian, *supra* note 21, 26-27.
[25] The expert witness has the right to check the files and materials related to the case, he could also take part in checking on the spot, inquiring of the parties and requesting people involved to give explanations.
[26] J. Wang and J. Wu, *supra* note 8, 86.

running simulated experiments, and so forth.[27] Meanwhile, the expert witness should also take on corresponding responsibilities. He should appear in court, present an 'Expert Witness Certificate', state clearly the authentication procedures, explain the conclusions, subject himself to cross-examination, and answer the questions with objectivity and disinterest.

Obviously, the status of expert witness in the Chinese legal system is different from those of the Anglo-American law system and of the continental law system. Following conscience and sense, the expert witness in the Chinese legal system should hold an objective and disinterested stand and apply professional knowledge to help the judge determine the facts. The independent status of the expert witness is a premise of the efficient operation of the judicial authentication system.

5 CONCLUSION

In conclusion, in litigation for marine pollution damages compensation, judicial authentication and the authentication report play vital roles in determining the extent and degree of the damages and in clarifying the facts. However, the authentication report should not be equated with scientific truth and be trusted blindly. Both the responsible party and the aggrieved party in a marine pollution accident should be aware of the importance of judicial authentication and use it to protect their own legal rights. Owing to the suddenness of marine pollution accidents, in the special geographic environment of the sea, once the pollutants enter the marine environment, they will be influenced by the particular temperature, wind direction, wind speed, tides, waves and ocean currents, in being diluted, extended and transported. Since the conditions of oceans sometimes change, it becomes difficult to investigate and obtain evidence following an accident, bringing obstacles to damage evaluation. Therefore, in marine pollution accidents, both parties should try to take and protect the evidence by way of recording and taking pictures at the time, for the purpose of presenting adequate and reliable materials for a future judicial authentication to draw an objective and just authentication conclusion.

---

[27] Civil Procedure Act of People's Republic of China, no. 72; Regulations for Expert Witnesses Management, art. 28.

LIST OF REFERENCES

H. Huang, 'Attempt to probe the Judicial authentication System Reform in China', (2001) 10 *People's Procurator* 20-22.

F. Jian, 'Researches on Some Issues in Judicial authentication', (2001) 4 *Law and Medical Science* 235.

Z. Lin and J. Tang, 'Assumptions on Judicial authentication System Reform and Legislation Improvement of China', (1995) 5 *Chinese Law* 22-27.

D. Shen, *British and American Acts of Testimony* (Zhong Xin Press, Beijing, 1996).

P. Tian, 'On the Expert Authentication Conclusions', (2000) 6 *Modern Law* 26-27.

J. Wang, 'Researches on Some Issues in Judicial authentication (1)', (2001) 3 *Law and Medical Science* 168-169.

J. Wang, 'Researches on Some Issues in Judicial authentication (2)', (2001) 4 *Law and Medical Science* 234-235.

J. Wang and J. Wu, 'Reviews on Basic Theories of Judicial authentication (2)', (2002) 5 *Forum for Law* 84-92.

J. Wu, 'Researches on Some Issues in Judicial Authentication', (2001) 4 *Law and Medical Science* 234-235.

J. Wu, 'Researches on Some Issues in Judicial Authentication (1)', (2001) 8 *Law and Medical Science* 168-169.

L. Yang, 'On the Effect of Expert Authentication Conclusions in Procedures', (2001) 4 *University of Electronics and Technology Journal (Social Science)* 66-69.

F. Zhang, 'Looking at the Ownership of Expert Authentication Commission from the Comparison between the two types of Testimonies', (2000) 7 *People's Procurator* 19-20.

THE NECESSITY OF INCORPORATING MARINE ENVIRONMENTAL TORTS INTO THE MARITIME LEGAL SYSTEM

Li Zhiwen and An Shouzhi

1 INTRODUCTION

Economic development in the 21st Century depends, to a great extent, on the sea. With further study of the resources of the sea and of inland rivers, the variety of economic activities which relate to the sea are increasing. Besides the traditional industries like transportation, the fishing trade and fisheries, there are other new industries like exploitation of oil and benthic mineral resource which are ballooning in size. Nonetheless, marine environmental torts at times occur in the utilization of marine resources. Some activities, such as discharge of large amounts of sewage and industrial wastes into the sea, in addition to other activities destructive of marine environments, result directly in frequent ocean disasters and the gradual deterioration of marine environment. This is especially so along the shores of China. Therefore, preventing and reducing pollution of the marine environments becomes an urgent matter for legislators to resolve. For that particular purpose, both the Civil and the Marine Environment Protection Laws have been regulating certain marine environmental torts from different aspects. However, since such marine environmental torts may involve various victims and cause serious damages, the lack of a relevant compensation system in either the Civil or the Environment Protection Laws means that, in actuality, they cannot achieve full, or even sufficient, compensation for damage caused by such torts. This insufficiency can be remedied, particularly through further regulatory measures and a compensation system for marine tort law.

2 DEFINITION AND TYPES OF MARINE ENVIRONMENTAL TORTS

A marine environmental tort is one which occurs on account of, for example, industrial activities or from other non-natural sources, causing the marine environment to be polluted or otherwise damaged, and potentially resulting in damage to public or private property, personal injury and/or harm to general environmental interests. Human activities have numerous consequences, many of which may result in a decline in the quality of the marine environment and a worsening of the normal existence and evolution of both human and other forms of life.[1] The ultimate difference between marine environmental

---

[1] M. Cao, *Tort Law of Environment* (Publishing House of Law, Beijing, 2000), p. 9.

torts and other torts is that the former prejudices the rights of unspecified persons in the specific area of environmental interests.[2]

2.1 CHARACTERISTICS OF MARINE ENVIRONMENTAL TORTS

Specific differences between marine environmental torts and other general tort are the inherent explanation for the need to cover marine environmental torts within the marine tort law system. The torts characteristics include:

1. Immensity of harmful consequences. With the development of economy, the exploitation and utilization of the ocean has continually expanded in scale, with the carriage of petroleum, the construction of benthic tunnels, the laying-down of benthic optical cables and so forth growing daily as technologies become ever more advanced. In the case of any accident happening during such economic activities, not only are the rescues difficult, but the scope of damage may also extent beyond effective control. Personal injuries, tremendous property losses and endangering of the marine environment are all potential consequences. An example of such extensive damage can be seen in the 1989 grounding of the oil tanker '*Exxon Valdez*' at Prince William Gulf in Alaska. It leaked a mass of petroleum, leading to losses of USD 8 billion. Furthermore, the deleterious effects caused to the local environment take many years to fully eradicate.[3]

2. Long-term effects of damage. Pollution of the marine environment often has impacts on the continued exploitation and utilization of the ocean and the securities of the marine environment. For instance, almost half of the sea area of the Bo Ocean in China has been polluted, and the benthic biological resources have been reduced to a tenth of their level in the 1950s. Ocean experts have warned that the environmental pollution in the Bo Ocean has now reached a critical point. If decisive countermeasures are not taken immediately, the Bo Ocean may well become the first 'dead ocean' on the earth within ten years.[4]

3. Combined private and social effects. Marine environmental torts are complex, emanating from proliferate sources. Pollution, then, may prejudice several kinds of rights and interests enjoyed by numerous non-specific persons, albeit in specific areas and at the same time. Thus, it can be difficult or even impossible to distinguish the infringer from the victim. Occasionally the infringer himself is, at the same time, the victim. For example, in 2002, red tides occurred nearly ten times in the sea areas of Dapeng and Shenzhen gulfs in Shenzhen. The reason for the high frequency of such tides was the large discharge of sewage into the Shenzhen gulf, the discharge point for the wider sea. The sewage, produced by millions of residents in Shenzhen, was rich in phosphorous, and coupled with the slowness of the water exchange,[5] such tides continued to occur.

---

[2] *Ibid.*, p. 19.
[3] <http://www.epa.gov/oilspill/exxon.htm (31 May 2004)>.
[4] <http://www.riel.whu.edu.cn/news.asp?sort=2&npage=7> (24 May 2004).
[5] <http://www.cenews.com.cn/news/2004-05-21/35128.php> (30 May 2004).

2.2 CATEGORIZATION OF MARINE ENVIRONMENTAL TORTS

Considering different categories of causal behaviour and contamination, the diverse forms of marine environmental torts can usually be divided into (1) a tort of oil pollution damage from ships, like the spillage, leaking and/or discharge from a ship due to perils of the sea during its operation; (2) pollution from ocean exploitation, such as pollution of the marine environment caused during the reconnaissance and exploration of seabed mineral resources, or the related disposal of rubbish or other substances produced during the process on the sea; (3) pollution from the discharge of land-sourced pollutants, including discharge of sewage and industrial waste products; (4) pollution through dumping of wastes, including residual waste from human activities both inland and on the sea; (5) the pollution by coastal construction projects, including that caused by exploitation and utilization of the coastal zone, as well as of ports and (6) the act of damaging the sea in the course of its exploitation and utilization, namely, acts resulting in variance of the natural environment of the sea areas, coastal evolution and ecology-damaging actions such as dredging and sand mining which can aggravate the erosion of the coast.

3 ANALYSIS OF THE DEFICIENCY OF THE STATUS IN THE CIVIL REMEDIES FOR MARINE ENVIRONMENTAL TORTS IN CHINA

3.1 THE STATUS OF CIVIL REMEDIES FOR MARINE ENVIRONMENTAL TORTS REGULATED BY LAW IN CHINA

In China, the applicable civil remedies for marine environmental torts which are regulated by law include: the General Principles of the Civil Law of PRC, the provisions of pollution prevention and control, as in the Environment Protection Law, the Marine Environment Protection Law, the Air Pollution Prevention and Control Law, the Water Pollution Prevention and Control Law, the Solid Waste Pollution Prevention and Control Law, and regulations regarding natural resources, like the Fishery Law, the Water Law, the Water and Soil Conservation Law and the Sea Areas Employment and Administration Law. Besides these, the Civil Procedure Law and The Supreme People's Court Regulation on Civil Trial Evidence (hereinafter called the Evidence Rules) provide procedural direction.

The Civil Law provides the principles of civil liability for a wide variety of torts, including marine environmental torts. Any person who pollutes the environment and causes damage to others in violation of the provisions for environmental protection and the prevention of pollution are civilly liable regardless of fault. The main remedies under the liability for environmental torts are cessation of infringements, removal of obstacles, clearance of dangers, return of property, restoration to the original condition, compensation for losses and so on.[6]

---

[6] § 2 and 3 of art. 106; art. 123; art. 124, and art. 134 of the General Principles of Civil Law.

Among the environmental pollution prevention and control laws, *the Marine Environment Protection Law* provides that the doctrine of no-fault liability applies to all environmental pollution cases.[7] Namely, one who has caused or may cause pollution damage to the marine environment must pay for the expenses incurred in connection with the eradication of such pollution, as well as compensate any losses to the government as well as individual or individuals who suffer as a result of the pollution of the marine environment. Under Article 92, infringers are not liable for pollution damage to the marine environment caused by act of war, unavoidable natural disasters or on account of the negligence (or other wrongful acts) of the departments responsible for the maintenance of beacons and other navigational aids in exercising those functions. Article 55 of the Water Pollution Prevention and Control Law provides that a tortfeasor who has caused an environmental pollution hazard has the obligation both to eliminate it and to compensate those who have suffered direct losses, except losses as may have been caused by the victim's own actions. Both the Environment Protection Law and the Solid Waste Pollution Prevention and Control Law provide that persons or interests that suffer losses from pollution are entitled to claim for damages. In addition, the Evidence Rules stipulate that if any person causes damage to other people as well as the environment by engaging in operations that are greatly hazardous, the burden of proof in the course of any proceedings shall be reversed.[8]

3.2 ANALYSIS OF CHINA'S CURRENT CIVIL REMEDY SYSTEM FOR MARINE ENVIRONMENTAL TORTS, AND ITS DEFICIENCIES

1 Article 124 of the Civil Law adopts no-fault liability for torts of environmental pollution, meaning that no subjective fault is required of the person made liable. However, the tortfeasor is not held responsible for compensation for the loss unless he violates an environmental protection or pollution prevention regulation. Nevertheless, not only are such regulations contrary to the general approaches to tortious liability (e.g., in the rules of Japan, Germany, America and the Taiwan Prefecture of China there may be civil liability, regardless of compliance with public law standards), but they also conflict with certain regulations in China's own environmental legislation.[9]
2 There is a general lack of a regulated compensation liability guarantee system, for example to establish limitation on amounts of compensation liability; financial securities or warrants; compulsory insurance; or a compensation or indemnity fund. In China, the general principle of total compensation in civil liability is applicable to marine environmental damage compensation. However, as for a compensation guarantee system, although the Regulation on Management of Offshore Petroleum Exploration and Exploitation and Environmental Protection provides that operators should set up certain financial guarantees in order to ensure compensation for potential pollution damage, neither limitation amounts of such guarantees nor limitation

---

[7] Art. 91 of the Marine Environment Protection Law.
[8] See art. 4 of The Supreme People's Court Regulation on Civil Trial Evidence.
[9] M. Wang, *Legal System of Relief of Tort of Environment* (China Legal System Publishing House, Beijing, 2001), p. 279.

amounts for compensation are covered.[10] As a result, when the infringer is incapable of payment or is unknown, victims are unable to obtain a full or prompt compensatory remedy.

3 The legislation which relates to marine environmental torts is too general to allow for victims to seek civil remedies. The Marine Environment Protection Law makes it clear that the legislation purpose is to protect and improve marine environment, protect marine resources, prevent pollution damage, maintain an ecological balance, protect the health of population and promote sustainable development of the economy and society. The law, though, merely regulates administrative liability for dumping wastes, like orders to improve within a defined and limited time and associated penalties, it does not include regulations on civil damage compensation, meaning victims of marine environmental torts are unable to get fully effective legal remedies.

4 Although infringers have had the responsibility to cease such infringement, remove any obstacles, eliminate any danger of causing environmental torts or any associable torts imposed upon them, the law falls short of providing any relevant supporting procedural regulation. Therefore, the judicial remedies are not perfect.

Above all, it is submitted, the deficiencies of present civil legal remedies in response to marine environmental torts in China can mainly be embodied in the fact that such remedies are still rooted in traditional tort law, and that recent developments are limited to aspects of no-fault liability, the inversion of the burden of proof, causation and responsibility, which do not present any breakthroughs compared with regulation of the general tort law. The reason for this is that legislation of civil remedies in environmental tort law mainly stipulates adherence to the individualized 'polluter-pays' principle, which does not take the social effects of marine environmental torts or any externalization of marine environmental tortious liability into account. The ideological reason behind such a deficiency is a trend towards concentration on economic growth, setting economic development against environmental protection, meaning that parties to marine economic activities are unable to get either an overall or prompt protection from damage. As a result, creating the sustainable development of China's marine economy has been greatly (negatively) affected.

4 THE NECESSITY TO INCORPORATE MARINE ENVIRONMENTAL TORTS INTO THE MARINE TORT LAW

4.1 THE TREND OF EXPANDING THE SCOPE OF MODERN MARINE TORT LAW

Marine tort law consists of the generalized legal norms to regulate tortious behaviour that infringes on the rights of persons and property at sea. Traditional marine tort law limits

---

[10] Art. 9 states simply 'An enterprise, an institution and operators shall have civil liability insurance or other financial securities'.

the torts 'at sea' to torts emanating from a vessel's maritime transport.[11] However, with increasingly active use and exploration of the seas, the scope of such application of marine tort law has expanded. In addition to ships' torts, other sources have been included. This has included, for example, sources such as production and operation at sea, which, directly or indirectly, put the material into the ocean, thus causing the natural environment and oceanic ecosystem to be polluted or destroyed and resulting in damage to the personal or property rights of third parties, marine environmental rights or to public properties.

Firstly, it can be seen that in modern times, activities at sea have become gradually more diverse and complicated. In addition to traditional transportation by sea, activities such as mining for minerals in the seabed, marine oil and gas exploration development and dumping of wastes in the sea are also prevalent. The equipment from artesian boring, and chemical materials used in these activities; poisonous materials, oil and other wastes from the drilling process to explore at sea; leakage from oil pipelines and well sprays; abandonment of equipment and constructions at sea as well as other forms of abandoned waste and other materials into the sea will all result in the pollution of the marine environment, and seriously infringe on rights of persons, both collectively and individually. These harmful activities have the same characteristics as the traditional 'ships' tort': great harm, complexity of causation, the special principle of imputation and compensation mechanism.

Second, the Certain Provisions of the Supreme People's Court Concerning the Scope of Cases Accepted by Maritime Court (hereinafter called the Provisions on the Case Scope) has definitely brought marine environmental torts within the competence of Maritime Courts.[12] In judicial practice in China, Maritime Courts not only take up the marine torts that involve ships, but also types of cases that involve marine pollution as well, including pollution of the sea caused by coastal projects, prospecting and exploitation of marine petroleum and the dumping of wastes, and so forth.[13]

It is obvious that the realistic demands for utilization of marine resources by humans and thus increased oceanic activities does help to extend marine tort law beyond the former category of ships' torts, towards objective coverage of marine environmental torts.

4.2 THE AVOIDANCE OF CONTRADICTION IN THE SYSTEM

In China, there are currently several kinds of civil law mechanisms used to settle marine environmental issues in terms of civil relief for common torts, meaning that the indemnification system for marine environmental torts has become different from the marine tort law, and resulted in conflicts of laws. For instance, the liable party in marine environmental law is most commonly the person(s) who *caused* pollution damage to the marine environment. However, if such damage is wholly due to actions of a third party,

---

[11] Y. Si, *New Maritime Law* (Dalian Maritime University Publishing House, Dalian, 1999), p. 4; R. Deng, *Research on basic theoretical issues of law of tort on vessels* (Publishing House of Law, Beijing, 1999), p. 13.

[12] Certain Provisions of the Supreme People's Court Concerning the Case Scope of Maritime Courts Accepting, arts. 5; 33-34; 36.

[13] <http://www.china.com.cn/chinese/2002/Sep/209009.htm> (31 May 2004).

who acts with intent or is at fault for causing damage to the marine environment, then the third party will be civilly liable. This can be compared with marine tort cases in which damage is caused by ships spilling oil, or drainage and so forth, when, according to the International Convention on Civil Liability for Oil Pollution Damage 1969 (and its subsequent protocol), it is the shipowner who is civilly liable. Furthermore, the Protection and Indemnity Clubs or insurance companies, which insure against civil liability for oil pollution damage, may also face civil liability for compensation for such oil pollution damage. According to the provisions of the International Convention on the Establishment of an International Fund for Compensation for Oil Pollution Damage 1971, the fund is also subject to civil liability for compensation of oil pollution damage. Nevertheless, a third party is only subject to liability when he acts with intent to cause such damage. Obviously, in the current law of China the same types of marine environmental torts may attract different legal approaches, and thus the law is not being applied uniformly. Therefore, bringing marine environmental torts into the field of marine tort law would effectively avoid such conflicts of the legal systems which currently concern the various marine environmental torts.

4.3 PROTECTION OF THE LEGAL RIGHTS AND INTERESTS OF THE PARTIES CONCERNED

4.3.1 The marine tort legal system is useful for the development and utilisation of economic activities at sea

General tort law applies the so-called 'personal liability' principle, namely that the infringer undertakes responsibility by himself. However, the special nature of marine torts means that the person liable may be separate from that person whose conduct was at fault.

This is most clearly manifested in the example of employer's liability and liability insurance systems in marine tort law, which result in the transfer of liability so that employer, having a special connection with the nominal 'tortfeasor', or otherwise insurers, assume the liability. Moreover, in some oil pollution liability instances, oil pollution insurers or financial guarantors may still undertake direct liability for compensation.[14]

Again, compared with the general tort law, the advantage of the system in which the subject bearing liability may be separate from the party whose actions caused the damage, is not only that it can promote innovation in employees' economic activities at sea, but it can also ensure that the victim of a marine accident is able to obtain full and prompt compensation.

4.3.2 The special system of compensation in marine tort law contributes to sufficient compensation, and to the development of the ocean

Many special risks exist in the carriage of goods by sea and operations at sea, and so, objectively, a special system of tort compensation to enforce limited compensation, different from the common civil compensations for loss, must be established, so as to promote the development of the economy of the sea. To deal with the problem that marine

---

[14] See the Special Maritime Procedure Law of the People's Republic of China, art. 97.

torts may cause serious damage that is beyond the ability of an infringer to compensate, and to settle compensation with in reasonable delay, marine tort law has established a unique system of compensation which joins limitation of liability, compulsory insurance and reserve funds. Thus, on the one hand, it effectively encourages operators to exploit the ocean while, on the other hand, it can effectively compensate any losses suffered by the injured, and it reasonably distributes the risks of operation at sea. The incorporation of marine environmental torts into this field of marine tort law, which has the special system of compensation, will surely help to protect the rights and interests of all parties, while maintaining the interests of the country, as well as the utilisation and protection of the ocean's resources.

4.4 FACILITATION OF JUDICIAL REMEDIES OF MARINE ENVIRONMENTAL TORTS

Until the 1980s, understanding of justice and the environment was still confined to the settlement of disputes caused by environment pollution. However, in 1992, the United Nations Conference on Environment and Development passed several important resolutions, including the 'Declaration of Rio de Janeiro' and the '21st Century Agenda', and since then sustainable development has become a focus in countries around the world. Sustainable development requires countries to expand the scope of judicial action, and to establish a perfect legal system for civil liability in the field of environmental resources, as well as to enlarge and refine civil actions. Nowadays, then, there is an urgency for the environmental justice and sustainable development in China to solve the following two problems: (1) little observance of the law, slack law enforcement and lack of penalties imposed on the infringers in the environmental justice due to the localism of protection and excessive administrative intervention, and (2) the reinforcement and perfection of a system of environmental justice and sustainable development[15] from the central to local authorities.

Today, maritime courts accept a large number, and various types of, cases of marine environmental torts, and also achieve good results. However, due to the effect of the attribution of powers regarding marine environmental torts, the existing judicial remedies still have problems. For instance, Articles 4 and 5 of the 'Provisions on the Case Scope' only clarify that the maritime courts have jurisdictional power over environment pollution cases in which the source of pollution is vessels, ports, oceans and/or the operation of installations in navigable waters. Chinese law states that the jurisdictional power of maritime courts is only in disputes related to shipping and sea-going ships, and thus it causes confusion as to the competence and jurisdiction of the maritime courts. In judicial practice, some disputes regarding marine environmental torts are dealt with by other courts that often take local protection into consideration when considering them. The main reason for this is that the enterprises which have caused the pollution also simultaneously contribute much to the local economy and finances. As such, though, there is a sharp and dramatic contradiction between partial interests of localities where the pollution occurs and the general interests of sustainable development and ecological protection. Maritime

---

[15] See S. Cai, *Sustainable Development and Legal System Construction of Environment Resources* (China Legal System Publishing House, Beijing, 2003), p. 557.

courts have many advantages in terms of their set-up and functioning, and have a sound ability to resist the localism of protection. Only by incorporating marine environmental torts into marine tort law, thereby defining the jurisdictional power of maritime courts, can the localism of protection be eliminated, and the justice element in the trial of such cases be guaranteed.

In addition, although there *are* enactments regarding eliminating damage caused by infringers of environment torts and other torts that can be linked to the environment in current Chinese law, there is still no procedure for settlement in civil actions. Maritime injunctions, as set up in the Special Maritime Procedure Law, are helpful in eliminating dangerous behaviour and marine torts. The maritime claimant may apply to a maritime court located in the place where the dispute takes place, or indeed the court in which he brings an action, for an order for either the tortfeasor's action or his inaction before commencing an action or during proceedings.[16] As such, the legal interests of the plaintiff would be fully protected.

5 CONCLUSIONS

With the development of society, humans have utilized the oceans and engaged in activities related to them using increasingly different means, causing an conflict between protection of environmental resources and economic development. In accordance with the international treaties which China has concluded or has acceded to, domestic laws, and realities of practice for the protection of ocean resources, there should be a new understanding of marine tort law, which expands the traditional 'vessels, shipping and port operation' to 'protection of the marine environment and exploration and utilisation of the marine resources'. Marine environmental torts should be incorporated into the scope of the marine tort law, which is also the basis of the jurisdiction of the maritime courts. Maritime courts should be given the jurisdiction in cases of marine pollution both specifically and generally. Thus, not only would the protection system of marine environment be perfected, but there would also be an effective measure to overcome the localism of protection, prevent marine pollution, and compensate the victim. Moreover, it would harmonize the economic development of our country as well as promote the safety of the marine environment, while promoting sustainable development.

---

[16] See Special Maritime Procedure Law, arts. 51-53.

LIST OF REFERENCES

S. Cai, *Sustainable Development and Legal System Construction of Environment Resources* (China Legal System Publishing House, Beijing, 2003).

M. Cao, *Tort Law of Environment* (Publishing House of Law, Beijing, 2000).

R. Deng, *Research on basic theoretical issues of law of tort on vessels* (Publishing House of Law, Beijing, 1999).

Y. Si et al., *New Maritime Law* (Dalian Maritime University Publishing House, Dalian, 1999).

M. Wang, *Legal System of Relief of Tort of Environment* (China Legal System Publishing House, Beijing, 2001).

THE LEGAL PROTECTION OF THE MARINE ENVIRONMENT IN CHINA: CURRENT SITUATION AND CHALLENGES

Guo Ping and Zhao Lujun

1 INTRODUCTION

Since the 1980s, the environmental protection of marine and inland waterways has drawn the attention of the Chinese government and relevant authorities daily. As a result, special laws and regulations have been enacted on marine environmental protection, prevention and control of pollution, and prevention of polluting the ocean by exploration and development, ships and from offshore projects and land-source pollutants. The legal system of marine environmental protection in China has already principally been set up. However, the problems are still so serious that pollution accidents occurring in different waters are dealt with according to different laws and regulations, even the type of a ship being possibly a decisive element of application of the law. These not only cause the confusion of application of the law in practice, but also make the legal enforcement authorities shift responsibility onto or interfere with each other.

This paper introduces the current situation about the legislation on the marine environmental protection in China, and analyzes the existing problems and proposes solutions as well as a train of thought.

2 CHINESE LEGISLATION ON THE MARINE ENVIRONMENTAL PROTECTION

2.1 LAW OF ENVIRONMENTAL PROTECTION (LEP)

In order to protect and improve the living environment and ecological environment, prevent and reverse the effects of pollution and other public hazards, ensure the health of people, promote the development and modernization of society, the 11th Meeting of the Standing Committee of the Seventh National People's Congress adopted the LEP, which was promulgated on 26 December 1989, implemented by the No. 22 Decree of the President on the same day. This is the basic legislation on environmental protection in China.

'Environment' as defined by Article 2 of the LEP refers to a totality of various kinds of natural and artificial factors influencing human survival and development, including atmosphere, water, ocean, land, mineral deposits, forest, grassland, wild living things, natural traces, trace of humane race, nature protection areas, scenic spots, cities and villages, and so forth. Article 3 of this Law provides that its scope of application includes coastlines and other sea areas under the administration of the People's Republic of China. It also elucidates and defines the supervision and administration and the protection and improvement of the environment, the prevention of pollution of the environment, and other public hazards as well as legal liability. Article 7 provides that the environmental protection administrative authority under the State Council is responsible for the integrated supervision and administration of the implementation of national environmental protection.[1] In addition, this State Council authority, the authority of harbour superintendence, the authority of superintendence in fishing administration and fishing ports, military departments of environmental protection, competent authorities of public security at all levels and the authorities of traffic, railway, civil aviation exercise supervision and control over the prevention and control of pollution of the environment in accordance with the relevant laws.

In fact, prior to the LEP, some regulations had already been promulgated successively, such as the Law of Marine Environmental Protection of 1982, the Law on Prevention and Treatment of Water Pollution of 1984 (hereinafter referred to as 'LPTWP'), as well as the Law on Air Pollution Prevention of 1987 (hereinafter referred to as 'LAPP').[2] After the promulgation of the LEP, China promulgated the Law of Preventing and Curing Environmental Pollution by Solid Wastes.[3] These pieces of legislation are branches of and supplements to the basic legislation of the LEP.

2.2 LAW OF MARINE ENVIRONMENT PROTECTION (LMEP)

On 23 August 1982, the 24th meeting of the Standing Committee of the Fifth National People's Congress adopted the first piece of legislation on marine environmental protection in China, that is, the LMEP.[4] Article 2 stipulates clearly that it only applies to domestic waters,[5] territorial waters, contiguous zones, the exclusive economic zone, the continental shelf, and other waterways under administration. Obviously, the LMEP only applies to the sea area. The question of prevention and control of pollution in waterways like rivers, lakes, canals, channels, reservoirs and so forth, shall be adjudicated by other laws and regulations.[6] Meanwhile, Article 5 contains stipulations on the division of duties

---

[1] The State Environment Protection Administration (SEPA) is directly in charge of the supervision and administration of environment protection according to the reform plan approved by the first meeting of Ninth National People's Congress.

[2] The LAPP was adopted by the 22nd meeting of the Standing Committee of the Sixth National People's Congress on 5 September 1987, amended by the 15th meeting of the Standing Committee of the Eighth National People's Congress on 29 August 1995. It came into force from 1 June 1988. It provides that if the ship causes the air pollution, it should be under the separate supervision of traffic authorities and fishery authorities.

[3] It was adopted by the 16th meeting of the Standing Committee of the Eighth National People's Congress on 30 October 1995 and comes into force from 1 April 1996.

[4] It was amended by the 13th meeting of the Standing Committee of the Ninth National People's Congress on 25 December 1999, came into force from 1 April 2000.

[5] According to art. 95 of the LMEP, domestic water means the sea areas face to the main land from territorial datum line.

[6] According to art. 45 of the LPTWP, Implementing Rules on the Prevention and Treatment of Water Pollution of the People's Republic of China had been approved by State of Council on 12 July 1989, which came into force from 1 September 1989.

among the authorities responsible for marine environmental protection, namely, the national marine administrative authority (State Oceanic Administration: SOA) is responsible for the investigation, monitoring, observation, appraisal and scientific research of the marine environment, the Maritime Safety Administration (MSA) is responsible for the supervision and administration, investigation and treatment of accidents of ships other than military or fishery ships, the treatment of pollution accidents with foreign ships, and so forth within or outside the harbour water areas; the fishery administrative department is responsible for the supervision and administration of non-military ships within fishing ports and of fishing ships outside fishing ports and so on. If a ship-related pollution accident causes damage to fisheries, then the investigation will be conducted by the MSA and fishery administrative department together; military department of environmental protection is responsible for the supervision and investigation of pollution from military ships. The content chiefly includes general provisions, the supervision and administration of marine environment, marine ecological protection, prevention and reversal of effects of pollution of marine environment from land-source pollution, prevention and reversal of effects of pollution of marine environment from offshore projects, prevention and reversal of effects of pollution of marine environment from dumping wastes, prevention and reversal of effects of pollution of marine environment from ships and their related activities, legal liability and supplementary provisions.

According to the LPTWP,[7] the matters concerning prevention and control of pollution of waterways like rivers, lakes, canals, channels, reservoirs, and so forth in the territory of China, will be governed by it. Article 4 defines the duties of the environmental protection departments of the people's government at all levels in relation to supervising and managing organs to prevent and control water pollution; navigation administration offices of traffic departments at all levels are the organs that supervise and manage ship pollution; the competent authority of water conservancy, hygiene administrative department, geological and mining department, competent municipal authority river water protection department at central level should supervise and monitor prevention and control of water pollution in cooperation with the environmental protection departments. It also governs the formulation of the standards for water environment quality and for pollutant discharge, the supervision and administration of preventing and controlling water pollution, prevention of water pollution on the earth's surface, prevention of underground water pollution and the corresponding liability resulting from breaches.

---

[7] It was adopted by the 5th meeting of the Standing Committee of the Sixth National People's Congress on 11 May 1984, came into force from 1 November 1984. It was also amended on the 19th meeting of the Standing Committee of the Eighth National People's Congress on 15 May 1996.

3 RELATED ADMINISTRATIVE RULES AND REGULATION CONCERNING THE PROTECTION OF THE MARINE ENVIRONMENT

3.1 ADMINISTRATIVE REGULATIONS ON PREVENTING SHIPS FROM POLLUTING SEA AREAS (ARPSP)

In order to implement the LMEP for the purposes of preventing ships from polluting the sea and safeguarding the ecological environment of the sea, the State Council promulgated the ARPSP on 29 December 1983.

The ARPSP applies to shipowners and other persons, ships with Chinese and foreign nationality on the sea or in seaports under administration. The harbour superintendent[8] is the competent authority to prevent pollution of the sea environment by ships. This Regulation is the first administrative statute that specifically concerns preventing pollution of the sea by ships. The provisions of this regulation are detailed, including Chapter I 'General Rules', Chapter II 'General Provisions', Chapter III 'Documents and devices of antipollution for ships', Chapter IV 'Oil operation of ships and discharge of foul water', Chapter V 'Carriage of dangerous cargo', Chapter VI 'Other shipping sewages', Chapter VII 'Shipping waste', Chapter VIII 'Waste dumping by ships', Chapter IX 'Repair and building as well as salvage of ships above and under water and project of ship dismantling', Chapter X 'Compensation for damages by ship's pollution accident', Chapter XI 'Punishment and rewards' and Chapter XII 'Supplementary provisions'.[9] Concerning the competent authority, Article 54 provides that functions and powers mentioned under these Rules are exercised by the National Fishery Superintendent Office within the waters of fishing ports, Article 55 states that the military environmental protection department may provide otherwise detailed regulations on the matter of pollution administration of military ships as well as the military administrative area inside seaport according to the 'LMEP' and this regulation. In addition, Article 53 stipulates that as to foreign ships, pollution administration can also be made on the basis of 'equal principles' in addition to the ARPSP. If an accident occurs which causes or might cause serious marine environmental pollution damage, the harbour superintendent has the right to take strict measures to avoid or reduce damage, including measures of compulsory removal or towage. All expenses should therefore be assumed by the ship involved.[10]

3.2 ADMINISTRATIVE REGULATION OF PREVENTING POLLUTION FROM DISMANTLING AND REPAIR OF SHIPS

In order to prevent pollution of the environment from dismantling of ships, and to protect

---

[8] See Arts. 2 and 3 of the ARPSP for the detailed provisions. Harbour superintendence had been replaced by the MSA on the basis of the combination of original Harbour superintendence of the People's Republic of China and the original Ship Survey Administration of the People's Republic of China in 1998 according to the approval of the State Council.

[9] According to Art. 52, 'ship' means all types of motorized and non-motorized ships, however, it excludes fixed and movable platforms used for exploration and exploitation of ocean oil which is governed else by the Administrative Regulations on Environmental Protection in Ocean Oil Exploration and Exploitation.

[10] See Art. 7 of the ARPSP for details.

the ecological environment, the State Council promulgated a Regulation on 18 May 1988.

It only applies to units and individuals engaged in the activities of dismantling ships on shore or on water.[11] Article 4 provides that the environmental protection department of government bodies above county level is responsible for organizing and coordinating, supervising and inspecting the environmental protection work of the ship dismantling industry, and is responsible for the environmental protection work of offshore ship dismantling outside the harbour water areas; the harbour superintendence and port navigation administrations are responsible for the environmental protection work of ship dismantling on water and within the water areas of harbours, and should provide assistance to the environmental protection department to supervise the work of preventing and reversing pollution from offshore ship dismantling outside the waters of harbours. The competent fishery authority on fish policy and fishing ports is responsible for the environmental protection work from the ship dismantling process within a fishing port. The military is responsible for the environmental protection within the water areas of a military port. The above-mentioned departments are generally called 'the Competent Authority'. In addition, the competent national ocean authority and the water resource protection authorities for large rivers shall provide assistance to 'the Competent Authority' to supervise the work of preventing pollution from the ship dismantling process according to the LMEP and the LPTWP. This regulation mainly contains detailed rules for planning and establishing a shipyard for dismantling, supervision and inspection of ship dismantling activities, items that should be noted in ship dismantling work and so forth, so as to ensure reduction of the risk of pollution in the ship dismantling process to the lowest level.

3.3 ADMINISTRATIVE REGULATIONS ON ENVIRONMENTAL PROTECTION ON OCEAN OIL EXPLORATION AND EXPLOITATION (REPOE)[12]

The State Council promulgated the REPOE on 29 December 1983. It applies to enterprises, institutions, individuals engaged in ocean oil exploration and exploitation within the sea areas under administration, as well as fixed platforms, movable platforms and other relevant facilities used.[13] The competent authority on environmental protection of ocean oil exploration and exploitation is the SOA[14] and its agencies.

---

[11] Art. 3 of these Rules provides that ship dismantling by shore includes dismantling in berth, dismantling in dock as well as dismantling in grounding which excludes the accidental grounding. It came into force from 1 June 1988.

[12] It came into force on the same date on which the rule was enacted. The detailed rules were promulgated on 20 September 1990 by the State Oceanic Administration with the title Implementing Rules of The Administrative Regulation on the Environmental Protection on Ocean Oil Exploration and Exploitation.

[13] According to the provision of Art. 30, the fixed and the movable platforms include the drilling ship, platforms and other platforms mentioned in the LMEP. The exploration and exploitation on ocean oil means the operational activities in exploration, exploitation, development, storage and carriage by pipeline etc.

[14] On 10 March 1998, the standing committee of the Ninth National People's Congress adopted a decision on the reform of subsidiary agencies of the State Council. According to the decision, the Ministry of Land and Resources is set up with the combination with the Ministry of Geology and the Mineral, State Land

3.4 ADMINISTRATIVE REGULATIONS ON DUMPING WASTES AT SEA (RDW)[15]

In order to control strictly the dumping of wastes into the ocean and to prevent the marine environment from being polluted, the State Council promulgated the RDW based on the LMEP.

Article 2 of the RDW defines 'dumping' as the disposal of wastes and other substances into the sea by using ships, airborne vehicles, platforms and other means of transport; disposing of ships, airborne vehicles, platforms and other artificial marine construction into the ocean; disposing of wastes and other substances produced by the exploration and development of the sea-bed mineral resources or by the relevant procession at sea, but operational discharge of wastes by ships, airborne vehicles, other apparatuses of transport and facilities is excluded. Article 3 specifically provides that this regulation applies to the dumping of wastes and other substances into inland sea water, territorial sea, continental shelf, and other sea areas under administration; loading of wastes and other substances for the purpose of dumping at the ports of China; carriage of wastes and other substances through the inland sea water, territorial sea, and other sea areas under administration; burning and disposing of wastes and other materials in the sea areas under administration. As to the disposal of wastes produced in the course of ocean oil exploration and exploitation, the REPOE shall be applied. The competent authority responsible for the dumping at sea is the SOA and its agencies.

3.5 ADMINISTRATIVE REGULATIONS ON PREVENTING OCEAN POLLUTION FROM LAND-SOURCE POLLUTANTS[16]

In order to strengthen the supervision and administration of land-based sources of pollution, preventing and reversing the ocean environment pollution from land-source pollutants, the State Council promulgated these Regulations under the LMEP.

The land-based sources of pollution referred to in this regulation refer to places and facilities and so on, from which the pollutants are discharged from the land to the sea areas and cause or may cause marine environmental pollution damage. Any unit or individual within Chinese territory who discharges land-source pollutants into sea areas shall obey this regulation, but pollution in the ship dismantling process shall be governed by the Administrative Regulations on Preventing Pollution from Ship Dismantling and Repair. The competent authority for environmental protection under the State Council is responsible for the whole country's work of preventing and reversing pollution from land-source pollutants of the marine environment; coastal environmental protection administrative departments of local people's government above county level are in charge of preventing and reversing pollution from land-source pollutants of the marine environment within its administrative jurisdiction.[17]

Administration, the SOA as well as the State Survey and Drawing Administration. The last two Administrations have been kept as the subsidiary authorities under the Ministry of Land and Resources.

[15] It is promulgated by the State Council on 6 March 1985 and came into force 1 April of the same year.
[16] It was promulgated by the State Council on 22 June 1990 and came into enforce from 1 August 1990.
[17] See Arts. 2, 3 and 4 for details.

There are no special rules on the problem of pollution in other water areas besides sea areas from land-source pollutants in China. However, in order to prevent the pollution of the Changjiang River water areas from ship and solid waste along the river bank, the Ministry of Communications, the Ministry of Construction and the State Environmental Protection Administration jointly promulgated the Administrative Regulations on Preventing the Pollution of the Changjiang River Water Areas from Waste of Ships and Solid Wastes along River Banks[18] on 24 December 1997 in accordance with the LPTW, the Law of Preventing and Curing Environmental Pollution by Solid Wastes as well as the Administrative Regulations on Urban Appearance and Environmental Sanitation and so forth. These particular Regulations apply to a ship's sailing, berthing and operation in the Changjiang River water areas and its owner (operator) as well as owners (operators) of terminal loading and unloading facilities. Article 24 provides that the competent administrative department in the State Council is responsible for the supervision and administration work of preventing and reversing pollution of water areas of the Changjiang River from ship's garbage. When a pollution accident from waste disposal happens, the shipowner (operator) shall take measures immediately, control and eliminate the pollution, report to the port superintendent office and local environmental protection department, and wait for an investigation and ruling or penalty. For situations that have caused or may cause serious pollution, the port superintendent office and the local environmental protection department are entitled to take compulsory measures to eliminate the pollution, all the expenses produced must be borne by the party that caused the pollution.

3.6 ADMINISTRATIVE REGULATIONS ON PREVENTING AND CURING OCEAN POLLUTION FROM COASTAL CONSTRUCTION PROJECTS

The State Council promulgated these Regulations according to the LMEP 'in order to strengthen the environmental protection administration of the coastal construction project, control strictly on new pollutions, protect and to improve the marine environment'.

According to Article 2 of these Regulations, a coastal construction project refers to a capital construction project, a technological transformation project and a regional development project that influences the ocean environment in order to control sea water or utilize some or all of the functions the ocean lying in or connected with the coast. It includes mainly ports, quays, building or repair shipyards, coastal thermal power stations, nuclear power stations, coastal oil depots, coastal mining areas, chemical, papermaking, iron and steel industries, project for treatment and disposal of solid wastes, engineering of discharging urban waste water into ocean and other construction projects of discharging pollutants into the ocean, engineering of the channel project and navigation course at the mouth of rivers, engineering of tidal power generating, engineering sea area enclosures, fishery's engineering, engineering of bridge and tunnel across the sea, dike and dam projects, engineering protection of dikes and dams and any other construction project that will change the nature of coast and mud-flats. This regulation applies to all units and individuals who build coastal construction projects within the territory of China.

---

[18] It came into force from 1 March 1998.

However, the environmental protection of a ship-dismantling shipyard's construction projects shall be governed by the Administrative Regulations on Preventing Pollution from Ship Dismantling and Repair. The competent authority of the environmental protection department of the State Council is in charge of the environmental protection work of the coastal construction projects in the whole country. The competent authority of the administrative department in the local people's governing bodies above county level is responsible for the environmental protection work of the coastal construction projects within its administrative jurisdiction.

3.7 IMPLEMENTING RULES ON OCEAN ADMINISTRATIVE PENALTY

Although the above-mentioned laws and regulations on ocean environmental protection have defined the relevant legal liability for violations, the Ministry of Land and Resources promulgated the Implementing Rules on Ocean Administrative Penalty on 25 December 2002.[19] These are the first rules that have the title of ocean administrative penalty and involve the procedure of administrative penalty in China.

These Rules make clear that any legal entities and individuals who violate the laws, rules and regulations concerning the usage of sea areas, marine environmental protection, laying submarine cable pipeline, administration of marine scientific research concerning foreign affairs shall be punished by the executive authorities of ocean administrative penalty. The competent authorities of the marine administrative departments of the governing bodies above county level are the executive authorities of the marine administrative penalty. Where the executive organ has set up the China Ocean Supervising Organization, the organization is responsible for implementing the marine administrative penalty; if no sea supervising organization is set up, the competent authorities of the marine administrative departments at the same level are responsible for the implementation. The China Ocean Supervising Organization applies the marine administrative penalty in the name of the marine administrative departments at the same level.[20]

---

[19] Came into force from 1 March 2003.
[20] China Ocean Supervising Organization is under the direct supervision of the SOA, and is responsible for supervising sea areas under administration, investigation or violation of ocean right and pollution damage, enforcing the supervision on the basis of entrustment, and empowering, etc. See the website: <http://www.soa.gov.cn/jigou>, last visited 11 March 2003.

4 RELATIONS AND EXISTING PROBLEMS BETWEEN LAWS, ADMINISTRATIVE REGULATIONS AND RULES ON MARINE ENVIRONMENT PROTECTION

4.1 RELATIONS BETWEEN LAWS, ADMINISTRATIVE REGULATIONS AND RULES ON MARINE ENVIRONMENT PROTECTION OF CHINA

With over 20 years' constant effort and legislation, the framework for the administrative legislation concerning marine environmental protection has tentatively been set up. The relations between the aforesaid laws, administrative regulations and rules are as shown in the following chart.

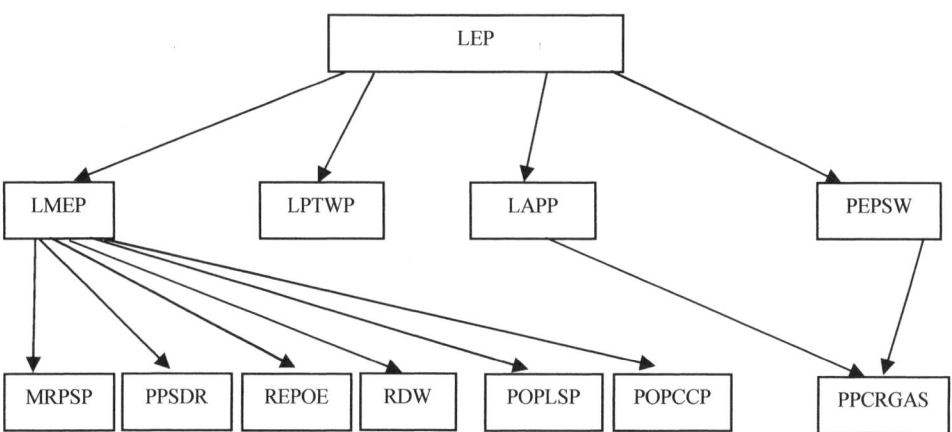

Note: Abbreviations that have been used herein are the following: the PEPSW refers to the Law of Preventing and Curing Environmental Pollution by Solid Wastes; the PPSDR refers to the Administrative Regulations on Preventing Pollution from Ship Dismantling and Repair; the POPLSP refers to the Administrative Regulations on Preventing Ocean Pollution from Land-Source Pollutants; the POPCCP refers to the Administrative Regulations on Preventing and Curing Ocean Pollution from Coastal Construction Projects; the PPCRGAS refers to the Administrative Regulations on Preventing Pollution of Changjiang River Water Areas from Garbage of Ships and Solid Wastes along River Bank.

4.2 PROBLEMS EXISTING IN LAWS, ADMINISTRATIVE REGULATIONS AND RULES ON MARINE ENVIRONMENT PROTECTION

Firstly, the system of law lacks unity on the applicable law.

From the chart 1, it may be concluded that the applicable laws, administrative regulations and rules vary greatly concerning marine pollution of the same nature but with different sources which potentially cause environmental pollution, for example, land sources, engineering dike and dam building, waste dumping, ship dismantling, ocean oil exploration and development, normal shipping operation, and so forth. Even on the matter of the marine environment pollution from the same pollution source, different administrative statutes or rules are put forward. For example, the Administrative Regulations on Preventing Pollution from Ship Dismantling and Repair applies to the marine pollution produced by ship dismantling and the engineering of ship-dismantling shipyard construction, regardless of whether it comes from the land sources or belongs to a coastal construction project, furthermore, the Administrative Regulations on Preventing Ocean Pollution from Land-Source Pollutants and Administrative Regulation on Preventing and Curing Ocean Pollution from Coastal Construction Projects shall not apply on this point. Waste produced during the process of marine exploration and development will be governed by the REPOE instead of by the RDW.

For this reason, the relevant competent administrative authority shall first confirm the sources of or reasons for the pollution before determining whether the pollution is within the jurisdiction of its competency or not, according to the applicable laws and regulations. If the source of the pollution cannot be ascertained or there are two or more sources of pollution, then it is not very clear by which administration authority and according to what law the administrative penalty should be applied. It complicates the application of the law, and renders it unpredictable and thus unfavourable to the efficiency of the enforcement of administrative law.

Second, the competent administrative authorities involved are manifold, and their duties as well as limits are unclear.

According to the above-mentioned relevant laws and administrative statutes, there are five to six competent authorities for marine environment pollution. If in addition the relevant provisions in the LEP and the LPTWP are considered, there as many as 14 or 15 administrative departments involved in environmental pollution. It is significantly difficult to decide how to divide the functions and powers and how to coordinate between each other.

Although the LMEP has defined the duties and functions of the SEPA, SOA and MSA, the precise competencies are not very clear yet. Other authorities, like the superintendent office of fishing policy, the fishing ports and the military environmental protection department, also play a role. The precise division of competencies among these authorities is not very clear yet. For example, it is held that the SEPA is responsible for the administration and supervision of the national environmental protection work and is in charge of the marine environmental protection work of preventing pollution from the land-source pollutants and coastal engineering of construction projects; however, the SOA is responsible for investigating, monitoring, observing, appraising and researching the marine environment; the MSA, the superintendent office of fishing policy and fishing ports, and the military environmental protection department are responsible for investigating and settling accidents within its jurisdiction. However, according to the REPOE, the RDW, the SOA and its agencies are also entitled to investigate the accidents and settle the disputes arising thereof. These concurrent competencies can cause problems. For example, if a certain pollution incident involves several pollution sources, how would the relationships between these administrative departments then be coordinated? If the administrative departments handling the accident make different decisions according to different laws and regulations, what responsibility should the polluter finally bear? There are still no answers to these questions.

For historical reasons China extends the administration mode of land to the ocean. As a consequence, the whole sea area is separated artificially according to the needs of the work of each functional department, and each administrative department is responsible for only one issue, without mutual interference. For example, the China Ocean Supervising Organization and its detachment are set up under the SOA and equipped with specific ships and personnel with the task of patrolling and monitoring at sea. The MSA is also equipped with similar facilities, ships and personnel. The direct consequence of this repeated investment is the increase in administrative costs and the decrease in administrative efficiency.

Third, there is too great an emphasis on the distribution of powers and on cooperation, and too little emphasis on coordination among the administrative departments.

Within the jurisdiction of every administrative department, the above-mentioned laws and regulations have clearly ruled on supervising powers such as the administration power, accident investigation rights, punishment and so forth. Because of the complexity of pollution, often many competent administrative departments are involved. However, there is no clear legislation on the relationships among the competent administrative authorities. In addition, whether the administrative departments are obliged to cooperate and help each other mutually is not very clear.

The administrative assistance refers to a legal mechanism according to which an administrative authority, based on its own condition and the needs of public affairs in the course of the implementation of its administrative power, is entitled to obtain the cooperation and help from another to implement the same administrative behaviour or to implement common administrative behaviours together.[21] If two different administrative activities are involved, administrative assistance cannot be constructed. It is not quite clear if the MSA has the right to deal with investigating a pollution incident that happened within the sea area under its administration. The law, however, does not govern the question of whether it is obligated to communicate with the SOA and ask for the latter to investigate and monitor the ocean pollution, and whether it is obliged to provide relevant information to the SOA in order to assess and analyze the environmental pollution damages. If the pollution accident involves fishery damages, the MSA shall allow the competent administrative authority of fishery to participate in the investigation according to Article 5 of the LMEP. There is at present no clear regulation in the law with respect to this administrative assistance.

Fourth, the party making the claim can often not be identified in case it is the entire State.

Who has the right to claim on behalf of the country when the country suffers large losses because of the marine pollution that causes damages to the marine ecology, marine aquatic resources, and the marine protection zone? Article 90(2) of the LMEP provides 'the department exercising the right of supervision and administration on the marine environment has the right to file the claim against the party responsible on behalf of the country.' However, combined with Article 5 of the LMEP, the SEPA, the SOA, the MSA,

---

[21] J. Hu, *Administration Law* (2nd edn, Law Press, Beijing, 2003), p. 184.

the superintendent office of fishing policy and fishing ports, the military environmental protection department as well as the departments of the coastal local governing bodies above the county level exercising the right of marine environmental supervision and administration and so forth each have the right to supervise and manage the marine environment. Hence there is not just one department that may exercise the right of supervision and management.

Fifth, legal liability is inconsistent; the standard of punishment is not unified and the strength of these sanctions is weak.

The ARPSP stipulates that the competent authority can impose punishment in the form of a warning or a fine as a result of the illegal activities. The highest fine that can be imposed on the shipowner is 100,000 RMB. The highest fine is only 1000 RMB if the conditions under Article 47 are satisfied. Fines are imposed upon the individuals directly responsible, but the maximum fine may not exceed 20 percent of the monthly salary of the perpetrator. The REPOE provides that the competent authority may issue the order to the party responsible to rectify within a time limit, or order it to pay fees for eliminating the pollution, or discharging the pollution, or issue a warning or fine; 100,000 RMB is the maximum fine for most enterprises and institutions; if the conditions of Article 27 are satisfied, the highest fines will be 1000 RMB, 5000 RMB respectively according to different scales; as to the individuals that are directly liable, the amount of fine is adjusted according to the specific circumstances.

The Administrative Regulations on Preventing Pollution from Ship Dismantling and Repair provide that, as regards illegal activities, it may be possible to order the punishment of rectifying illegal action within the time limit, the business being suspended or closed, being warned and fined a different amount; if the conditions in Articles 17 and 18 are met, the amount of fine will vary between 10,000 RMB and 100,000 RMB as well as below 10,000 RMB; imposed administrative sanctions can be on the party that is directly responsible for its entity or a higher competent authority. Article 20 of the RDW provides that a warning will be issued or a fine of under 2000 RMB will be issued for forging documents and not reporting in time; if it is serious in nature, the licence will be suspended or revoked and a fine of between 2000 and 5000 RMB will be imposed; a fine from 5000 RMB to 20,000 RMB will be imposed for unauthorized dumping; fines for unauthorized dumping without approval range from 20,000 RMB to 100,000 RMB; if criminal responsibility is involved, the criminal responsibility is imposed by a judicial authority. Articles 24 to 31 of the Administrative Regulations on Preventing and Curing Ocean Pollution from Coastal Construction Projects also provide punishments similar to the Administrative Regulations on Preventing Pollution from Ship Dismantling and Repair with the exception of a maximum fine of 200,000 RMB.

Obviously, punishments pursuant to the above-mentioned laws and regulations are weak, and the amount of the fine is too low to achieve the objective of protection of environment, which is even far lower than that defined in the relevant provisions of the LMEP. It cannot achieve the objective of effectively punishing and educating the party responsible, let alone to meet the expenses incurred to eliminate the pollution. As regards the problems involving dumping at sea, Article 87 of the LMEP provides for a maximum of 1,000,000RMB, while the RDW only provides a maximum fine of 100,000 RMB, one tenth that of the LMEP. One explanation for this phenomenon is that most of the above-mentioned laws and regulations were enacted during the second half of the 1980s and have not been adjusted and revised since the LMEP was revised in 1999. This incongruity among the relevant pieces of legislation on marine environment pollution in China is not favourable to the protection of the marine environment.

Sixth, the stipulations of marine administrative penalty procedure are not unified.

The Implementing Rules on Ocean Administrative Penalty was enacted 1 March 2003. It clearly provides that it applies to entities and individuals who violate marine laws, regulations and rules on the usage of sea areas, marine environmental protection, laying of submarine cable pipeline or administration of marine scientific research concerning foreign affairs and such violations will be penalized by the implementing administrative departments in accordance with the law.

However, except the SOA, other competent authorities such as the MSA, SEPA, the superintendent office of fishing administration and fishing ports, the military environmental protection department, and so forth, shall not apply these Rules when they impose administrative punishments in accordance with the law because these were promulgated by the Ministry of Land and Resources alone. However, the Ministry of Communications promulgated the Administrative Penalty Regulations on Safety Supervision on Water on 2 September 1998. Therefore, the MSA shall impose administrative penalties according to this particular regulation within the scope of its duty. Other departments can only punish according to the stipulations of the Administrative Penalty Law as well as the relevant provisions because there are no other specific rules on administrative penalty procedure.

5 PRELIMINARY CONSIDERATIONS AND METHODS TO SOLVE THE PROBLEM

Taking into consideration the current situation in China, this paper proposes tentatively the following points on considerations and methods in order to solve the existing problems.

5.1 PERFECTING THE RELEVANT LEGISLATION

This paper suggests that the relevant authorities in China revise and perfect the above-mentioned laws, regulations and rules as soon as possible according to the relevant provisions of the LMEP in order to ensure these regulations and laws are coordinated and connected with each other. The primary focus should be on unclear limits of enforcement authority, on the lack of clarity of administrative duty, on the lack of unity in applying the law, and especially on enhancing the strength of punishment according to the stipulations of the LMEP.

5.2 ORGANIZATIONS AND FORMS

This paper proposes the following steps: firstly, a special body or agency or the leading group in the State Council shall be established as soon as possible and given the responsibility for coordinating and negotiating with the above-mentioned administrative departments. This could help in avoiding and reducing the situations of shifting onto each other, interfering with each other and could solve the phenomenon of 'each department acting on its own'. Then the problem of administrative assistance could be solved before significant changes and adjustment among the above-mentioned organizations. Second, if it is possible in the future, one special body like the American Coast Guard could be set up and be provided special equipment and personnel to deal with problems related to marine environmental pollution and joint action in work to protect the environment. When marine pollution is found, it is the duty of this special agency to investigate the accident and take emergency measures for preventing the pollution from expanding. Then according to the division of functions and powers among the administrative departments mentioned above, the special agency will transmit the case to the relevant department(s) to handle, or will have the right to demand that every relevant department provide the necessary cooperation and assistance. If the sources of pollution are undetermined, there is a variety of sources or reasons for the pollution and there is no way to ascertain which administrative department has the jurisdiction, the special agency has the exclusive right to deal with directly or authorize and appoint one competent authority to deal with the pollution. Certainly, creating such a special agency and defining the limits of its authority, including the relationship with relevant administrative departments, should be carried out through legislation.

5.3 Promoting the Legal Effect Level of the Department's Regulations and Rules

As stated above, the existing laws on the administration of marine environmental pollution are mostly regulations and rules promulgated by different administrative departments, the legislative level is relatively low which results in restraints and limits on the scope of application and enforcement. Therefore, some measures shall be taken as soon as possible in order to promote those regulations and rules to administrative statutes in accordance with the LMEP and maintain the compulsory nature and seriousness of law.

6 CONCLUSIONS

With an increasing amount of attention being paid to the marine environmental protection, the administrative legal system necessary for the protection of the marine environment has already been set up in China. Meanwhile, for historical reasons, the existing laws on marine environmental protection create several problems. In order to protect China's marine environment more effectively, this paper suggests that these problems must be solved in the interest of China making the system of administrative protection of the marine environment more mature and effective.

LIST OF REFERENCES

J. Hu, *Administration Law* (2nd edn, Law Press, Beijing, 2003).

RISK ANALYSIS OF NOXIOUS LIQUID SUBSTANCES TRANSPORTATION BY WATER IN THE YANGTZE DELTA AND THE RESEARCH OF ITS COUNTERMEASURES

Wei Zhijie and Song Ruqing

1 INTRODUCTION

The Yangtze Delta is located at the meeting point of the two developed areas along the coast and the Yangtze River. Being one of the areas with the most flourishing economy and the most prosperous tourism industry, it is acknowledged by international economists as the sixth city group in the world.[1] Industrial economy and shipping economy are well developed in this area. In particular, the petrochemical industry, as one of the pillars of the regional economy, has recently made rapid progress. There consequently has been a great increase in transportation of petroleum and noxious liquid substances (NLS) by water. Because of the intrinsic risk of carriage by water, transportation of oil and NLS by water has brought about a great potential threat to the environment of water areas, to people's lives and to production activities, and this threat is becoming increasingly severe. Therefore, it is urgent to analyze and predict the risk of transporting oil and NLS by water and to investigate countermeasures like establishment of systems reacting to emergencies and compensation regimes for pollution damage.

Since there is experience concerning the management of oil substances, this article will focus its analysis on the risks involved in transporting NLS by water, combining the regional characteristics of the Yangtze Delta, and will discuss the mechanism of prevention, salvation and compensation.

2 ANALYSIS OF THE RISK IN NLS TRANSPORTATION BY WATER IN THE YANGTZE DELTA

Noxious liquid substances (NLS) are any non-oil substances that are likely to pose a threat to human health, damage biological resources and flora and fauna in water, reduce the advantages or prevent the proper use of other water resources if they are introduced into a water environment. Annex II of MARPOL 73/78, Regulations on Control over Pollution of NLS in Bulk, divides NLS into the four types (A, B, C, and D) according to the degree of threat to the marine environment. Substances in Annex III beyond these four types are

---

[1] Materials about economic and ecologic features of the Yangtze Delta, from <http://www.chinacsj.net>.

regarded as essentially innocuous.

Bulk NLS carried by vessels are numerous in variety and possess various physical and chemical properties. They usually have one or several characteristics, which mainly include strongly corrosive, toxic, flammable, explosive, self-reactive, heat-sensitive, and have temperature requirements to prevent solidification, instability, and sensitivity to impurities and so forth. These properties define the hazard of NLS as regards fire, health, water pollution, air pollution and reaction.

Characteristics in the natural environment and economic development of the Yangtze Delta together with the existing conditions of NLS transportation by vessels have created special risks. Pollution accidents in this area have the potential of causing great harm.

2.1 EXISTING CONDITIONS IN NLS TRANSPORTATION IN THE YANGTZE DELTA[2]

The transportation of NLS by water in the Yangtze Delta is quite substantial. According to statistics, in 2000, NLS carriage accounts for 54.9 percent of the total for the country. The domestic trade in NLS was 2,890,000 tonnes and that of foreign trade was 3,980,000 tonnes, the total thus being 6,870,000 tonnes. In 2001, transportation of NLS accounted for 57.1 percent of the total for the country. Domestic trade was 3,260,000 tonnes and that of foreign trade was 5,040,000 tonnes, the total thus rising to 8,300,000 tonnes. In recent years, the growth rate has been 20.8 percent.

Altogether, there are over 120 types of noxious liquid substances carried in bulk by vessels in the Yangtze Delta, including the four types of A, B, C, and D. Among the NLS transported in the Yangtze Delta, Types C and D account for a large proportion. There are 29 varieties of Type C substances, accounting for 23.4 percent of the total, and 36 varieties of Type D substances accounting for 29 percent. Among these four types, substances of Type D accounts for the largest share of transportation – about 50 percent of the total, and those of Type C account for 27.6 percent. Substances of Types A and B respectively account for 0.7 percent and 13.1 percent, and those of Annex 3 account for 6.4 percent.

The loading and unloading capability of NLS in the Yangtze Delta is quite high. There are nearly a hundred terminals, accounting for 40 percent of the total number throughout the country. Those terminals with the storage capability over 4,217,000 m^3 account for 63.3 percent throughout the country. The rates of distribution for Jiangsu, Shanghai and Zhejiang are respectively 50, 25, and 23.

Clearly the Yangtze Delta is the main distribution centre of noxious liquid substances in bulk.

According to the 2001 statistics, a total of 19,854 ships carrying NLS sailed in and out of ports throughout the country, the number of ships that sailed in and out of the Yangtze Delta harbours was 9,488, accounting for 47.8 percent of the total. Vessels carrying NLS of 1000-3000 m^3 in the Yangtze Delta are mostly older than 10 years. These vessels are in relatively poor condition and the amount of NLS transported in and out of harbours is relatively high.

---

[2] Data of this part are from A. Sun, X. Liu and R. Song, *Research on Emergent Countermeasures for Pollution Accidents by NLS Carried by Ships in Littoral Areas*.

2.2 STATUS OF NLS POLLUTION ACCIDENTS IN THE YANGTZE DELTA

From the 1990s to 2001, a total of 17 pollution accidents involving bulk NLS transportation occurred in the Yangtze Delta. The categories of accidents include pollution by accidental spills, pollution by operational spills, pollution during navigation and pollution in terminals. According to statistics, in terms of variety of bulk NLS in pollution accidents, there are 11 varieties. No pollution accidents have occurred involving Type A substances; six accidents occurred with Type B substances, accounting for 35.3 percent of the total, with a spill quantity of 769 tonnes, accounting for 60.8 percent; five accidents occurred with Type C substances, accounting for 29.4 percent of the total, with a spill quantity of 208 tonnes, accounting for 15.4 percent; six accidents occurred with Type D substances, accounting for 35.3 percent of the total, with a spill quantity of 340 tonnes, accounting for 23.8 percent.

Among these varieties, accidents involving styrene occurred most frequently (six times, accounting for 35.3 percent), and those of benzene (including xylene) occurred five times, accounting for 29.4 percent. The largest quantity of NLS that was emitted was styrene (869 tonnes, accounting for 60.8 percent of the total). The amount of animal/vegetable oil was 205 tonnes, accounting for 14.4 percent, and vitriol, 208 tonnes,

Accidents occurred most frequently with Type B substances, which is the most hazardous of all. The amount of spill is quite large and the damage from pollution is very severe and must be sufficiently attended to.

2.3 DEVELOPMENT TREND IN NLS CARRIAGE IN THE YANGTZE DELTA

The Yangtze Delta is economically solid. It has become the largest base for light industries, machine and electricity and steel and petrochemicals. With the strategy of economic integration, many international giants were attracted to invest here. In recent years, BP, DOW, and BASF have all established production bases in the Yangtze Delta. At the same time, logistics in the chemical industry has developed rapidly in this area. Compared with other means of transportation, shipping NLS by water has the features of low cost and simple technique for loading and unloading. Transportation of NLS by water is steadily rising.

1. NLS vessels navigating in and out of the Yangtze Delta are increasing in number and become larger.
   With the development of logistics resulting from the growth in the chemicals industry, NLS vessels are increasing in number. After the reconstruction of the channels in the Yangtze estuary, navigation lanes have been provided for large vessels. To pursue higher profits and lower costs, many enterprises construct vessels of large tonnage to transport NLS. Therefore, NLS vessels navigating in and out of the Yangtze Delta will become larger.
2. Diversity in varieties of NLS transported in and out of the Yangtze Delta.
   Before 1989, NLS transported in the Yangtze Delta were mainly animal and vegetable oils, substances resembling oil, and alcoholic substances, which are all Type C and D substances with over 40 varieties; in the 1990s, there were already 10 varieties of

Types A and B substances; by 2001, there were 120 NLS varieties, among which were 20 varieties of Types A and B substances(10 varieties of Type A substances and 12 varieties of Type B substances). Varieties of serious hazard have also increased in number.

3 The number of vessels used to transfer NLS on inland waterways increased rapidly.

With the implementation of the strategy of West Development, refined chemicals industry have also grown quickly in the west, which increases the demands in chemical materials and as a result leads to the development of NLS transportation using inland waterways. Vessels carrying NLS to the upper reaches of the Yangtze River through inland waters will also gradually increase in number.

4 With the increase in flow of vessels, accidents occur more frequently.

There are many islands in this area and the navigation channels are rather narrow. In particular, the Jiangsu part of the Yangtze River is a natural river, with abundant water and a large amount of mud and many islands. There are also shoals in many navigation channels and there is much change with the navigation lanes. Because the administrative monitoring means lie behind the development of transportation, the present monitoring devices and means of control are insufficient for organizing transportation.

The poor hydrology and climate conditions pose a significant threat to safe navigation of vessels in the Yangtze Delta. Tides in this area are irregular half-day tides. Daily tides can reach 0.5-1.5 meters, and the flow is about 3 kts (knots per hour). Severe weather conditions in the region include very low temperatures, tropical storms and high winds throughout the year. In addition, there is fog an average of 25 days per foggy.[3]

The Yangtze Delta is located in the Yangtze estuary, flowing into the East Sea, with nearly 1000 km of coastline and 600 km of shoreline along the Yangtze. It incorporates the 'Golden Coasts' with the 'Golden Channel', and it has become the largest group of coastal and riparian harbours within China with Shanghai harbour, Ningbo harbour, Zhoushan harbour, Zhangjiagang harbour and Nantong harbour. It connects the international market with the inland market. It boasts a large flow of vessels. The daily flow of vessels is 2500, among which up to 1200 medium and large vessels sailing in and out of the Yangtze River, including 200 NLS vessels of tonnage of over 500. With the development of regional economy, the flow of vessels into the Yangtze Delta will also increase rapidly.

If existing conditions of the navigation channels in the Yangtze Delta do not improve and the means of monitoring are not improved, the frequency of traffic accidents of vessels in this area will increase rapidly and the risk of pollution will become more severe.

2.4 RISK ANALYSIS OF POLLUTION ACCIDENT HAZARDS IN THE YANGTZE DELTA

Through analysis on the existing conditions of NLS transportation, vessels, and loading

---

[3] About the Management of Risk in the Carriage of NLS in Bulk in Water Areas in the Backward Position of the Yangtze River (Shanghai Maritime Safety Administration).

and unloading operations in the Yangtze Delta as well as of the conditions of NLS pollution accidents and the developing trend in NLS transportation, it is known that: the Yangtze Delta has become the centre for the chemical production industry and the distribution centre for chemicals; the amount of liquid chemicals carried in bulk is increasing dramatically, and the variety is also growing in number, especially in the Types A and B substances that pose a great hazard; those vessels are relatively old and are in poor condition; the complexity of the navigation channels and climate influences navigation security; there is a severe risk of pollution accidents with the carriage of NLS in the Yangtze Delta.

NLS have special physical and chemical characteristics, and accidents involving such substances have the potential of causing great harm to the state of the regional economy and the abundant ecology in the Yangtze Delta. This is evident in the following aspects:

1. Pollution accidents cause severe damage to ecological environment.

    The Yangtze Delta lies in the humid monsoon zone of the subtropics. It boasts a mild climate, luxuriant soil, and a well-developed agriculture since ancient times. It is nicknamed the 'Land of Fish and Rice' and 'Silk Town'. Aquaculture resources there are quite rich. The famous Zhoushan fishing grounds, Shengsi fishing grounds and the Yangtze estuary fishing grounds are all located in the nearby sea areas. There are over 180 varieties of fish, shellfish, and other creatures with local and seasonal features, rare creatures like *acipenser sinensis*, as well as 160 kinds of tidal land biological resources. Aquaculture and the fishing industry are among the principal industries for residents there.[4] With the thriving tourism, the Yangtze Delta has become world renowned as a tourist destination with many cities with special features for tourists. Those ecological resources have increased the quantity of environment-sensitive areas in the Yangtze Delta and the level of risk in pollution hazard is also heightened. NLS is very noxious. Once pollution accidents occur, great harm will be caused to the ecological resources of this area. As a result, rare animals might even die out. This has been borne out by the 'April 17' spill in the estuary of Yangtze in 2001.

2. Pollution accidents do great harm to human health and influence social stability.

    The Yangtze Delta is the belt of a water network. Rivers and the sea join here. The Yangtze River is the main source of water for life and for industrial and agricultural production in the riparian cities of Jiangsu and Shanghai. The population in the Yangtze Delta accounts for 6 percent[5] of the whole country and approximately 50,000,000 people drink the water from the Yangtze. Safe drinking water is connected with social stability. In 2002, the 'Woo Seok' styrene spill occurred in Zhenjiang. The watersource for Dagang Town was closed. Within a night, the price of bottled fresh water rose ten times the original price. If the 'April 17' styrene spill, in which hundreds of tonnes of styrene spilled into the water, had not happened in the Yangtze Estuary but rather in Jiangsu section of Yangtze River, the noxious substances would have diffused into the branches of the Yangtze with the water flow and entered into

---

[4] *Ibid.*
[5] For materials on the economic and ecological features of the Yangtze Delta, from <http://www.chinacsj.net>.

littoral water areas affected by tides and floodwaters, which would greatly affect fresh water for tens of millions of people. Industrial and agricultural production was also severely damaged and social stability was threatened.

The Yangtze Delta is an area with a dense population. For example, in Shanghai, there is an average of 2116 people per square kilometre.[6] Some NLS, for example, Type B styrene, will volatilize noxious gas shortly after its leakage. The winds will spread the gas and threaten human safety near the site of the accident. Thus, people must be evacuated. If the leakage is great, as many as millions of people will need to be evacuated. A great deal of manpower and materials will be involved, which will also lead to social instability.

3 Pollution accidents affect the investment environment of the Yangtze Delta and restrict the process of economic integration.

The Yangtze Delta is located in the industrial area at the point where the opening strip in the eastern littoral areas meets the industry-intensive strip along the Yangtze River to form a 'T'. Among the 35 cities with the strongest economy in China, ten are located in the Yangtze Delta. Among all the counties (cities) with the largest comprehensive strength, half are within the Yangtze Delta. Nearly 100 industrial zones with an industrial production value of over CNY 10 billion gather here. There are also over 1000 large enterprises. Four hundred of the 500 world top enterprises have settled here. It has become the area with the densest population, the fastest developing economy, and the largest scale of economic gross. The Yangtze Delta covers one percent of the area of the country. Its population accounts for 5.8 percent in the whole country. However, in 2002, it created CNY 1,910.5 billion, accounting for 18.5 percent of the total GDP for the country, it contributed financial income of CNY 404.6 billion, accounting for 22 percent for the whole country, and it accomplished exporting of USD 2.4 billion, accounting for 28.4 percent of the whole country.[7] It is evident that the economic status of the Yangtze Delta is important to the economic development of China.

Economic development calls for investment, a socially stable living environment and a good natural environment. The high risk of NLS pollution from transportation accidents in the Yangtze Delta would necessarily affect the economic development of this area.

3 COUNTERMEASURES (PREVENTION, SALVAGE AND COMPENSATION)

Through analysis of the risk in NLS carriage by water in the Yangtze Delta, the following conclusions may be drawn: NLS pollution accidents would severely influence the ecological environment, human health and economic development in the Yangtze Delta. Major pollution accidents will become great calamities to the Yangtze Delta. Active and effective countermeasures must be taken:

---

[6] *Ibid.*
[7] *Ibid.*

1 To establish active industrial policies, make reasonable arrangement and avoid risks.
   Loading and unloading operations of NLS in the Yangtze Delta mainly centre in densely-populated areas and environmentally sensitive areas of important ecological resources like the coastal areas along the Yangtze in Jiangsu, the Huangpu River and the Hangzhou harbour, which increases the risk of pollution. Governments of various regions should lay down reasonable plans in accordance with the characteristics of the Yangtze Delta and Regulations on the Control over Safety of Dangerous Chemicals. They should try to arrange the carriage of explosive, flammable and noxious substances in littoral water areas and the lower water area of the Yangtze in Jiangsu so as to reduce the occurrence of pollution accidents and mitigate the pollution in water areas resulting from accidents.
2 To establish investigation systems of liability for pollution.
   Such laws (regulations) as Marine Environmental Protection Law of the People's Republic of China, Water Pollution Prevention and Control Law of the People's Republic of China, General Principle of Civil Law and Criminal Law of the People's Republic of China have established systems to investigate liability for pollution, clarified the requirements and means of investigating the civil, administrative, and criminal liability that should be imposed on the party concerned for pollution accidents.
   Judicatory and administrative organs should follow the laws and regulations, investigate accurately the legal liability of the party concerned, punish the party concerned, urge the parties concerned to enhance their management and strengthen self-discipline.
3 To establish an effective mechanism for prevention, salvage and compensation following an accident, and to reduce and control the damage of accidents and resume the ecological environment.

3.1 MEANS TO CONTROL THE OCCURRENCE OF POLLUTION ACCIDENTS

Pollution accidents occur when certain factors comprising potential dangers are activated by accident conditions. As long as the factors and the conditions for accidents are controlled, accidents can be prevented from happening. Providing for a safe system of 'personnel, machinery, environment and control' and countering the accident factors like personnel, ships, terminals and navigation environment, highly efficient administration and professional self-discipline should be established; the operating and managerial quality of personnel should be enhanced; technical conditions of NLS vessels and terminals (devices) should be monitored and controlled; techniques in loading and unloading operations should be optimized; traffic organization and means such as forecasting bad weather should be improved so as to prevent pollution accidents from happening.

3.1.1 Control of the Factor of Personnel

The personnel is the first element in safe production. Professional techniques for operation personnel are directly connected with the effect of prevention and control over NLS pollution accidents. STCW (78/95) provides administrative standards in training, certification and watchkeeping for seafarers, and Regulations on Training of Seafarers for Tankers Carrying Liquid Substances in Bulk also stipulates requirements for such seafarers. Through effective management and means of control, it is expected that pollution accidents caused by human factors may be controlled effectively.

1. Activities of publicity and education should be carried out in various forms.
    Administrative organizations and enterprises may clearly publicize the laws and regulations in administration of safe transportation of NLS as well as compulsory regulations and standards through publicizing, training and other effective means, and continuously enhancing awareness of safety and professional standards of the personnel concerned.
2. System of entry for personnel should be established.
    According to the requirements of conventions and the law, there should be strict evaluation through a system of tests. Personnel working in the field of NLS transportation in bulk should be equipped with knowledge of safety and trained for safe operation. Entry should be strictly regulated.
    Before seafarers are engaged by shipowners or loading/unloading units assigned to related posts, there should be tests on the health and psychological conditions of the persons who will be operating the machinery and working on the ships and persons in good condition should be selected.
3. An evaluation tracking system should by established.
    Seafarers should be evaluated using a tracking system according to STCW so as to ensure that seafarers have the most up-to-date knowledge. Through the examination on seafarer operation, the qualification of seafarers should be ensured.
4. There should be strict system for responsibility investigation.
    Persons causing accidents and operators behaving illegally should be closely investigated. A marking system should be strictly implemented and those concerned should be encouraged to observe regulations.

3.1.2 Control over the Factor of Vessels

NLS is carried on vessels. Their technical condition is linked with the frequency of pollution accidents. MARAPOL 73/78, IBC, BCH and ISM have all provided standards for the seaworthiness and load qualification of NLS vessels.

1. Survey on vessels.
    The administration should carry out surveys on the technical condition of vessels carrying NLS in accordance with regulatory requirements to appraise the compliance of NLS vessels with regulations.
2. Control of port States and flag States.
    Through security survey, maintenance of technical conditions of vessels should be supervised and examined so as to encourage execution of the necessary maintenance on vessels in compliance with the requirements of conventions.

3 Establish an evaluation system of declaration on loading qualification for carriage of dangerous goods.
   Through the declaration and survey of vessels carrying NLS when entering harbours, passing through territory, and loading cargoes before leaving ports, a survey of the qualification for loading NLS should be completed.
4 To establish evaluation system for technical conditions of vessels.
   When hiring vessels, NLS cargo owners should carry out the necessary safety evaluation of the managerial ability of shipowners and technical conditions of vessels in accordance with common international practice and hire vessels with a high level of management and technical conditions sufficient to carry NLS.
5 To improve the requirements for types of vessel.
   For NLS to enter fresh riparian water areas in Jiangsu, the type of vessel should be better than that regulated in IBC CODE. Type III vessels should be forbidden to carry styrene into the Yangtze River.

3.1.3 Control over the Factor of Loading and Unloading Terminals (Facilities)

Loading and unloading operation is an important link in the carriage of bulk NLS. Analysis of accidents reveals that the frequency of pollution accidents in loading and unloading operations is the highest, accounting for 80 percent of all such accidents. Controlling terminals, devices and operations of loading and unloading should follow the regulations in Port Law strictly when carrying out administrative measures. Port administrative departments and maritime administrative organizations should form a joint force through division of work and cooperation basing on their own scopes of functions.

1 To implement the system of approval in building.
   Before terminals and devices are constructed, the design should be examined according to national standards to implement control before construction. Before terminals and devices are put into use, their compliance with national standards should be surveyed.
2 To implement the evaluation system of security and pollution prevention.
   Professional organizations should evaluate aspects like terminals for loading and unloading operations, safety of devices and equipment of devices of pollution prevention, regulations for operation, control systems, and operational techniques of the personnel to decide whether these aspects meet the conditions of safety and pollution prevention. Then port administrative departments issue Permits for Port Operation of Dangerous Cargoes.

3.1.4 Control over the Factor of Environment for Navigation

The navigation lanes of the Yangtze Delta are quite complex, and changes in hydrology and climate are drastic, elements that influence the safe navigation of vessels. Effective measures must be taken to strengthen administration of navigation order and provide information.

1. The routeing system should be implemented and navigation lanes should be defined. Accidents involving collisions could lead to major pollution accidents. To react to the complexity in channels and the large flow of vessels in the Yangtze Delta, advanced means in the world should be adopted. The routeing system should be implemented and navigation lanes should be stipulated so as to reduce the probability for vessels to meet transversally and avoid the occurrence of collisions.
2. VTS and AIS systems of full coverage should be established in the Yangtze Delta to execute traffic organization for NLS vessels and provide information service for navigation.
3. A weather-forecasting system for navigation should be established to provide weather service for passing vessels to avoid bad weather.

3.1.5 Control over the Factor of Administration

1. Compulsory pilot navigation and convoy should be implemented. For vessels carrying NLS, especially Types A and B NLS, pilot navigation and convoy should be compulsory;
2. A system of ship/bank survey should be implemented;
3. Shipowners and loading and unloading operations should be obligated to follow international conventions or standards. A safety and pollution prevention control system should be set up and the standard of scientific administration should be improved.

3.2 COUNTERMEASURES OF EMERGENCY RESPONSE TO POLLUTION ACCIDENTS

NLS pollution accidents are characterized by strong outburst, broad scope, and great degree of difficulty in providing assistance. Only when decisions are made without hesitation, that is people within the range of the accident are evacuated rapidly, salvage personnel and enough equipment to enclose and clean up pollution in enclosures are deployed quickly, and sensitive environmental resources are properly protected, can pollution damage be reduced to the most effectively. When a system of emergency response for vessels is established, effective emergency response can be taken when there is an accidental oil spill. An emergency response plan having the nature of regulation and technique should be established in advance, and a command system should be set up. To counter the possible pollution damage by emergency oil spill accidents, plans should be set in advance and should be organized for implementation. The practice of many countries has shown that establishing an emergency response system is effective in reducing pollution damage from accidental spills.

The system of emergency response should include:

1. Emergency plans;
2. Information system for emergency response
3. Guarantee system of emergency response
4. System of technical support for emergency response

3.2.1 Emergency Plans

In recent years, there has been rapid development in the research of countermeasures for emergency reactions to oil spills. Concern for emergency systems for oil spills and continuous work has turned the system into a mode and provided a theoretical base. It can also be applied to accidental chemical spills, which happen less frequently. NLS emergency plans resemble emergency plans for oil spills in aspects like organization structure, requirements of emergency plans, survey and report, clear operating procedure, training and maneuver, finance and responsibility. The similarities are beyond the scope of this paper and will thus not be considered in any greater detail.

1. Levels of plan for emergency response

Because of the special geographic features of the Yangtze Delta, the connection with the river and the sea, the effects of tides and the broad scope of expansion of pollution accidents, emergency plans of regional cooperation must be established.

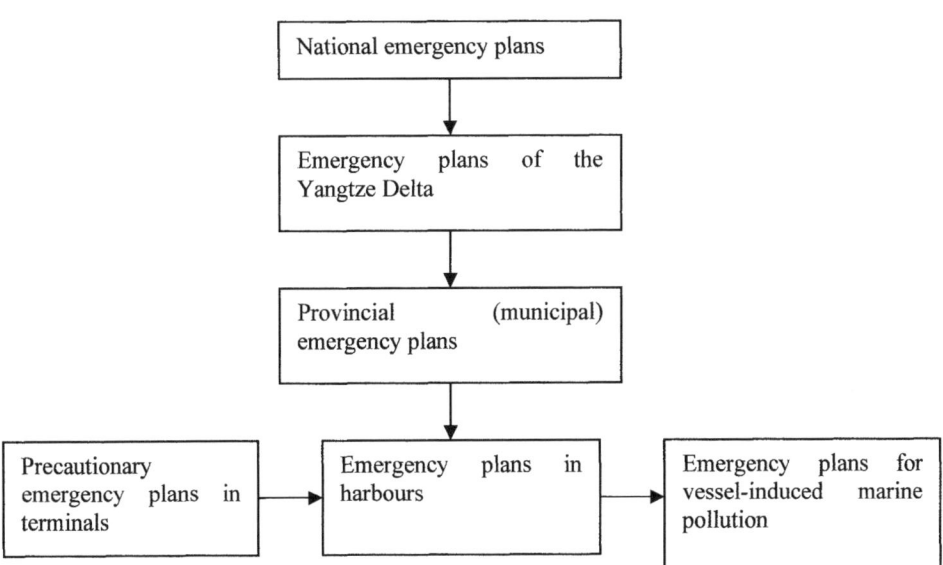

2. Organizations

The Safe Production Law of the People's Republic of China provides, 'governments of all levels above the county level should organize related departments and set down emergency assistance precautionary plans for particularly large accidents of safety in production within the jurisdiction of their own administrative region and systems for emergency assistance should be established.' According to Article 33 of the Environment Protection Law of the People's Republic of China, environmental protection administration of local people's government above the county level should report in a timely manner to the local government bodies when the environment is severely polluted or the pollution threatens the safety of residents' lives and property. The local governments should carry out effective measures so as to relieve or reduce the damage. Emergency plans should consider all levels of governments responsible and organizing bodies should be set up.

The emergency NLS organizing bodies can be established by referring to the reacting system for oil spills. Coordination organs should be set up in the Yangtze Delta to organize implementation of regional emergency plans based on the organization institution of Jiangsu, Zhejiang and Shanghai.

3.2.2 Information System for Emergency Response

Based on the GIS (geographic information system) platform, emergency information management of NLS pollution accidents, coordinating direction and functions of technical support system may be designed. Computer software should be developed for the management of emergency devices and teams, direction in emergency handling, evaluation of pollution damage and technical support should be provided for the handling of pollution caused by NLS. Assisted by such outside devices as VTS, CCTV and WEB, related departments should use their best efforts to actualize real-time delivery of the scene of the accidents to the command centre.

1. Input of basic information of ships. Using the facilities of VTS and AIS, information of vessels should be input in time;
2. The indication and revision of geographic information of the area under jurisdiction: including the channels, sensitive zones, and the location of devices for salvage and emergency response;
3. Actions of emergency response should be input, sent and searched in time;
4. Management of salvage information as well as the team and devices for emergency response;
5. Contact with members of the organization and experts;
6. Statistics on the condition with emergency teams and use of devices.

3.2.3 Guarantee System for Emergency Response

The guarantee system mainly includes: emergency team, emergency facilities and financial preparations. It is crucial for the reaction to emergencies of pollution accidents.

The government should establish corresponding financial and tax policies, and cultivate professional pollution cleaning companies as the subject of emergency guarantee.

1. Emergency team: this team should chiefly comprise pollution cleaning companies with professional training and volunteers in cleaning up pollution may be the complement; when major pollution accidents happen, the army and police should be employed for emergency handling;
2. Facilities: the government should establish facility storerooms in environmentally sensitive areas, important environmental protection areas and high-risk areas. Through financial and tax policies, manufacturers of pollution cleaning devices may be regarded as storerooms. Laws should be made to require enterprises of loading and unloading operations to supply the necessary appliances.
3. Financial preparations:
   Sufficient financial preparations are the guarantee of proper operation of emergency plans. The government should establish financial preparations through appropriating funds, compulsory insurance and establishment of funds.

3.2.4 System of Technical Support for Emergency Response

The system of technical support for reaction to emergencies is an important technical guarantee for emergency response. The functions of this system should include:

1. Marking parameters of harmful and noxious substances and inquiry of hazard
2. Inquiry of means in dealing with various kinds of accidents
3. Simulation of diffusion and efflorescence
4. Evaluation of pollution damage

The complex chemical and physical properties of chemicals determine that chemicals should differ from oil in emergency means and techniques. Oil usually floats on the water's surface, while only a small number of chemicals do so; oil spills are visible, while a spill of many kinds of chemicals is not easy to see; reaction techniques in oil spill are quite advanced, while reaction techniques in NLS are still developing; to skim oil from the water surface will be very successful, while skimming does not usually apply to NLS.

There are chiefly three types of means selected for emergency response based on the physical characteristics of NLS and according to the need of emergency NLS accidents. These are: forecasting means, monitoring techniques and the final systematic retrieve.[8]

1. Spill of volatile chemicals – to monitor the presence of gas in the air and make sure whether vulnerable people and the general public have been evacuated.
2. Spill of floating substances
   - Diffusion forecast of floating substances on the water

---

[8] 'Problem assessment and response arrangements', 1999, edition IMO.

The calculation of diffusion and floating of substances can be achieved by universal vector pictures used in oil spills. Such means apply to spills of most noxious and harmful substances. It is important to consider related factors like wind and flow, because these will determine the location of spilled substances.
- Reaction of spilled floating substances
There are usually several methods of dealing with spilled floating substances. It is essential to reduce vaporizing by covering with foams, using absorbent or other chemicals, reclaiming those substances by such devices as oil-skimming machines, or combining the above methods.

3   The reaction of spilled soluble substances
A substance will take the form of mist or can feather when it dissolves in the water. The substance will gradually dissolve in the water. Through on-site monitoring and simulation of diffusion as well as tracking the diffusing and floating substances, threats to the environment can be evaluated (to fisheries, leisure and fresh water areas). Sometimes, the spill of noxious substances can be treated with various chemicals in order to eliminate the harmful influence on human beings and the environment.

4   The reaction of spilled sediment
Spilled sediment will settle to the lower level of the sea. Although it seems to dissolve only slightly, there is solution, which is determined by the other substances in the water. The relative density in the water should be carefully monitored and the harm to environment, fisheries, leisure areas, and fresh water areas should be evaluated.

The spilled substances on the sea floor can be collected using various digging devices or machines, but not all the digging equipment are appropriate for reclaiming noxious substances from the sea floor.

3.3 COMPENSATION REGIME FOR POLLUTION ACCIDENTS

Considering the great threat to life and property from damage of caused by NLS transportation accidents and the ill effects on marine environment, the IMO held an international conference at its headquarter on 3 May 1996 and adopted the 1996 International Convention on Liability and Compensation for Damage in Connection with the Carriage of Hazardous and Noxious Substances (HNS 1996). HNS is based on the 1992 International Convention on Civil Liability of Oil Pollution Damage (CLC92), which applies to damage caused by tankers carrying persistent oil substances, and FUND 92. HNS combined the two conventions. It provides sufficient, timely and effective compensation for damage to human beings, property and environment caused by shipping HNS. The HNS has developed over a long period of time and it has not come into force until recently. However, it does provide experience for China as concerns what kind of compensation mechanism should be established.

Research on special topics has been launched in China aimed at establishing a proper compensation mechanism for oil pollution damage that can work in the circumstances of China. Much progress has been made in the research, but there has been little concern for research regarding compensation for pollution damage of harmful and noxious substances.

For the Yangtze Delta, because of its special economic and social position as well as its ecological environment, research should be carried out in advance. A mechanism for compensating pollution damage should be set up through legislation so as to ensure the damaged resources are renewed and compensated for. A perfect compensation mechanism for pollution damage should include the following elements (this list is not exhaustive):

1. Compulsory insurance systems for pollution. A compulsory insurance regime should be imposed on vessels carrying NLS and depend on the load; a maximum compensation amount should also be stipulated;
2. Establishment of special fund. It should include:
   - Starting capital based on national investment.
   - Contributions from domestic cargo owners. Whether it is carriage by sea or on inland waters, cargo owners should contribute according to the quantity of cargo they actually receive.
   - Fine imposed on ships. Fines collected in accordance with related laws and regulations can be considered as a source for the fund.
3. Scope of compensation provided by the law;
4. Means of damage calculation and principle of compensation provided by the law;
5. Compensation claimant. It must be clarified who has the right to claim civil compensation on behalf of the country.

4 CONCLUSIONS

The Yangtze Delta is the area with the most developed and most active economy in China. The chemical industry, as one of the important pillars of the regional economy, has played an important role in the economic construction as well as social development of this area. However, through analysis of the existing conditions of NLS transportation, of vessels, and of loading and unloading terminals, as well as of pollution accidents by NLS and the developing trend in NLS transportation, it is known that with the increasing number of vessels carrying liquid chemicals in bulk in the littoral areas of China and the Yangtze Delta and the increase in varieties and quantities of NLS, especially the more hazardous Types A and B substances, the risk associated with carrying liquid chemicals especially NLS in bulk by water increases daily. The potential threat to the water environment, water resources, human life and production activities is also becoming greater. Therefore, there is an urgent need to strengthen the research analysis and precautions against risk in vessel-related pollution accidents, to take practical action in legislation and specific measures, to speed up establishment and improvement of the emergency mechanism for vessel-related pollution and a compensation mechanism for damage and to enhance the ability in neutralizing risks.

LIST OF REFERENCES

A. Sun, X. Liu and R. Song, *Research on Emergent Countermeasures for Pollution Accidents by NLS Carried by Ships in Littoral Areas.*

FINANCIAL CAPS FOR OIL POLLUTION DAMAGE: CHINA AND THE INTERNATIONAL CONVENTIONS

Michael Faure and Wang Hui

1 INTRODUCTION

Marine pollution, especially caused through oil, has been a source of increasing concern around the world. As early as the 1960s, the first serious incidents with oil tankers gave rise to legal actions at the international level. Many international conventions emerged aiming on the one hand at measures for the prevention of oil pollution damage, and, on the other, at compensation for damage caused by marine (oil) pollution.

An important feature of the conventions, like the civil liability convention for marine oil pollution from ships, is that the liability of the tanker owner is limited to a specific amount. These financial limits (also referred to as caps) originate in maritime law where there has long been a tradition of limiting the carrier's liability. However, more recent incidents with oil tankers have demonstrated that the original caps as laid down in the conventions were largely insufficient. Hence, new conventions emerged through the International Maritime Organisation to increase the financial limits, but the incidents of the Erika and the Prestige showed once more that the amounts were insufficient. As a result, new amendments were adopted in 2000 to substantially increase the limit and again in 2003 a Supplementary Fund was adopted to provide the third tier compensation.

Hence, the goal of this paper is to examine critically the financial caps for marine oil pollution and to examine this more particularly within the context of Chinese law and its relationship with the international conventions. Hence, this paper will be structured as follows:

After this introduction, in a second section, the international conventions will be critically examined. Attention will be paid to the 'old' CLC convention, but also to alternative compensation mechanisms that once existed, such as TOVALOP and CRISTAL. Attention will also be given to the revision of the CLC Convention and the task of the International Oil Pollution Compensation Fund today. Also the most recent evolutions of the international regime in 2000 and 2003 will be examined. The question will always be asked why the financial limits on liability still exist and how these financial caps are assessed.

A third section will address the limitations of liability for marine oil pollution in Chinese law. In that section, the relationship between the international conventions and Chinese law will be discussed and an overview will be provided of potential limits in Chinese legislation.

Section four will critically discuss potential problems that may arise under financial caps which are set too low. It will be argued that financial caps cannot only lead to undercompensation, but may potentially also affect the preventive effect of liability systems.

After discussing these potential problems with (too low) financial caps some policy recommendations will be formulated in section five. Hence, The objective of this paper is to draw some lessons from the experience with financial caps from the international conventions for marine pollution law in China.

2 INTERNATIONAL REGIMES

The increasing sea borne transport of oil constitutes a growing pollution risk to the seas and the oceans. Among all the contaminants from ships, oil has the highest public profile. Today the compensation of damage caused by oil pollution is mainly regulated through international conventions like the International Convention on Civil Liability for Oil Pollution Damage (the CLC Convention) and the International Convention on the Establishment of an International Fund for Compensation for Oil Pollution Damage (the Fund Convention).

2.1 PRE-1969 REGIME

Until 1969, there were no international conventions which specifically addressed the liability for ships which caused marine pollution. Liability for oil pollution damage was generally limited to the vessel's liability tonnage with amounts limited under the 1957 International Convention Relating to the Limitation of Liability of Owners of Sea-going Ships or subject to different national law concerning tort and liability.

The Torrey Canyon disaster in 1967 constituted a turning point. The maritime world realized that it was not prepared for such a major pollution incident. The existing rules on limitation of liability were not adequate for a major pollution incident.

This led to intense activities at national and international levels. As a result, the International Maritime Consultative Organization (IMCO, now IMO) produced two international conventions – the Civil Liability Convention in 1969 and the Fund Convention in 1971, which turned out to be the basis of an oil pollution liability system in the following years, even to present. In addition, the shipping industry and the oil industry produced two private schemes the TOVALOP in 1969 and CRISTAL in 1971. This paper will first take a closer look at the contents of these regimes and more particularly at the amounts of liability.[1]

---

[1] Since the financial caps are the main focus of this paper, other aspects of the conventions will not be addressed here. For information on those other aspects we refer the reader to the literature listed in the footnotes.

2.2 1969/1971 REGIME

2.2.1 CLC Convention 1969 & Fund Convention 1971

2.2.1.1 CLC Convention 1969

Although most major shipping countries are now parties to the CLC 1992,[2] the 1969 CLC Convention set up the basic principle for oil pollution damage compensation which still applies today, the present discussion will begin with the 1969 CLC Convention.[3]

The CLC Convention 1969 set up strict liability for the owner of a tanker that caused the pollution and compulsory insurance for the tanker owner as basic principles, and under certain circumstances, the tanker owner enjoys a limitation of liability.

Strict Liability

The shipowners are held strictly liable whenever an oil spill occurs. Apparently, the introduction of strict liability for oil pollution damage was not accepted that easily at the discussions preceding the CLC Convention. Many countries with a large tanker fleet first opposed strict liability, but finally the delegates accepted the description proposed by the IMO. The principle of strict liability was accepted, but it was weakened by the limitation of the liability. Moreover, in some specific cases the tanker owner will not be strictly liable under certain conditions, for instance if the damage is the consequence of an act of war or if the damage is exclusively caused by the intentional act or negligence of a third party.

As Chen rightly notes, the replacement of the traditional fault-based liability by a form of strict liability may now sound common place after the development of environmental law, but could certainly be considered as extraordinary and innovative when the CLC came into being in 1969.[4]

Compulsory Insurance

The owner of a tanker carrying more than 2,000 tonnes of persistent oil as cargo is obliged to maintain insurance to cover his liability. The victims may in addition bring legal action directly against the insurer. The convention stipulates that the vessel should prove the

---

[2] As of 1 March 2004, there are 44 States party to the 1969 CLC Convention, while 92 States are party to the 1992 CLC Convention. Statistics from the website of the IOPC Fund: <http://www.iopcfund.org>.

[3] For a comment on this convention see *inter alia* D. Abecassis, 'I.M.O. and liability for oil pollution from ships: a retrospective', (1983) *Lloyds Maritime and Commercial Law Quarterly* 45-46; E.H.P. Brans, *Liability for damage to public natural resources. Standing, damage and damage assessments* (Kluwer Law International, The Hague, 2001), pp. 315-316 and E. Gold, *Handbook on marine pollution* (Gard, Arendal, 1985), pp. 44-47; O. Ozçayirz, *Liabibility for oil pollution and collisions* (Lloyds of London Press, London, 1998), pp. 211-218.

[4] See X. Chen, *Limitation of liability for maritime claims. A study of US law, Chinese law and international conventions* (Kluwer Law International, The Hague, 2001), p. 140 and see M. Jacobsson, 'The international conventions on liability and compensation for oil pollution damage and the activities of the international oil pollution compensation fund', in C. De La Rue (ed.), *Liability for damage to the marine environment* (Lloyds of London Press, London, 1993), p. 41.

availability of insurance or another form of financial security to cover its liability. As proof of the availability of financial security, the vessel should carry a certificate.[5]

As will be argued below, compulsory insurance can increase the utility of the risk averse person; it can lead to a protection of the insured – the tanker owner in this case, since the larger risk of oil pollution damage compensation is removed from the insured tanker owners by the payment of a relatively small premium; and it can also reduce the transaction cost[6] in the sense that it is already decided *ex ante* through the conditions of the insurer who should intervene and bear the costs when an accident happens.[7]

The goal of the CLC Convention is to ensure that tanker owners have appropriate levels of insurance to cover their potential liability for pollution damage.[8] However, how to set the optimal level of insurance remains a critical issue.

Limitation of Liability

The shipowner whose ship caused pollution was entitled to limit his liability in any incident to 133 Special Drawing Rights (SDR) for each tonne of the ship's gross tonnage with a maximum liability of 14 million SDR (around USD 18 million for each incident). During the discussion at the IMCO conference, the level of compensation available from shipowners was the cause of great concern for the insurance industry. The UK, having a large P&I club industry, was opposed to high limits of liability as it believed that high limits would dry up the funds which were available in the world insurance market.[9]

For a potentially liable party, the argument was that a limitation of liability clearly provided protection against the consequences of a catastrophic accident. For the victims, however, a limitation of liability has no positive effects. They remain unprotected to the extent the damages exceed the limitation of liability, unless these damages are compensated by other means, for example, through a compensation fund. If the liability is unlimited, or is limited to an amount higher than the insurance coverage, the liable party will have to use his personal assets for at least part of the losses. This could provide an incentive for prevention. However, the personal solvency of the defendant in this case will limit the victim's chance of obtaining compensation.[10]

The argument is that the limitation provided makes this insurance coverage possible.[11] If the limitation were not predictable or if it were even unlimited, the insurers would not

---

[5] The contents of this certificate are set in Art. VII.2 of the Convention. For further details *see* E. Gold, *supra* note 3, 45 and X. Chen, *supra* note 4, 141. *See also* O. Ozçayirz, *supra* note 3, 215-217.

[6] G. Skogh, 'The Transaction Cost Theory of Insurance: Contracting Impediments and Costs', (1989) *Journal of Risk and Insurance* 726-732.

[7] The economic rationale of compulsory insurance will be further discussed below in 4.4.

[8] *See* K. Le Couviour, 'Responsabilité pour pollutions majeures résultant du transport maritime d'hydrocarbures. Après l'Erika, le Prestige... L'impératif de responsabilisation', (2002) *La Semaine Juridique* (JCB) 2271.

[9] Official record of the IMCO conference, LEG 1.

[10] Paper prepared by a Consultant, Financial Limits of Liability and Compulsory Insurance under the Draft Protocol on Liability and Compensation for damage resulting from transboundary movement of hazardous wastes and their disposal, Conference of the Parties to the Basel Convention on the Control of Transboundary Movements of Hazardous Wastes and Their Disposal, fifth meeting, Basel, 6-10 December 1999, United Nations Environment Programme.

[11] *See* L. Le Couviour, *supra* note 8, 2271.

be able to assess the magnitude of the risks and set an appropriate premium, and it would also be difficult for the shipowner to shift the risks to their liability insurance.

These financial caps are, at least towards third parties, often considered inefficient since they lower the incentives to take care and may thus contract the insurer with a moral hazard problem. So the financial limits are paradoxical, since they, on the one hand, are introduced to make the risks insurable, but on the other could also lead to a moral hazard. Both aspects of the caps will be addressed further in section 4 where a critical analysis is provided.

2.2.1.2 Fund Convention 1971

Although the 1969 CLC Convention provided a useful mechanism for ensuring the payment of compensation for oil pollution damage, it did not deal satisfactorily with all the legal, financial and other questions raised during the IMO conference in 1969.

Since the level of the ceiling for compensation under the CLC 1969 was considered insufficient to compensate the victims in cases of serious oil spills, the 1971 Convention was thus adopted to address the question of compensation by vesting obligations on oil cargo owners. It imposed liability on a fund to which the contributors are the oil receivers. This Fund was intended to cover the damage not covered by the CLC Convention, thus to relieve the shipowners from the additional financial burden imposed by the CLC Convention.[12] Many scholars hold that the combination of the CLC Convention and the IOPC Fund lead to a sharing of the costs of oil pollution damage between the vessel owners and the oil industry.[13]

2.2.2 Voluntary Schemes

In the wake of intense discussions after the Torrey Canyon disaster, while waiting for the international conventions to be implemented worldwide, the industries decided to take action since the negotiations between governments at international level would last long, while the urgency of an adequate and prompt compensation regime was highlighted by the Torrey Canyon accident. The shipping and oil industries respectively set up, as stopgap measures, two voluntary regimes that comprise two steps.[14] In the first step, the individual tanker owner whose vessel causes a spill is responsible. In the second step, the responsibility of the cargo owner is triggered where the tanker owner's applicable limit of liability is exceeded. This division was worked out in two separate schemes. These industry schemes are often known as TOVALOP and CRISTAL, which are discussed below.

---

[12] *See* X. Chen, *supra* note 4, 142-143 and *see also* M. Jacobsson, *supra* note 4, 39-56 and E.H.P. Brans, *supra* note 3, 317-318.

[13] *See* D. Abecassis, *supra* note 3, 47; J. Bongaerts and A. Debièvre, 'Insurance for civil liability for marine oil pollution damages', (1987) *The Geneva Papers on Risk and Insurance* 148 and E. Gold, *supra* note 3, 115.

[14] *See* C. White, 'The voluntary oil spill compensation agreements- TOVALOP and CRISTAL', in C. De La Rue (ed.), *Liability for damage to the marine environment* (Lloyds of London Press, London, 1993), pp. 57-70.

2.2.2.1 TOVALOP

On 7 January 1969, seven oil company shipowners signed the Tanker Owners Voluntary Agreement concerning Liability for Oil Pollution (TOVALOP). This applies only when the CLC does not apply.[15] It can hence be seen as a residual scheme. One of its main advantages is that it allows for reimbursement of the costs of preventive measures taken in case of a threat of pollution. It also includes tankers in ballast, which are excluded from the CLC. Moreover, the procedure of settlement of claims is quick and inexpensive.

2.2.2.2 CRISTAL

On 14 January 1971, with the anticipation that the Fund Convention would come into force, the oil industry adopted its voluntary agreement, the so-called CRISTAL, being the Contract Regarding an Interim Supplement to Tanker Liability for Oil Pollution. CRISTAL was designed to compensate oil pollution victims by supplementing TOVALOP when its compensation proved insufficient.

The industry scheme whereby tanker owners and oil importers undertook voluntary liability to pay compensation for oil pollution damage not only had an impact on world opinion, but also relieved pressure on national governments to introduce their own unilateral solutions.[16]

Originally, TOVALOP and CRISTAL were created as interim measures before the 1969 CLC and the 1971 Fund Convention could be fully implemented. When the entire international oil pollution compensation regime came into force in 1978 (1969 CLC came into force in 1975, and the 1971 Fund Convention came into force in 1978), the question arose whether the industry schemes were still necessary.

However, the situation then was that many countries had not yet become party to the international conventions, while a substantial part of the world's tanker fleet and oil companies were involved in the voluntary industry schemes. In order to avoid the application of different national laws and maintain consistency in all cases, these two voluntary schemes were modified in 1978 to stay in line with their counterparts, being the international conventions.

2.3 1984/1992 REGIME[17]

2.3.1 1984 Protocols

Since the adoption of the 1969 CLC Convention and the 1971 Fund Convention, the gross volume of spilled oil declined in the 1970s. However, the serious damage caused by

---

[15] See J. Bongaerts and A. Debièvre, *supra* note 13, 147-148 and E. Gold, *supra* note 3, 25-26.
[16] Cristal was considered an important precedent in the literature since it showed that the oil industry also could, under the pressure of the public opinion, take its responsibility for oil pollution damage (see E. Gold, *supra* note 3, 26).
[17] See M. Joransson, 'The 1984 and 1992 Protocols to the Civil Liability Convention 1969 and the Fund Convention 1971', in C. De La Rue (ed.), *Liability for damage to the marine environment* (Lloyds of London Press, London, 1993), pp. 71-82.

Amoco Cadiz[18] in 1978 showed that the old regime was ineffective and inadequate in the event of major oil spills. Hence, the financial limits under the old regime were faced with the urgent need of upgrading.

Consequently, soon after entry into force of the 1971 Fund Convention (in 1978), informal consultations commenced aiming at increasing the amount of compensation under the two conventions. In 1984, two protocols were adopted to achieve this objective. The protocols indeed increased the financial limits. These protocols, however, were designed to fail, since the conditions for their entry into force could not be met without American participation.[19] As a result, these 1984 Protocols never entered into force.

The US is one of the world's largest consumers of oil and one of the largest shipping nations. Therefore, its participation would have greatly changed the situation of the international system, especially from a financial point of view – the US would have been the most important contributor to the IOPC Fund. US approval would have helped establish a uniform international oil pollution liability regime. However, the country did not join the international conventions but enacted its own regime instead – the Oil Pollution Act was adopted in 1990, effectively closing the door to US participation in the international scheme.

The changes, particularly the increase in compensation amounts, established in the 1984 Protocols were urgently needed, as demonstrated by a series of further incidents, notably, the 1991 Haven accident off the coast of northern Italy. Accordingly, there was a growing sentiment, especially among the European members of the IOPC fund, that something had to be done to bring the substance of those instruments into force in order to keep the international system up to date and avoid the threat of further regional schemes. The 1992 Protocols thus came into force.

2.3.2 1992 Protocols

The 1992 Protocols are almost identical in substance to the 1984 Protocols but there was a change to the entry into force provision, which eliminated the need for US participation. The new protocols also provided for an increase in the financial limits. The new protocols entered into force on 30 May 1996.[20]

The compensation limits are those originally approved in 1984:

- for a ship not exceeding 5,000 gross tonnage: liability is limited to 3 million SDR (about USD 3.8 million)

---

[18] The CLC Convention entered into force on 19 June 1975, after the incident following the accession of France. Claims arising from the incident were litigated in the US against Standard Oil of Indiana, which actually controlled the operations of the tanker, but which was not the registered owner. In this way the restraints of the 1969 CLC, including its inadequate limits, were avoided. For an instructive account, *see* E. Fontaine, *The French Experience of Tanio and Amoco Cadiz, Incidents compared*, Genoa Seminar, 21-23 September, CMI.

[19] *See* X. Chen, *supra* note 4, 141.

[20] For an excellent overview of the contents and importance of the 1992 Protocols *see* E. Brans, 'Liability for ecological damage under the 1992 Protocols to the Civil Liability Convention and the Fund Convention, and the Oil Pollution Act of 1990', (1994) *Tijdschrift voor Milieuaansprakelijkheid* (TMA) 61-67 and 85-91 and E. Brans, *supra* note 3, 344-360.

- for a ship 5,000 to 140,000 gross tonnage: liability is limited to 3 million SDR + 420 SDR (about USD 538) for each additional unit of tonnage
- for a ship over 140,000 gross tonnage: liability is limited to 59.7 million SDR (about USD 76.5 million)

The international system had encountered many problems during the 20 years' application. In addition to the problem of insufficient compensation, two other problems are intrinsic to the international system: firstly, the refusal by the IOPC Fund to cover environmental damages; second, the lack of effective sanctions to the pollution prevention system. The refusal of the US to take part in the international regime has also posed an obstacle to the highlighted concerted international efforts to control oil pollution by ship.

France played an important role in the revision process because it had suffered the greatest exposure to the risks of oil spills. The French delegation proposed an increase in the amount of compensation available from the IOPC Fund. Another problem raised was the need to extend the field of application of the conventions: some of the terms of the CLC and the Fund Convention needed to be redefined to give a broader scope to the conventions.

The 1992 Protocols thus came into force with a substantial increase in limitation of liability in comparison with the old regime of 1969 CLC and 1971 Fund. The amounts available after the entry into force of the 1992 Protocols were thus substantially increased, as indicated above. However, the basic principle of a joint contribution by the oil industry and the shipping industry as well as a system of strict liability of the tanker owner with financial caps on liability remained in existence.

2.3.3 Disappearance of TOVALOP & CRISTAL

For many years, the industry schemes TOVALOP and CRISTAL provided a worldwide voluntary scheme of compensation for pollution from tankers. However, these voluntary schemes are considered to have fulfilled their intended purpose as most of the countries have joined the international conventions, the voluntary schemes were thus abrogated on 20 February 1997.

2.3.4 Winding up of the 1969/1971 Regime

From 16 May 1998 on, Parties to the 1992 Protocols ceased to be parties to the 1969 CLC Convention and the 1971 Fund Convention due to a mechanism for compulsory denunciation of the old regime established in the 1992 Protocols. However, for the time being, the two regimes co-exist, since there are a number of States still party to the 1969 CLC Convention that have not yet ratified the 1992 CLC Convention.

The 1971 Fund Convention ceased to be in force in May 2002 as most of the States with major contributors had left the 1971 Fund to join the 1992 Fund. Thus, the former regime lost its financial basis and became financially vulnerable. Member States who remained in the 1971 Fund would have been confronted with serious financial disadvantages, since the financial burden was spread over increasingly fewer contributors. Should an incident occur involving the 1971 Fund, the 1971 Fund would not be able to pay compensation to victims since there would be few or no contributors in the remaining Contracting States, or the remaining contributors would have been called upon to provide a much greater proportion of the compensation than previously. In order to avoid these damaging consequences, the IMO and the IOPC Fund Secretariat have been working actively to encourage the governments who have not yet done so to accede to the 1992 Protocols and to denounce the 1971 regimes.[21] As a result, after some denunciations of the 1971 Fund Convention, it ceased to be in force on 24 May 2002.[22]

2.4 THE 2000/2003 REGIME

2.4.1 The 2000 Amendments

2.4.1.1 Background

The legal regime concerning the damage caused by oil pollution has been rapidly evolving and still is. This is partially due to new incidents, showing the inadequacy of the latest regime adopted and making yet another change necessary. Following the Erika incident in 1999, a 50 percent increase in the amounts available under the CLC and the Fund Conventions was adopted by IMO Legal Committee in October 2000. This latest change took effect from November 2003 under the tacit amendment procedure.

The adoption of the increased limits comes in the wake of, again, two major incidents: the Nakhodka in 1997 off Japan and the Erika disaster off the coast of France in December 1999.[23] The European Commission has played a central role in the process of updating the international regime of oil pollution compensation.[24] Several issues were on the agenda.

---

[21] A. Blanco-Bazan, *The Erika casualty, Legal Issues from the IMO View*, IUMI 2000 Conference London, 10-13 September 2000, Liability Workshop.

[22] Under Art. 43.1 of the 1971 Fund Convention, it would have ceased to be in force when the number of Member States fell below three. Although many States denounced the 1971 Fund Convention, it was unlikely that the number of Member States would fall below three in the foreseeable future. For this reason, at the September 2000 Diplomatic Conference, a Protocol was adopted to modify the termination conditions for the 1971 Fund. The modified Protocol stated that the 1971 Fund would cease to be in force on the date on which the number of Member States fell below 25, or 12 months following the date on which the Assembly (or any other body acting on its behalf) noted that the total quantity of contributing oil received in the remaining Member States had fallen below 100 million tonnes. As the number of Member States fell below 25 on 24 May 2002, the 1971 Fund Convention ceased to be in force from that day.

[23] See on the incident with the Erika and the consequences for further (European) action towards the prevention of oil spills C. Roche, 'Après l'Erika: la prevention de la pollution des mers par le renforcement de la sécurité maritime en Europe (Erika I)', (2002) 3 *Revue Juridique de l'Environnement* 373-392. And see K. Le Couviour, *supra* note 8, 2269-2272.

[24] For some examples of Europe's activity in this domain, *see e.g.* Directive 2001/150/EC of 19 December 2001 amending Council Directive 94/57/EC on common rules and standards for ship inspection and survey organizations and for the relevant activities of maritime administrations, *OJ* L19/9 of 22 January 2002, Council Decision of 19 December 2002 authorizing the Member States in the interest of the community, to sign, ratify or accede to the international convention on civil liability for tanker oil pollution damage, 2000 (the Tankers Convention) *OJ* L256/7 of 25 September 2002 and Directive 2002/84/EC of the European Parliament and of the Council of 5 November 2002 amending the directives on maritime safety and prevention of oil pollution from ships, *OJ* L324/53 of 29 November 2002, just to mention a few of the

Firstly, concerning the inadequate limits of liability, the increases in compensation limits have been given urgency by moves within the European Commission to establish a separate European Community compensation fund, the so-called COPE (Compensation for Oil Pollution in Europe) Fund, a third tier of compensation (up to one billion euros), in response to the perceived shortfall of the international oil pollution liability funds.[25]

Second, concerning the inappropriate balance between responsibilities of different players and their exposure to liability, and the shipowners' unbreakable limitation rights, the European Commission has criticized the central role played by private organizations in the area of maritime safety, developing rules more directly to regulate classification societies (responsible for verifying the seaworthiness of vessels) and recommending amendments to CLC 1992 to weaken the right of liability limitation of the shipowner.

2.4.1.2 2000 Amendments

The influence of the European Commission on the IMO had its effect. At the IMO Legal Committee meeting in October 2000, resolutions were passed which increase the limits of the 1992 CLC and Fund Convention by 50 percent, with effect from November 2003. (The proposal of a third tier of compensation originally developed by the European Commission was taken up by IMO as well.) The compensation limits set by the 2000 amendments are therefore as follows as of November 2003:

- For a ship not exceeding 5,000 gross tonnage: liability is limited to SDR 4.51 million, approximately USD 5.78 million USD(under the 1992 Protocol, the limit was SDR 3 million)
- For a ship 5,000 to 140,000 gross tonnage: liability is limited to SDR 4.51 million (USD 5.78 million) plus SDR 631 (USD 807) for each additional gross tonne over 5,000 (under the 1992 Protocol, the limit was SDR 3 million plus SDR 420 for each additional gross tonne)
- For a ship over 140,000 gross tonnage: liability is limited to SDR 89.77 million, approximately USD 115 million (under 1992 Protocol, the limit was SDR 59.7 million)

2.4.2 2003 Supplementary Fund Protocol

The evolution of the international regime makes clear that limits were increased every time following new incidents. When the new 2000 amendments had just been adopted, another major disaster, in 2002, with the tanker 'Prestige' again caused new challenges to

---

European initiatives that were published in 2002, but one could point at many other initiatives showing European activism in this domain.

[25] *See* in this respect, COM(2000) 802 final, the Communication from the Commission to the European Parliament and the Council on a second set of Community measures on maritime safety following the sinking of the oil tanker Erika, Brussels, 6 December 2000, 51-83; and the amended proposal for a regulation of the European Parliament and of the Council on the Establishment of a Fund for the Compensation of Oil Pollution Damage in European Waters and Related Measures, *OJ* C227 E/487 of 24 September 2002.

the international regime. The European Commission was once more arguing that the new regime of the 2000 amendments was even outdated before it entered into force in November 2003. Higher limits or even unlimited liability were again on the agenda.[26]

The current international regime, by providing a combination of channelling of liability and setting almost unbreakable limits, provides an effective shield for unscrupulous operators. Yet, strict channelling and a stringent test for breaking the limitation have undoubtedly contributed to quick settlement of claims. This especially benefits claimants who do not have the financial means to sustain long and costly litigation to establish who is liable and whether, once the responsible party is identified, that person can maintain the right to limit liability.[27]

So far no specific proposals have been suggested for dramatic changes to the current regime. The course today is probably not to tamper with this financial aspect of the international scheme. A preferable course might be to focus on appropriate modifications to international conventions aimed at maritime safety, such as MARPOL and SOLAS, as well as promoting greater vigilance under the various port State control agreements, to address the problem of substandard ships and their operators. These are issues that will be high on the political agenda in the future. Typical in this respect is the reaction of the European Commission after the Prestige accident, which issued immediately a communication.[28] The Commission suggests that Member States should support proposals aimed at restricting the right of shipowners to limit their financial liability if they are at fault. Again, the channelling of liability is criticized and hence the Commission pleads in favour of 'removing the *de facto* immunity of other key players, in particular the charterer, operator or manager of the ship from compensation claims (other than from recourse claims by the registered owner)'. Finally the Commission also points to the fact that the international regime does not provide for adequate compensation for damage to the environment as such, this being more particularly ecological restoration.[29]

The threat of the establishment of a separate European fund apparently had its influence on the IMO. During a diplomatic conference in London between 12 and 16 May 2003, a new protocol to the Fund Convention was accepted that established a supplementary fund bringing the total amount of compensation for oil pollution damage to approximately EUR 920 million or USD 1 billion (the limits have been increased by about 50 percent). The establishment of this new fund substantially improves the capacity of compensation to pollution victims, compared to the previous situation. The new convention establishing the supplementary fund will only enter into force three months after eight States, representing a total of at least 450 million tonnes of contributing oil, have become parties to it. Effectively, this would mean that if the EU Member States alone were to contract, they could bring the supplementary fund protocol into operation. Thus the European Commission proposed to the Council to take a decision urging the European Member States to ratify the supplementary fund convention as soon as

---

[26] *See* G.J. Van der Ziel, 'De olieramp met de 'Prestige'', (2003) *Tijdschrift voor Milieuaansprakelijkheid* 7.
[27] *Ibid.*
[28] COM(2002) 681 final, Communication from the Commission to the European Parliament and The Council on improving safety at sea in response to the Prestige accident, Brussels, 6 December 2002.
[29] *See* note 26 at 7.

possible.[30] So far, there are two countries that have ratified the Supplementary Fund Convention (Denmark and Norway). In sum, as a result of these latest changes, the international regime now provides for a maximum available amount of compensation of USD 1 billion.

3 CIVIL LIABILITY FOR OIL POLLUTION DAMAGE IN CHINA

China acceded to CLC69 and later CLC92, but the Fund Convention 1992 applies only to the Hong Kong Special Administrative Region (SAR). However, there is no specialized legislation in China that is comparable with, for instance the Oil Pollution Act 1990 in US, which regulates the oil pollution damage as such.[31] In the China Maritime Code, and the Marine Environment Protection Law, there is not even a separate Chapter concerning the liability for oil pollution damage. Hence, the compensation for oil pollution damage in Chinese practice remains a complex issue.

3.1 INTERNATIONAL CONVENTIONS EFFECTIVE IN CHINA

3.1.1 CLC Convention

China acceded to the CLC Convention 1969 on 30 January 1980, which entered into force in China on 29 April 1980, and applied to Hong Kong SAR with effect from 1 July 1997. Due to the compulsory denunciation procedure, China has denunciated the 1969 CLC and joined the 1992 CLC Convention, which has been effective in China since 5 January 2000. As the 2000 Amendment to the CLC Convention took effect, the Ministry of Communication in China published the announcement on 11 December 2003. As a result, this Amendment is effective in China as well.

3.1.2 Fund Convention

With respect to the Fund Convention, the Chinese government has declared that it only applies to the Hong Kong SAR. Mainland China is to date not yet a party to the Fund Convention. This might be because most large oil companies in China are State owned, while according the Fund Convention, the oil industry that has large import volumes must contribute to the Fund. Hence, effective lobbying by special interest groups – more particularly the (largely government-owned) oil companies – is thus the reason why China so far did not accept the Fund convention.

However, as mentioned above, there is so far neither a specialized legislation nor even a separate Chapter in the China Maritime Code or Marine Environment Protection Law concerning oil pollution damage compensation, which resulted in the different

---

[30] COM(2003) 534 final of 8 September 2003.
[31] For an overview of China's practice in the law of the sea, see J. Greenfield, *China's practice in the law of the sea* (Clarendon Press, Oxford, 1992).

interpretation by various local courts. Hence, the amount of liability today in China may largely depend on the court's decision concerning the applicable law.

3.2 DOMESTIC LEGISLATIONS IN CHINA

There are divergent opinions concerning the applicable law in cases of oil spills in China. One prevailing opinion holds that in such cases, when there is foreign element involved, the international conventions to which China is a party shall apply.[32] This complies with the provisions on foreign-related matters in the General Principles of the Civil Law, and the China Maritime Code, which states that in case of existing conflict with international conventions to which China is a party, the international conventions shall prevail unless the Chinese government has made a reservation. Moreover, the Regulation of 1983 on the Prevention and Control of Marine Pollution from Vessels, stipulates in Article 13 that 'ships carrying 2,000 tons or more of bulk oil cargo navigating on international sea lanes' shall comply with the requirements in the CLC Convention. Hence, all Chinese ships engaging in international oil transportation must buy insurance as required by the CLC Convention, and their liability in case of oil pollution incident is limited to that provided for in the CLC Convention. This might be an issue since according to this theory, those not navigating on international lanes, being offshore and inland water navigating ships, Chinese national law shall apply. As will be seen below, in the current legislation in China, the financial limits for the shipowner is lower than those under the CLC Convention, and the compulsory insurance is not effectively implemented as there is so far only one general provision in the Marine Environment Protection law concerning the State's responsibility to establish the compulsory insurance. Moreover, there is not yet a compensation fund in China despite the fact that the revised Marine Environment Protection law has included such requirement on the State to perfect the compensation system for oil pollution and to establish such fund.

Another opinion states that the CLC Convention shall apply to all cases of marine oil pollution incidents as there is no provision in any relevant domestic legislation that stipulates otherwise.[33] This may seem to contribute to the harmonization and uniformity of international and domestic law by avoiding conflicts of law, however, this does not seem so practical for China in its current situation, since China as a developing country bears the dual tasks of economic development and environmental protection. So the uniform application of international conventions to cases even where no foreign elements are involved might not be helpful to facilitate the economic development in China.

In all, the difference between the domestic law in China and the international conventions leads to complications in the practice. This will be discussed below. Here

---

[32] Z. Lujun, *Legal Issues in Marine Oil Pollution* (original text in Chinese: Hai Shang You Wu Chu Li Zhong De Ruo Gan Fa Lu Wen Ti), International Seminar on Compensation Regime for Ship-source Pollution Damage, Shanghai, June 2001.

[33] Y. Xiaohan, Case on Limitation of Liability for Oil Pollution Damage Compensation, Guangzhou Maritime Court Cases, 24 March 2003.

relevant provisions in Chinese law concerning liability and compensation for oil pollution will be considered.[34]

3.2.1 China Maritime Code[35]

Chapter XI of China Maritime Code contains provisions on Limitation of Liability for Maritime Claims.[36] In Article 208, it is specifically provided that the provisions of this Chapter shall not apply to 'Claims for oil pollution damage under the International Convention on Civil Liability for Oil Pollution Damage to which the People's Republic of China is a party'. Hence, all the internationally navigating ships are subject to the limitations provided under the CLC Convention, irrespective of their nationalities, Chinese or foreign.

Where the CLC is held not applicable in certain oil spill incidents (mainly tankers navigating in offshore and inland waters), the indemnity for oil pollution damage is considered a restricted credit right in Article 207, the limitation of which shall be calculated according to the approach provided in Article 210. Article 210 distinguishes between the liability for loss of life or personal injury, and loss other than this, which mainly includes property damage. It is interesting to note that the limit for claims for personal injury is almost twice that for property damage.

A lower standard for the limitation is set forth in the China Maritime Code compared to that in the CLC Convention. The difference in standards seems contrary to the protection of victims of oil pollution and is not very helpful for the development of the shipping business.[37]

3.2.2 Marine Environment Protection Law

The revised Marine Environment Protection Law took effect on 1 April 2000. The most important revision in respect of marine oil pollution liability is Article 66, which stipulates that:

> 'The state shall make perfect and implement the system of compensation for civil liability for pollution damages caused by vessels, establish principles of liability to be shouldered jointly by shipowners and cargo owners in accordance with liability for pollution damage caused by vessels, and establish marine insurance for pollution and a system of compensation funds for oil pollution damages.'

---

[34] *See* on the procedure to assess the damage to the marine environment in China, W. Mao Shen, L. Shu Jian, and S. Man Tang, 'The normal procedure of assessment of damage to the marine environment in Chinese judicial practice', in C. De La Rue (ed.), *Liability for damage to the marine environment* (Lloyds of London Press, London, 1993), pp. 29-31.

[35] Maritime Code of the People's Republic of China, adopted at the 28th Meeting of the Standing Committee of the 7th National People's Congress on 7 November 1992 and effective as of 1 July 1993. An English translation of this Maritime Code was first published in 2000 through the Foreign Languages Press, Beijing.

[36] For a discussion of the limitation of liability for maritime claims in the international conventions as well as in Chinese law, *see* X. Chen, *supra* note 4.

[37] *See* X. Chen, *supra* note 4, 148.

Thus, the establishment of a compensation fund and compulsory insurance now has a legal foundation through the revision of the Marine Environment Protection Law. However, the principle of joint liability between the shipowner and the cargo owner remains at a theoretical stage, there has been no compensation fund set up in China so far and the compulsory insurance is not fully implemented in practice.

3.3 PRACTICAL ISSUES

3.3.1 Negative Effect on Clean-up

Since China has not established a complete system for oil pollution compensation that corresponds to the international regime (sharing of liability of shipowners and oil receivers, compulsory liability insurance for shipowners and compensation fund), there is no reliable financial source for oil pollution damage. Thus, pollution damage is often inadequately compensated, and thus great social losses are incurred. As a result, the clean-up activities and preventive measures were not encouraged.[38]

Moreover, among the ships carrying oil on domestic lines, a large number is small privately-owned tankers. Some of these shipowners have only one single vessel registered under their names, which lowers their financial capability in case of liability. Some of these tankers are poorly maintained old tankers, which increases the potential accident risk. Under this situation, the shipowner is often insolvent so he is unable to pay the full compensation; when the pollution damage exceeds the shipowner's liability, the surplus part cannot be paid.

3.3.2 No Compulsory Insurance for Cabotage Tankers & Low Financial Limits under Chinese Law

According to the CLC Convention, tankers carrying more than 2,000 tonnes of oil for international transportation are obliged to take insurance, while under the China Maritime Code, there is no such requirement. Moreover, for tankers of less than 2000 tonnes, there is no insurance requirement even under the Convention. However, these ships are often employed by private owners in China for coastal water shipping. Under the revised Marine Environment Protection Law of 2000, Article 66 stipulates that the State shall ensure the establishment of insurance for pollution. This is, however, only a general provision which is not effectively implemented in practice.

Therefore, the problem arises in cases of ships engaged in coastal and inland water shipping and tankers under 2,000 tonnes, where there is no effective legal duty to obtain compulsory insurance. In practice, in those cases only a letter of credit is needed, following the instructions from the Ministry of Communications. This discrepancy affects China's oil pollution compensation as a whole. On the one hand, the absence of

---

[38] L. Hong and Z. Zhengming, *Setting up the Chinese Characteristic compensation system for oil pollution damage from vessels* (original text in Chinese, Jian Li You Zhong Guo Te Se de Chuan Bo You Wu Sun Hai Pei Chang Ji Zhi), International Seminar on Compensation Regime for Ship-source Marine Pollution Damage, Shanghai, June 2001.

requirements considering the proof of financial security for cabotage vessels is inefficient as far as oil pollution damage compensation is concerned and, on the other hand, the financial limits set by domestic legislation (mainly the China Maritime Code) might be too low.

3.3.3 No Compensation Fund in China

Although in the newly revised Marine Environment Protection Law joint liability between the shipowner and the cargo owner is established, there has been no compensation fund set up in China yet.

At present, spills from small ships seem, in practice, to be the more problematic issues, but there is of course an increasing risk, given the huge increase in volumes of imported oil and the growth of the shipping industry, that a major spill will exceed the owner's limits provided for in the CLC Convention, giving rise to problems of inadequate compensation under the CLC Convention as well.

Thus, it is necessary to set up a compensation fund in China. However, what type of compensation fund is suitable for the particular Chinese situation is still under discussion.

4 ANALYSIS

4.1 INTRODUCTION

The history sketched thus far of the international conventions demonstrates that the conventions are apparently reactive, in the sense that they are every time adapted a new incident again showed that the previously agreed limits were insufficient to meet – again – greater oil spills. Hence, the CLC Convention, adopted in 1969 and entered into force in 1975, was amended with a protocol in 1976,[39] in 1984 (although those protocols did not enter into force), in 1992 and recently in 2000. This constant changing of the Convention, whereby every time the limits were adapted, raises of course the more fundamental question whether there should be a limitation of liability at all. In the economic analysis of law, liability rules are intended to have a deterrent effect. Economists have, therefore, often claimed that a financial cap on liability could negatively affect the incentives to take care. This argument is also used precisely within the context of oil pollution damage. Hence, this paper will analyze some of the economic effects of financial caps from an economic perspective, assuming that no insurance is available (4.2). However, the introduction of financial caps was in part motivated by referring to the fact that an unlimited liability of the tanker owner would be uninsurable. The question, thus, arises whether the availability of insurance, as usually provided through the Protection and Indemnity clubs (P&I clubs) can justify financial caps (4.3). However, one may argue that without financial caps and without insurance, potentially insolvent tanker owners may be insufficiently motivated to prevent oil pollution. However, that obviously raises the

---

[39] Although that change was mainly necessary to change the currency from the old point carré franc to the special drawing rights (SDR) as used by the International Monetary Fund (EMF).

question as to whether the appropriate remedy should not be the introduction of a duty to seek financial coverage (as provided in the CLC convention) rather than a financial cap (4.4). Finally, the question of course arises of what the consequences of this analysis are for the particular situation of China which have just been sketched above (4.5).

4.2 FINANCIAL CAPS ON TORT LIABILITY: AN ECONOMIC PERSPECTIVE

4.2.1 Financial Caps Cause Underdeterrence

In the literature it has been indicated that there may be good reasons to favour a strict liability rule for environmental damage, like oil pollution, the main reason being that only a strict liability rule would lead to a full internalization of those highly risky activities.[40] Only with strict liability the potential injurer would also have an incentive to adopt an optimal activity level. This full internalization is obviously only possible if the injurer is effectively exposed to the full costs of the activity he engages in and is therefore in principle held to provide full compensation to a victim. An obvious disadvantage of a system of financial caps is that this will seriously impair the victim's right to full compensation. However, if the cap is indeed set at a much lower level than the expected damage, this would not only violate the victim's right to compensation, but the full internalization of the externality would not take place either. From an economic point of view, a limitation of compensation therefore poses a serious problem since there will be no internalization of the risky activity. Indeed, if one believes that the exposure to liability has a deterrent effect, a limitation of the amount of compensation owed victims poses another problem. There is a direct linear relationship between the magnitude of the accident risk and the amount spent on care by the potential polluter. If the liability therefore is limited to a certain amount, the potential injurer will consider the accident as one with a magnitude of the limited amount.[41] He will, thus, take care to avoid the occurrence of accidents with a magnitude equal to the limited amount and he will not take the care necessary to reduce the total accident costs. Obviously, the amount of care taken by the potential injurer will be lower and a problem of underdeterrence arises. The amount of optimal care, reflected in the optimal standard, being the care necessary to reduce the total accident costs efficiently, will be higher than the amount the potential injurer will spend to avoid an accident equal to the statutorily limited amount.

4.2.2 Strict Liability versus Negligence

This section was begun with the remark that according to the economic literature a rule of strict liability might be favoured in case of major industrial accidents, but that such a rule only works optimally if the injurer is fully exposed to liability. This outcome, however,

---

[40] S. Shavell, 'Strict Liability versus Negligence', (1980) *Journal of legal Studies* 11 and S. Shavell, *Economic Analysis of Accident Law* (Harvard University Press, Cambridge, 1987), p. 8.

[41] *See* M. Faure, 'Economic Models of Compensation for Damages caused by Nuclear Accidents: some lessons for the division of the Paris and Vienna conventions', (1995) *European Journal of Law and Economics* 21-43.

changes when financial caps are introduced. Landes and Posner have shown that in case of insolvency (defined as a situation where the magnitude of the expected loss might exceed the injurer's assets) a negligence rule might provide better incentives. The same applies obviously to situations where the expected loss exceeds the amount to be paid by the injurer as a result of a statutory limitation. Negligence has the advantage that the injurer can avoid being held liable if he follows the due care standard required by the legal system. Therefore, under a negligence rule, the injurer will have an incentive to avoid negligence, even if financial caps apply, as long as the costs of taking care (to avoid negligence) are lower than the financial cap. By abiding by the standard of due care, he can avoid having to pay compensation. Strict liability, on the other hand, always leads to underdeterrence as soon as the expected loss is larger than the amount of the financial cap, whereas this is only the case under negligence when the costs to take care are higher than the financial cap.[42] From this it follows that when financial caps are used by the legislator this seems to be an argument in favour of negligence and against strict liability.[43]

4.2.3 Subsidization Effect

Another effect of protecting a certain industry through a statutory limitation is that this constitutes an indirect subsidization of that particular industry.[44] This point was also raised in the Netherlands during the parliamentary debate preceding a statutory change of the nuclear liability statute.[45] The amount of guarantee provided by the Dutch State was increased to the exceptional amount of NLG 5 billion. It was mentioned in parliament that the ministry of finance would have to charge the licensee of a nuclear power plant for this guarantee provided by the State. If this were not the case, nuclear energy would remain too cheap, since the energy price would not reflect the true costs of the nuclear risk. Evidently, this problem of 'over-consumption of nuclear power' would be reduced if other energy producers enjoyed a limitation of liability as well. In that case a second-best solution could be achieved. However, it seems that nuclear energy producers are the only ones enjoying this benefit of the limitation of liability.

Finally it should be mentioned that so far it has been assumed that the financial cap is set at a lower level than the level of expected loss (otherwise the mentioned problem of underdeterrence would obviously not arise) and that the injurer's wealth exceeds the amount of the financial cap. Thus, the insolvency of the injurer and resulting judgement proof are not yet considered problems.

---

[42] This point has been made by W. Landes and R. Posner, 'Tort law as a regulatory regime for catastrophic personal injuries', (1984) *Journal of Legal Studies* 421-422 and by S. Shavell, 'The Judgement Proof Problem', (1986) *International Review of Law and Economics* 74.
[43] See also M. Faure and R. Van den Bergh, *Objectieve aansprakelijkheid, verplichte verzekering en veiligheidsregulering* (Maklu, Antwerp, 1989), pp. 186-187.
[44] M. Radetzki, 'Private Arrangements to cover large-scale liabilities caused by nuclear and other industrial catastrophes', (2000) *Geneva Papers on Riks and Insurance* 180-195.
[45] M. Faure, 'De verzekering van het nucleaire risico', in *In Volle Verzekerdheid, Opstellenbundel ter gelegenheid van het afscheid van Prof. Mr. O. van Wassenaer van Catwijck* (Tjeenk Willink, Zwolle, 1993), pp. 247-260.

4.2.4 In sum

Applying this analysis to the limitations provided for in the CLC Convention, it can be argued that as long as the financial limits are set at a lower level than the real potential magnitude of the damage, a risk of underdeterrence might emerge: tanker owners may lack sufficient incentive to take the necessary care to reduce the risk caused by oil pollution damage. This limitation, moreover, may in fact constitute a subsidy to the oil industry (assuming that the price of an increased liability by the tanker owner can be passed on to the oil industry) since the injurer is not fully exposed to the risk caused by oil pollution damage. From an economic perspective, this constitutes a market failure since the relative prices of oil will be too low. They will indeed, given the financial caps, not reflect the true social costs caused by oil pollution damage. Moreover, economic analysis also shows that if an insolvency problem arises (caused by capping the liability), negligence would be a preferred rule. In that case, the potential injurer (the tanker owner in the case of oil pollution) would still have incentives to follow the due care standard. By following the due care standard, he would not be held negligent and hence could avoid liability. This therefore leads to the somewhat surprising conclusion that if financial caps were to be introduced, they should preferably be combined with a negligence rather than with a strict liability rule. Economic analysis can thus shed a critical light on the combination of strict liability and financial caps as provided for in the CLC Convention. However, this analysis so far assumes risk neutrality and unavailability of insurance.

4.3 FINANCIAL CAPS TO INCREASE INSURABILITY?

4.3.1 Capacity as Insurability Problem

If the situation where injurers are risk averse but liability insurance is available is now considered, the question should be asked whether a statutory cap on liability is a necessary tool to guarantee the insurability of risks. This argument is often advanced in the context of compulsory insurance and is also raised as an argument to cap the liability for oil pollution. Often the legislator introduced compulsory insurance (as a result of international conventions) and consequently argued that the amount of compensation in tort liability should be limited to make the particular risk insurable.

Generally, one can argue that within liability insurance it is usually not the amount of the expected damage that causes uninsurability of risks, but more often the unpredictability of certain risks.[46] The insurability question is indeed analyzed by looking both at the probability and the magnitude of the risk. The amount is not necessarily the main problem since competitive insurance markets have arranged a variety of devices to cope with large risks as well. Reinsurance, co-insurance, or pooling of risks are well-

---

[46] An event is defined as insurable when insurers can set a premium which reflects the risk and enables them to make a profit and subsequently a market emerges (M. Faure and P. Fenn, 'Retroactive liability and the insurability of long-tail risks', (1999) *International Review on Law and Economics* 487-500). *See* generally on insurability problems the papers published in (1995) *The Geneva Papers on Risk and Insurance* 407-462 and M. Faure, 'The limits to Insurability from a Law and Economics Perspective', (1995) *Geneva Papers on Risk and Insurance* 454-462.

known phenomena that allow insurers to provide large amounts of insurance coverage. All of these instruments are well known in the area of oil pollution as well. The high level of the risk itself therefore does not make certain industrial accidents uninsurable *per se*. Moreover, as will be discussed below, by adjusting the policy conditions the insurer can limit the amount for which he is willing to provide coverage.

Usually the problem of insurability of major environmental disasters like oil spills refers to the 'hard to predict' character of those risks which may make insurers both ambiguous and averse towards these risks. They might respond to insurer ambiguity[47] by requiring an additional risk premium. The insured, however, may not be willing to pay the additional risk premium if they do not recognize the ambiguity facing an insurer.[48]

4.3.2 Limitation of the Duty to Insure

More principally, one can also argue that even in cases where there is a limited availability of insurance coverage (which is already difficult for the legislator to judge, if possible at all) this should not necessarily lead to a limitation of the liability of the injurer. If it indeed appears that the possibilities to obtain liability insurance coverage are limited to a certain amount there is no reason to limit the liability itself to that same amount. A clear alternative would be to introduce a duty to insure up to the available amount of insurance coverage, but to keep the liability of the injurer unlimited. This will, on the one hand, have the advantage that the duty to insure is limited to realistic amounts, whereas, on the other hand, the incentives for the injurer to take care remain at least partially in existence because the injurer is still exposed to risk where the magnitude of the harm was higher than the insured amount.

To summarize, from an economic point of view there are, in unilateral accident situations, very few convincing reasons to limit the amount of compensation due the victim. If insurability problems exist, these can be resolved by limiting the duty to insure. Recent examples have also shown that with respect to the nuclear liability conventions some countries have introduced a duty to insure up to a limited amount, but have left the liability of the licensee of the nuclear power plant itself unlimited. This has been done for instance in Austria,[49] Germany, Japan, Switzerland and Sweden.[50] Perhaps this constitutes an interesting example for the case of oil pollution as well. The advantage to this approach is that in those cases where injurers have assets at stake that outweigh the limited amount for which they had to purchase insurance coverage, they will still have incentives to

---

[47] See for the basic literature on insurer ambiguity H. Kunreuther, R. Hogarth and J. Meszaros, 'Insurer Ambiguity and Market Failure', (1993) *Journal of Risk and Uncertainty* 71-87.
[48] These problems have been discussed extensively in M. Faure and P. Fenn, *supra* note 46.
[49] See M. Hinteregger, 'La nouvelle loi autrichienne sur la responsabilité civile pour les dommages nucléaires', (1998) 62 *Bulletin de Droit Nucléaire* (BDN) 27-34.
[50] See M. Trebilcock and R.A. Winter, 'The economics of nuclear accident law', (1997) *International Review on Law and Economics* 221.

further reduce the accident risk. A generalized limit on liability does not take into account the differing financial possibilities of injurers and their insurers.[51]

4.3.3 Contractual Limitations

Although there are, generally, very few arguments in favour of a generalized statutory limitation of the liability, this does not mean that there may be no reasons for a contractual limitation in insurance policies. In many insurance policies, these limitations already exist since an insurer will hardly ever provide unlimited coverage for the liability risk.

Thus, an insurer could consider lowering the amount of liability coverage generally. This is, by the way, what is done in most environmental liability policies in practice: almost no policy will provide coverage to a potential polluter without a financial limit.

4.3.4 Unlimited Liability to Control Moral Hazard

There is, finally, one general argument related to insurance which can be put forward against financial caps introduced in legislation. Statutory limitations could be contrary to the insurer's interests, since they eliminate one way of reducing the moral hazard problem, which is to expose the insured party to risk for the uninsured top slice of liability.[52] It should apparently be in the insurer's interest to have a system of unlimited liability, where, on the one hand, a partial exposure to risk may be used by the insurer as a device to control moral hazard and where, on the other hand, the insurer may put contractual limitations on the amount of coverage (in the absence of a duty to insure up to a certain amount) depending upon the demand for insurance of the particular injurer and the willingness to provide coverage of the insurer. Thus, contractual limitations seem to be a better device that allows for an optimal differentiation of risk, thereby providing optimal control of moral hazard.

4.3.5 Financial Caps to increase the Insurability of Oil Pollution?

In sum, the insurability argument cannot fully justify the introduction of financial caps for oil pollution damage. The high magnitude of the damage may be a problem in cases of oil pollution as many incidents have shown. However, the insurer can limit its liability by contractual limitations. Moreover, if a duty to insure were introduced (and below it will be argued that there are sound arguments to do so) the (conventional or statutory) duty to insure could be limited to an insurable amount. The advantage of keeping the potential injurer, more particularly the tanker owner, fully liable for the remaining risk, is that this may provide additional incentives for prevention. This of course assumes that funds are indeed available in addition to the insured amount that can be exposed to risk. This may

---

[51] *See* T. Vanden Borre, *Efficiënte preventie en compensatie van catastroferisico's. Het voorbeeld van schade door kernongevallen* (Intersentia, Antwerp, 2001); OECD, *Liability and Compensation for Nuclear Damage, An International Overview* (OECD, Paris, 1994), pp. 312-354.

[52] *See* generally on the devices to control moral hazard, S. Shavell, 'On Moral Hazard and Insurance', (1979) *Quarterly Journal of Economics* (QJE) 541-562. See also M. Faure and T. Hartlief, 'Remedies for Expanding Liability', (1998) 18 *Oxford Journal of Legal Studies* (OJLS) 681-706.

be the case for some tanker owners, but definitely not for all. It seems, therefore, more appropriate to look at a mechanism which allows differentiation between injurers (and their available assets) and the potential accidents (and the corresponding magnitude of the loss). The generalized legislative caps may be too unbalanced and hence have possibly negative influence on the incentives towards deterrence.

4.4 A DUTY TO SEEK INSURANCE COVERAGE

In this paper, financial caps for oil pollution damage are analyzed. One implicit argument underlying the financial caps is that injurers might in any event face insolvency if they must pay an amount of damages exceeding their individual assets. The literature has dealt with this so-called 'judgement proof' extensively. Obviously, if the injurer had no assets that exceed the limited amount of liability, a liability exceeding the amount of the financial cap will not provide additional incentives. One significant disadvantage, however, of a generalized statutory financial cap is that it neglects the differences between injurers. Some might indeed have fewer assets than the limited amount; others may have more. By introducing a generalized financial cap, one excludes the possibility of providing additional incentives to these 'wealthy' injurers.

Moreover, as indicated above, a risk-averse injurer can protect himself against his potential insolvency by purchasing liability insurance. A disadvantage of financial caps in this respect is, as indicated above, that with caps the injurer's only duty is to purchase liability insurance to guarantee that he can pay the statutorily limited amount. The cap has the disadvantage that incentives for further reduction of damage are then diluted. Moreover, the insurer, in controlling the moral hazard problem, will only require the injurer to take care to avoid accidents that have a maximum magnitude of the statutorily limited amount and not to avoid accidents with the total expected amount of damages. This is once again an argument against a financial cap in combination with liability insurance. By leaving liability unlimited, the insurer can provide optimal insurance coverage, taking into account the expected damage.

Insolvency, however, may pose a problem in this respect. If the expected damages largely exceed the injurer's assets, the injurer will only have incentives to purchase liability insurance up to the amount of his own assets. He is indeed only exposed to the risk of losing his own assets in a liability suit. The judgement proof problem may therefore lead to underinsurance and thus to underdeterrence. Jost has rightly pointed to the fact that in these circumstances of insolvency, compulsory insurance might provide an optimal outcome.[53] By introducing a duty to purchase insurance coverage for the amount of the expected loss, better results will be obtained than with insolvency whereby the magnitude of the loss exceeds the injurer's assets. In the latter case, the injurer will indeed

---

[53] P.J. Jost, 'Limited liability and the requirement to purchase insurance', (1996) *International Review of Law and Economics* 259-276. A similar argument has recently been formulated by M. Polborn, 'Mandatory Insurance and the Judgement-Proof Problem', (1998) *International Review of Law and Economics* 141-146 and by G. Skogh, 'Mandatory insurance: transaction costs analysis of insurance' in B. Bouckaert and G. De Geest (eds.), *Encyclopedia of Law and Economics, II, Civil Law and Economics* (Edward Elgar, Cheltenham, 2000), pp. 521-537. Skogh has also pointed out that compulsory insurance may save on transaction cost.

only consider the risk as one where he could at most lose his own assets and will set his standard of care accordingly. When he is, under a duty to insure, exposed to full liability, the insurer will have an incentive to control the behaviour of the insured. Via the traditional instruments for the control of moral hazard, the insurer can ensure that the injurer will take the necessary care to avoid an accident with the real magnitude of the loss. Thus, Jost and Skogh argue that compulsory insurance can, provided the moral hazard problem can be cured adequately, provide better results than under the judgement proof problem. This is probably one of the explanations, for instance, for the introduction of compulsory insurance for traffic liability. Uninsured and insolvent drivers who have little money which they may lose compared to the possible magnitude of accidents they may cause may have little incentive to avoid an accident. Insurers might better be able to control this risk and could force the injurer to take care under the threat of closing the option of the insurance for him. Thus, the insurer acquires a duty to insure the licensee of the activity.

An additional advantage is that this compulsory insurance will lead to victim compensation. Victims are protected against the potential insolvency of the injurer through the compulsory insurance scheme. Obviously, this only works if moral hazard can be controlled adequately and insurers have appropriate incentives to do so. One condition is therefore that the insurance market should be competitive; another is that regulation should not inhibit inappropriate control of moral hazard (e.g., prohibit to exclude risky drivers). In sum: if it is the judgement proof problem that the legislator is worried about as regards oil pollution incidents, the appropriate answer at a policy level is clearly not to introduce legislative caps, but more likely to introduce a duty to insure if the conditions mentioned by Jost have been met.

In sum, the approach chosen in the CLC convention, this being to impose a duty to seek insurance coverage, corresponds well with the lessons from economic analysis. Such a duty may constitute an appropriate answer to the insolvency risk and may hence remedy the danger of underdeterrence caused through insolvency.

4.5 CHINA

It is fair to say from the above analysis concerning the (in)efficiency of financial caps and the description of the legislative situation in China that the inefficiencies inherent in the international system are even more serious in the situation of China. The financial caps are, as described above, far lower than even the caps provided for under the CLC and, more seriously, there is not yet an effective compulsory insurance system in Chinese national law. Given the insolvency risk the first priority for Chinese legislation should probably be to specify the originally too general provision on compulsory insurance to such an extent that it can be implemented effectively in practice according to the international model.

Moreover, the question arises as to whether in case of major oil spills in China sufficient funds will be available to provide compensation for the damage caused by oil pollution. As the shipping industry in China is growing, it is becoming imperative to set up a national compensation mechanism to deal with the growing ship-source oil pollution. China is a State party to the CLC Convention 1992, but mainland China has not acceded to the Fund Convention, nor set up a domestic fund that can be called upon as a second layer to protect the victims. Therefore, it is confronted with an unfavourable situation in which spills cannot be effectively cleaned up, and economic losses cannot be sufficiently compensated.

China's accession to the Fund Convention therefore seems inevitable and imminent. However, some Chinese scholars[54] have argued that the accession to the Fund might not seem an efficient solution to the central government of China at the current stage, as the actual costs would exceed the potential benefits. Chinese oil companies would constitute asignificant contributor to the Fund due to the increasing volume of oil import into China, and this large contribution would be an excessively heavy burden for the oil industry and thus will be detrimental to China's economic development. Therefore, further evidence is needed to prove whether accession to the Fund Convention is the optimal solution for China to strike the balance between environmental protection and economic development.

The relevant government departments have gained adequate understanding of the issue and have conducted studies on whether China should accede to the Fund Convention and how China can establish a compensation mechanism for ship oil pollution. Through extensive investigation, it is suggested a Chinese-style compensation mechanism be set up for ship oil pollution, which sounds more practical and more suitable for the present situation in China.[55]

The domestic-line vessels are often inferior to the international-line vessels in terms of ship age and technical conditions have thus increased the risks to port safety and the marine environment. It was suggested that such a Chinese-style fund should take account of the Chinese economy, the financial capacity of the shipowner, the compensation demand and the insurance supply, the proper proportion of the insurance and the fund. This domestic fund will solve problems like payment of the cleanup fee and the compensation if the ship at fault cannot be identified or has escaped; the advance payment for large-scale, long-term cleaning operation; insufficient compensation for the pollution damage, and so forth.

5 CONCLUDING REMARKS

The starting point for this analysis was the limitation of liability in the international conventions regulating oil pollution damage. An attempt was made to sketch the evolutions in the relevant international conventions and more particularly the relationship with Chinese law in this respect. These evolutions were described and critically analyzed from an economic perspective. Of course international conventions that regulate the compensation regime for oil pollution damage address a large number of issues, whereby the authors recognize that they only selected one of them, this being the financial caps on liability. The reason that focus was on financial caps is that this feature constitutes an

---

[54] L. Hui (from the Environment Protection Centre of the Ministry of Communication), *On China Oil Pollution Compensation System* (original text in Chinese: Guan Yu Wo Guo You Wu Sun Hai Pei Chang Zhi Du Wen Ti), International Seminar on Compensation Regime for Ship-source Marine Pollution Damage, Shanghai, June 2001.

[55] L. Hong and Z. Zhengming, *Establishing a Chinese Compensation Mechanism for Ship oil Pollution* (Institute of Scientific Research under the Ministry of Communications, Beijing, 2002).

important deviation from the general rule in tort law that a tortfeasor should in principle be fully liable for the damage he causes, thus enabling the victim to receive full compensation for the damage suffered. The question of course arises as to what justifies, specifically for the area of oil pollution damage, a deviation from this general rule. The issue of financial caps merits specific attention because one could notice many differences and variations in the amount of the caps. When, at the occasion of every new incident, it became clear that the amount of the previously adopted caps was too low, protocols and amendments were introduced to raise the amount to more realistic levels. That obviously raised the question as to whether liability should be limited at all. Moreover, one can increasingly observe criticism of the international oil pollution compensation regime, not only from economists, but also from environmentalists, arguing that the regimes mainly provide protection to tanker owners and the oil industry, but do not guarantee an effective compensation for damage. Even today, after the many changes the CLC Convention has undergone, the regime is still subject to criticism, not only because the question arises as to whether the financial limits will be adequate to provide compensation for recent catastrophes like the Prestige, but also because the conventions traditionally did not provide compensation for so-called ecological damage.

It is now possible to attempt to formulate a few modest lessons from the analysis in this paper at various levels.

Applying the economic analysis of tort law to financial caps teaches us that economics holds that there are very few reasons to introduce financial caps for third party liability. This outcome is certainly true when the parties involved in an accident are risk neutral, but it also applies in case of risk aversion, at least when insurance is available. The better option seems to be to agree on the optimal amount of liability via contract, which is obviously only possible when transaction costs are low, for example, when victim and injurer stand in a relationship to each other via the price mechanism.

The principal argument against financial caps assumes that the injurer has assets at stake which exceed the amount of the financial cap and that the expected amount of the damage will also be greater than the cap. In case of a judgement proof problem, the appropriate answer is clearly not to limit liability to the amount the injurer has at stake, but to seek insurance coverage. Through diversified contractual arrangements between the insurer and the injurer, an optimal amount of coverage can be determined in an individual case. The incentives can then be controlled via the risk differentiation in insurance and unlimited liability can apply for the excess, if the expected loss is (ex post) higher than the insured amount and the injurer still has assets at stake. In some cases, compulsory insurance might be an appropriate mechanism to control the judgement proof problem, but even compulsory insurance is no reason to introduce financial caps in legislation. The duty to purchase insurance coverage can be limited to a certain amount, but liability could remain unlimited.

This economic idea that an exposure to liability will provide adequate incentive for accident prevention has also been stressed by the European Commission, which recently argued that one of the criteria for deciding whether a compensation regime was satisfactory was whether it discourages operators from using vessels in less than satisfactory condition. This shows that the European Commission is also well aware of the fact that there is a relationship between an operator's exposure to liability of and the incentives to take adequate preventive measures.

However, one should proceed with some caution when drawing too soon the lesson from economic analysis at the normative level that financial caps should be absolutely abolished. This will as such not guarantee that sufficient coverage will be available and, hence, that incentives for an appropriate prevention will be met. Tanker owners might even (but this is also the case under financial caps) be persuaded to organize their own insolvency, for example, by limiting the available assets in one corporation.[56] However, the mere fact that the corporate world will always seek ways to organize their own insolvency (which may thus cause underdeterrence) should, within the scope of liability law, not be a reason for financial caps. Appropriate remedies might be sought in corporate law (as in some cases the piercing of the corporate veil, allowing victims to seize assets of a parent company in cases of damage caused by a subsidiary). Moreover, incentives for appropriate incentives and for responsible behaviour by tanker owners could also be provided via a code of good conduct for corporations. This may fit precisely within the modern tendency to stress the importance of socially responsible governance by corporations.

China can undoubtedly learn a few important lessons from the international regime. The most important lesson is probably that appropriate incentives in the fight against underdeterrence can be provided by implementing a duty to insure. Compulsory insurance for oil tankers, in cooperation with insurance companies, establish a method for oil pollution insurance that is applicable to small oil carriers of less than 2000dwt. The establishment of the oil pollution compensation fund expands the scope of its accession to the Fund Convention.

However, one could also argue that China should perhaps learn from the mistakes of the international regime whereby every time new amendments and protocols had to be accepted to adopt the financial limits to new major oil pollution incidents. If China were to introduce a national oil pollution damage regime (and this may be the case after the introduction of a new paragraph on this issue in the China Maritime Code), it should probably immediately either set relatively high limits capable of dealing with major incidents or, following economic logic, introduce unlimited liability, but limit the duty to insure to an insurable amount.

Economic analysis of law stresses the importance of exposing tanker owners to full liability, based on the belief that this will provide them with incentives to invest in preventive measures. However, the oil pollution compensation regime is clearly not only an issue of prevention and efficiency, but also of distribution, more particularly if one views the interests of the victims. In that respect the International Fund Convention provides an interesting and effective example where other parties (oil receivers) provide additional compensation.

In that respect China can learn from the successful experience abroad and develop a complete system of compensation, which includes compulsory insurance for oil tankers, establishment of a fund for oil pollution compensation, and determination of the compensation standards. When the time is appropriate, there should be immediate

---

[56] In China there have already been practices setting up so-called one-ship corporations after the financial limits of liability were raised. For critical comments concerning these single ship companies see K. Le Couviour, *supra* note 8, 2273.

accession to the Fund Convention. Only in this way will it be possible to maintain a sustainable development of the shipping industries and the maritime economies and solve the thorny issue that has long beset China's anti-pollution work.

There are, incidentally, interesting parallels between evolutions in Europe and in China. Europe has apparently, more particularly after the Erika incident, not been very satisfied with the compensation regime provided through the IMO-made international conventions and is now considering the establishment of a fund for the compensation of oil pollution damage in European waters.[57] A similar evolution whereby national or regional systems come in addition to the international regime is observable in China as well where the establishment of a domestic fund is seriously considered.[58]

Finally, in sum, it should be stressed once more that only 'one view of the cathedral' has been provided here by discussing only one aspect of the international compensation regime for oil pollution damage, this being the financial caps. Other issues, such as the channelling of liability and the question of whether ecological damage can be compensated, are obviously central as well, but beyond the scope of this paper. Moreover, the compensation issue was addressed from the belief that the exposure to liability of a potential injurer will provide efficient incentive for prevention. However, it is well documented in literature that the prevention of accidents cannot only be achieved through liability rules, but obviously through regulation as well. Hence, the effectiveness of the international regime aiming at the prevention of oil pollution spills is probably even more important. Indeed, prevention is always better than cure.

Moreover, it should again be stressed that law and legal remedies can only to some extent hope to influence the behaviour of social actors. A true change towards ecological behaviour of the oil industry will only be achieved if the norms are truly internalized and if the oil industry becomes aware that prevention of oil spills and providing adequate compensation is an essential part of their duty towards socially responsible governance. However, the public at large should equally be aware of the fact that increased preventive measures and increased compensation always come at a price: these (necessary and efficient) steps will undoubtedly lead to (probably even small) increases in oil prices which the public at large must be willing to pay.

---

[57] *OJ* C227E/487 of 24 September 2002.
[58] For an overview of regional solutions *see* E. Gold, *supra* note 3, 40-44 and *see also* M. Harjono and E. Leemans, 'Vervuilen loont. Organisatie van internationale scheepvaart stimuleert milieuvervuiling op zee', in (2003) 2 *Justitiële verkenningen*, (criminaliteit op zee) 68-69.

LIST OF REFERENCES

D. Abecassis, 'I.M.O. and liability for oil pollution from ships: a retrospective', (1983) *Lloyds Maritime and Commercial Law Quarterly* 45-46.

A. Blanco-Bazan, *The Erika casualty, Legal Issues from the IMO View*, IUMI 2000 Conference London, 10-13 September 2000, Liability Workshop.

J. Bongaerts and A. Debièvre, 'Insurance for civil liability for marine oil pollution damages', (1987) *The Geneva Papers on Risk and Insurance* 145-187.

E. Brans, 'Liability for ecological damage under the 1992 Protocols to the Civil Liability Convention and the Fund Convention, and the Oil Pollution Act of 1990', (1994) *Tijdschrift voor Milieuaansprakelijkheid* (TMA) 61-91.

E.H.P. Brans, *Liability for damage to public natural resources. Standing, damage and damage assessments* (Kluwer Law International, The Hague, 2001).

X. Chen, *Limitation of liability for maritime claims. A study of US law, Chinese law and international conventions* (Kluwer Law International, The Hague, 2001).

M. Faure, 'De verzekering van het nucleaire risico', in *In Volle Verzekerdheid, Opstellenbundel ter gelegenheid van het afscheid van Prof. Mr. O. van Wassenaer van Catwijck* (Tjeenk Willink, Zwolle, 1993), pp. 247-260.

M. Faure, 'The limits to Insurability from a Law and Economics Perspective', (1995) *Geneva Papers on Risk and Insurance* 454-462.

M. Faure, 'Economic Models of Compensation for Damages caused by Nuclear Accidents: some lessons for the division of the Paris and Vienna conventions', (1995) *European Journal of Law and Economics* 21-43.

M. Faure and T. Hartlief, 'Remedies for Expanding Liability', (1998) 18 *Oxford Journal of Legal Studies* (OJLS) 681-706.

M. Faure and P. Fenn, 'Retroactive liability and the insurability of long-tail risks', (1999) *International Review on Law and Economics* 487-500.

M. Faure and R. Van den Bergh, *Objectieve aansprakelijkheid, verplichte verzekering en veiligheidsregulering* (Maklu, Antwerp, 1989).

E. Fontaine, *The French Experience of Tanio and Amoco Cadiz, Incidents compared*, Genoa Seminar, 21-23 September, CMI.

E. Gold, *Handbook on marine pollution* (Gard, Arendal, 1985).

J. Greenfield, *China's practice in the law of the sea* (Clarendon Press, Oxford, 1992).

M. Harjono and E. Leemans, 'Vervuilen loont. Organisatie van internationale scheepvaart stimuleert milieuvervuiling op zee', in (2003) 2 *Justitiële Verkenningen,* (criminaliteit op zee) 68-69.

M. Hinteregger, 'La nouvelle loi autrichienne sur la responsabilité civile pour les dommages nucléaires', (1998) 62 *Bulletin de Droit Nucléaire* (BDN) 27-34.

L. Hong and Z. Zhengming, *Setting up the Chinese Characteristic compensation system for oil pollution damage from vessels* (original text in Chinese, Jian Li You Zhong Guo Te Se de Chuan Bo You Wu Sun Hai Pei Chang Ji Zhi), International Seminar on Compensation Regime for Ship-source Marine Pollution Damage, Shanghai, June 2001.

L. Hong and Z. Zhengming, *Establishing a Chinese Compensation Mechanism for Ship oil Pollution* (Institute of Scientific Research under the Ministry of Communications, Beijing, 2002).

L. Hui (from the Environment Protection Centre of the Ministry of Communication), *On China Oil Pollution Compensation System* (original text in Chinese: Guan Yu Wo Guo You Wu Sun Hai Pei Chang Zhi Du Wen Ti), International Seminar on Compensation Regime for Ship-source Marine Pollution Damage, Shanghai, June 2001.

M. Jacobsson, 'The international conventions on liability and compensation for oil pollution damage and the activities of the international oil pollution compensation fund', in C. De La Rue (ed.), *Liability for damage to the marine environment* (Lloyds of London Press, London, 1993), pp. 39-55.

M. Joransson, 'The 1984 and 1992 Protocols to the Civil Liability Convention 1969 and the Fund Convention 1971', in C. De La Rue (ed.), *Liability for damage to the marine environment* (Lloyds of London Press, London, 1993), pp. 71-82.

P.J. Jost, 'Limited liability and the requirement to purchase insurance', (1996) *International Review of Law and Economics* 259-276.

H. Kunreuther, R. Hogarth and J. Meszaros, 'Insurer Ambiguity and Market Failure', (1993) *Journal of Risk and Uncertainty* 71-87.

W. Landes and R. Posner, 'Tort law as a regulatory regime for catastrophic personal injuries', (1984) *Journal of Legal Studies* 421-422.

K. Le Couviour, 'Responsabilité pour pollutions majeures résultant du transport maritime d'hydrocarbures. Après l'Erika, le Prestige... L'impératif de responsabilisation', (2002) *La Semaine Juridique* (JCB) 2269-2277.

Z. Lujun, *Legal Issues in Marine Oil Pollution* (original text in Chinese: Hai Shang You Wu Chu Li Zhong De Ruo Gan Fa Lu Wen Ti), International Seminar on Compensation Regime for Ship-source Pollution Damage, Shanghai, June 2001.

W. Mao Shen, L. Shu Jian, and S. Man Tang, 'The normal procedure of assessment of damage to the marine environment in Chinese judicial practice', in C. De La Rue (ed.), *Liability for damage to the marine environment* (Lloyds of London Press, London, 1993), pp. 29-31.

OECD, *Liability and Compensation for Nuclear Damage, An International Overview* (OECD, Paris, 1994).

O. Ozçayirz, *Liabibility for oil pollution and collisions* (Lloyds of London Press, London, 1998).

M. Polborn, 'Mandatory Insurance and the Judgement-Proof Problem', (1998) *International Review of Law and Economics* 141-146.

M. Radetzki, 'Private Arrangements to cover large-scale liabilities caused by nuclear and other industrial catastrophes', (2000) *Geneva Papers on Riks and Insurance* 180-195.

C. Roche, 'Après l'Erika: la prevention de la pollution des mers par le renforcement de la sécurité maritime en Europe (Erika I)', (2002) 3 *Revue Juridique de l'Environnement* 373-392.

S. Shavell, 'On Moral Hazard and Insurance', (1979) *Quarterly Journal of Economics* (QJE) 541-562.

S. Shavell, 'Strict Liability versus Negligence', (1980) *Journal of legal Studies* 1-25.

S. Shavell, 'The Judgement Proof Problem', (1986) *International Review of Law and Economics* 43-58.

S. Shavell, *Economic Analysis of Accident Law* (Harvard University Press, Cambridge, 1987).

G. Skogh, 'The Transaction Cost Theory of Insurance: Contracting Impediments and Costs', (1989) *Journal of Risk and Insurance* 726-732.

G. Skogh, 'Mandatory insurance: transaction costs analysis of insurance' in B. Bouckaert and G. De Geest (eds.), *Encyclopedia of Law and Economics, II, Civil Law and Economics* (Edward Elgar, Cheltenham, 2000), pp. 521-537.

M. Trebilcock and R.A. Winter, 'The economics of nuclear accident law', (1997) *International Review on Law and Economics* 215-243.

T. Vanden Borre, *Efficiënte preventie en compensatie van catastroferisico's. Het voorbeeld van schade door kernongevallen* (Intersentia, Antwerp, 2001).

G.J. Van der Ziel, 'De olieramp met de 'Prestige'', (2003) *Tijdschrift voor Milieuaansprakelijkheid* 3-8.

Y. Xiaohan, Case on Limitation of Liability for Oil Pollution Damage Compensation, Guangzhou Maritime Court Cases, 24 March 2003.

C. White, 'The voluntary oil spill compensation agreements- TOVALOP and CRISTAL', in C. De La Rue (ed.), *Liability for damage to the marine environment* (Lloyds of London Press, London, 1993), pp. 57-70.

PART IV: COMPARATIVE CONCLUSIONS

COMPARATIVE CONCLUSIONS

Michael Faure and James Hu

1 INTRODUCTION

In the introduction, the reasons for this book and for including the particular contributions in it were outlined. The result is an overview of papers dealing with various aspects of prevention and compensation related to marine pollution damage. Attention in that respect was paid to both the international level itself (mainly the Civil Liability Convention and the Fund Convention) and evolutions in Europe, in the US, in other countries (like Indonesia) and in China. The main conclusions of the contributions to this book will be summarized here both as far as the approach is concerned and with respect to the contents. There will be no attempt to summarize or repeat the contents of the papers. Focus will therefore be on a few major points and tendencies.

This book contains contributions from a wide variety of contributors with very different backgrounds. Thus there where contributions Indonesian (see e.g. the contribution by Agoes), Chinese (see the many contributions in part III), European (see e.g. the contributions by Heine, Huybrechts and Van Damme, and Richter) and American (more particularly by Boyd and Cohen) academics. The contributions were not only submitted by lawyers, but also by economists who equally addressed the economic approach to the legal remedies to oil pollution damage. In addition to contributions from academics, the book also contains contributions from practitioners, for instance, from judges (Cheng Pingping and Guan Zhengyi) and attorneys (e.g. Chen Xuebin), and from employees of the Maritime Safety Administration (Song Ruqing and Wei Zhijie). This substantial input, also from practitioners, had the chief advantage that a valuable contribution could be provided by people who deal with marine pollution problems in practice.

This great variety in the contributors from various countries and various backgrounds (academics and practitioners) had the principal advantage that the legal aspects of the marine pollution problem could be addressed from all of its various angles, which was borne out by the contributions. The mix of speaker for instance led to a Chinese lawyer discussing European law (Wang Hui) and a German Lawyer discussing Chinese environmental criminal law (Thomas Richter). These examples show the truly comparative nature of the book, which was precisely the approach chosen.

2 PREVENTION

A few of the papers in part I dealt with the important issue of preventing marine pollution damage. In her paper, Wang Hui sketches the complicated evolution at the EU level concerning the prevention of marine pollution damage. In that respect, as her contribution showed, the so-called port State control, the notification system and the role of classification societies, *inter alia*, play important roles. It is striking to note in this respect that the European Union seems, as Wang Hui describes, to wish to run faster than the international regime. Europe being confronted with many major oil spills, like the Prestige and the Erika, apparently did not wish to wait for the international regime to implement further measures. Thus, Europe expedited the phasing out of single hull tankers, for instance, after the US had also banned single hull tankers from its waters. In this respect, an interesting 'race for the top' seems to have taken place in the sense that Europe took more stringent measures earlier, particularly because it feared that the phasing out of single hull tankers by the US may precisely create an increasing pollution danger for Europe.

In addition, some of the (Chinese) case studies address the prevention issue. For instance, the case study with respect to the transport of noxious liquid substances by Wei Zhijie and Song Ruqing stresses how through increased controls and the implementation of a risk analysis, pollution by noxious liquid substances could be prevented in the Yangtze delta. The importance of (emergency) plans, an efficient organization and information system is strongly stressed in that respect.

Although most of the legal literature today deals with the issue of how to compensate victims after marine pollution has occurred, it may be clear that the prevention of marine pollution damage is of course always better than the cure. Additionally, the international regime as developed by the IMO is increasingly imposing safety standards on ships all with the aim of preventing marine pollution damage from occurring. Most scholars do not question the effectiveness of tanker design regulations aimed at preventing accidental oil spills. Questions still arise with respect to the prevention of deliberate oil spills and with respect to the monitoring and enforcement of the regulations aiming at prevention.

3 MONITORING AND SANCTIONS

Another set of papers deals with these monitoring and enforcement issues. Marc Cohen provided an overview of empirical (mostly economic) research with respect to the effectiveness of various prevention and enforcement measures as they have been used in the US. These studies found that increased monitoring and enforcement can indeed reduce both the frequency and the size of oil spills. However, Cohen equally indicates that not all enforcement activities have a similar effect. Within the US context especially targeted monitoring and random patrols of ports and coastal waterways looking for spills seem to have been quite effective. An interesting suggestion made by Cohen is to make greater use of private law enforcement. Especially in cases where public polluters (e.g. governments) are the source of pollution, public agencies may be less inclined to enforce effectively. In those cases private enforcement may well be worth investigating. Finally, Cohen also stresses the importance of non-legal remedies like social norms, community pressure and reputation. Thus it may well be beneficial to have a large amount of publicity when enforcement actions are taken, since such a 'shaming' of polluters could well lead to additional deterrence.

Similar conclusions are reached in a set of papers addressing the way in which marine pollution damage should be criminalized. Günter Heine strongly argues that the criminal law should not only focus on individuals, but equally on legal entities. After having sketched some shortcomings of the international framework, Heine sketches that in case of major pollution incidents a criminal law approach that is based mainly on an individual responsibility of Human beings will have significant shortcomings. Thus, he pleads strongly (and in line with an international trend) in favour of corporate criminal liability. Of course, it is argued that the sanctions should then also be adapted to the corporate actors and measures should be taken to avoid situations where an actor could escape criminal responsibility, for example, by erecting one-ship corporations.

Richter focuses on the well-known problem of the relationship between regulation (administrative law) and criminal law (also referred to as the administrative dependency of the criminal law) in the context of Chinese environmental law. Richter indicates that it is striking that in the field of marine oil pollution, there does not appear to be a true connection between criminal law and oil pollution regulation. He, thus, points to the fact that today Chinese law may be inadequate in providing protection to the marine environment. This is, by the way, a conclusion that is also reached by Guo Ping and Zhoa Lujun who address the legal protection of the marine environment in China and equally plead in favour of regulations providing a more direct and adequate protection of the marine environment.

An important issue as far as the application of regulations and especially the criminal law is concerned (and also dealt with in the paper by Heine) is to what extent States have inspection and monitoring powers (and can thus apply their criminal law) also outside their territory. This issue is also explicitly addressed with respect to the situation in China concerning pollution in the exclusive economic zone (EEZ) by Chen Xuebin. The author argues that Chinese law must be amended in accordance with the United Nations Convention on the Law of the Sea (UNCLOS) in order to give Chinese authorities the ability to exercise jurisdiction and thus also to control vessels in the exclusive economic zone.

4 THE INTERNATIONAL COMPENSATION REGIME AND THE NATIONAL APPLICATION

Many contributions to this book also critically review the compensation of victims of oil pollution damage. Many critically discuss the regime as far as its ability to compensate victims of oil pollution is concerned. Of course the case studies discussed in Part III also make clear the weaknesses of the international regime, this being that in case of catastrophic tanker incidents (like the one with Erika or the Prestige) compensation will be inadequate. Maria Paz Garcia Rubio makes clear in her description of the *Prestige* case that the scope of the incident was so large that neither the international regime nor Spanish law were adequately able to provide compensation to the victims. The incident with tankers like the Prestige and the Erika also led the European Union to develop its own initiatives as far as the compensation regime is concerned. Thus, Wang Hui described in her paper that the European Union proposed a European compensation fund which would provide higher amounts than (at the time) did the civil liability and fund conventions. Because of this European initiative, Wang Hui explains that the IMO changed the Convention in 2003 in order to raise the limits substantially. These financial limits on liability were also critically reviewed by Michael Faure and Wang Hui, also from an economic perspective, arguing that financial caps for oil pollution damage leads both to underdeterrence and to inadequate compensation for victims.

In this respect, James Boyd sketches an interesting alternative, as it is constituted by the American example. The American example is not only interesting in that the Oil Pollution Act provides a separate regime (since the US did not join the international conventions). The definition of pollution damage seems also larger than the one contained in the international regime and the same is the case for the amounts of compensation. Thus Boyd argues that the American model is at least capable of providing adequate compensation in a majority of the cases and thus the American example may constitute an interesting model for other countries or, as Boyd presents it, perhaps even for global solutions.

Many of the Chinese papers contained in part III also critically review the international conventions and the application of the conventions in China (e.g., the papers by Han Lixing/Guan Zhengyi and James Hu/Yang Bo). As all of these papers stress, China has adopted the Civil Liability Convention (CLC 1969), but not the Fund convention. This leads some authors also to examine whether there should be a national oil pollution compensation fund in China, in order to provide additional compensation in the same way as the Fund Convention provides this as a supplement to the CLC. Another interesting issue mentioned in these studies is that, according to Chinese law, the CLC Convention applies apparently only when an international element is included in the tanker incident. Thus the CLC convention does not, for instance, apply to oil pollution damage caused by coastal vessels in China. In that case, Chinese domestic law applies which, however, provides a lower amount of compensation than the international regime and no specific rules for oil pollution damage.

An important conclusion from many of these studies is therefore that either the international regime should be entirely (thus including the Fund Convention) applicable to tanker incidents in China (including with coastal vessels) or a separate regime should be introduced into Chinese domestic law to guarantee adequate compensation to victims, which is not the case today.

5 COMPENSATING MARINE POLLUTION DAMAGE IN CHINA

Many of the papers in Part III dealt with the various aspects of the legal regime with respect to compensation of marine pollution damage in China. Already an adequate description of the law in this respect seems valuable since studies in the English language describing the legal regime with respect to marine pollution damage in China are rare. The contributions in Part III provide valuable insight into the legal doctrinal discussions that take place in China as well. A result of many of these contributions is that, strikingly, the legal regime is not at all that clear and that, as far as the situation is clear, most of the contributors argue in favour of serious revisions. For instance, Han Lixin and Guan

Zhengyi again discuss the debated issue whether the Civil Liability Convention (CLC 1969) should apply in China to oil pollution damage cases without a foreign element. The case law seems divided in that respect. As was mentioned, James Hu and Yang Bo clearly state in their contribution that the Civil Liability Convention does not apply to oil pollution damage caused by Chinese coastal vessels and they therefore argue strongly in favour of a separate legal regime to deal with this damage.

Zhao Yuelin, James Hu and Wang Hua deal with an interesting question that has led to a great deal of debate in other legal systems as well, this being the precise legal nature of the clean-up costs under Chinese law. The issue is indeed heavily debated since administrative authorities often proceed to a clean-up based on regulatory law, which raises the question of whether the claim they have on the polluter is of a civil or an administrative nature. They argue that in fact the administrative authorities clean up rather than the polluter and enjoy a competence to do so on the basis of administrative law. However, the costs incurred can, they argue, be reclaimed from the polluter based on civil liability.

Similar issues are also addressed in various papers that analyze the difference between environmental damage (and tort) versus maritime law. The question can indeed be asked to what extent, for example, oil pollution damage should be treated differently than other environmental torts. It is an issue that has received attention in many countries. For instance, in Belgium Huybrechts and Van Damme argue that the protection of the marine environment receives specific attention under environmental law (as a result of a specific statute), apparently independent of maritime law. A similar issue apparently also arises in China as one can learn from the contribution by Li Zhiwen and An Shouzhi. They argue that, nowadays, maritime pollution issues are dealt with under traditional tort law, without sufficient linkage with maritime law. These authors argue that marine pollution incidents like oil pollution are better dealt with by maritime tort law since this would enable victims to use the special maritime procedure law which may be helpful for the claimant.

6 DAMAGE ASSESSMENT, CLAIMS HANDLING AND INSURANCE

Many papers also deal with important practical aspects with respect to the compensation of marine pollution damage. One important issue in this respect is, for instance, how environmental damages should be assessed, if it is possible to compensate the injured parties.

This is an issue that has received a great deal of attention in the US, as one can read in the contribution of James Boyd. There, various methods to assess ecological damage have been successfully applied and Boyd shows how authorities and the judiciary in the US have, notwithstanding substantial difficulties, been able to put a monetary figure on the environmental damage caused by oil pollution, for example. Although this may, within the context of a country like China, sound like a luxury problem, the paper by Wu Lijing and Zhang Min shows that similar issues arise in China as well. Indeed, under Chinese law, procedures exist for the judicial authentication of marine pollution damage. However, the authors argue that, in practice, there may be problems with this authentication report;

hence, they formulate various suggestions for increasing the adequacy and quality of those reports.

Some Chinese papers also deal with the important issue of claims handling by victims. Thus, Ma Jing-jing and Du Jiang propose setting up a national claim mechanism for oil pollution damage. In essence, they point to the different mechanisms that victims could use to obtain compensation for damage suffered and to the different roles of, on the one hand, private victims and, on the other hand, administrative authorities like a maritime safety administration.

Finally, many papers also deal with a highly important aspect of the compensation mechanism, this being insurance. As in many other legal systems, in China the liable tanker owner will also often have insurance coverage. Many contributors of course point to the fact that within the context of the civil liability convention, the tanker owner is even under a duty to have insurance coverage. Interesting questions arise in that respect (mostly under national law, since this is regulated in the Convention) with respect to the division of liability between, on the one hand, the liable tanker owner and, on the other hand, the insurer. A question can in that respect arise as to whether the victim may have a direct right of action against the insurer. That is precisely the issue addressed by Judge Chen Ping Ping and she argues, *inter alia*, in the interest of victim protection for a joint and several liability of both the liable tanker owner and his insurer whereby the victim could basically choose at whom he addresses his claim.

7 RESULTS

From the summary of the contributions to the book presented here, it follows that the research of which the results where presented pointed to a few important results both for the international regime, for enforcement strategies and for developments at the regional and national levels. Indeed, many contributions formulated criticism with respect to the international regime as it has been developed within the context of the International Maritime Organization (IMO). The case studies presented in this book and the critical (also economic) analysis show that in fact the conventions were apparently not able to deal with major pollution incidents. The amounts provided for in the civil liability and fund convention may be adequate for an average pollution incident. When, however, catastrophic incidents like the Prestige and Erika occurred, the conventions proved inadequate to provide victim compensation.

In this respect is also striking that some regions have developed separate regimes. The US for instance has intervened intensely in the development of the international regime, but have not actually joined it themselves. After the Exxon Valdez incident, the US, to the contrary, developed its own oil pollution act whose features (as the contribution by James Boyd showed) to a large extent differ from those developed under the international regime of the civil liability and fund convention. For a moment, it seemed as though Europe (as shown in the contribution by Wang Hui) also went its own way. Indeed, the European Commission for a while proposed a European compensation fund that would merely apply to tanker incidents leading to pollution damage in European waters whereby higher amounts of compensation would be made available. Interestingly this European initiative led in turn to the most recent amendment of the international regime as a result of which today almost one billion US dollars can be made available to victims of oil pollution damage. This example shows that these initiatives at the regional (EU) level may have an significant influence on the development of the international regime.

The various contributions showed other interesting features as far as the interaction between the international and the national levels is concerned. Of course, many contributors examined to what extent the conventions were applicable in China, for example. However, other contributors (e.g., Huybrechts and Van Damme) showed that to some extent the national legal systems also continue to develop their own solutions for the protection of the marine environment, also irrespective of the international regime or with a different scope and goal.

Notwithstanding the criticism that was formulated on the international regime in many contributions, it was also shown that for a country like China the conventions could lead to a substantial improvement of the situation of the victims, for example. It was in this respect striking to note that the legal situation in China is still uncertain even as far as the applicability of the Civil Liability Convention is concerned. It is undoubtedly applicable when a foreign element is included, but when this is not the case, the case law appears unsettled and the same is, as the contributions in this volume show, the case for legal doctrine. It is, however, not debated that China has not joined the Fund Convention and therefore the compensation for victims in China will be lower than in the international regime.

Thus, many authors still examine to what extent current domestic law is able to provide compensation to victims of marine pollution damage. Also in that respect it is striking that still many uncertainties exist and still a great deal of regulation is lacking. If one conclusion may be drawn from the many contributions in this book dealing with the situation in China, it is undoubtedly that almost all Chinese contributions end with a call to the legislator to issue regulation both with respect to prevention and with respect to compensation.

However, as other contributions made clear, it is not only in Chinese law that much is left to be desired as far as the legal situation is concerned. For instance, as far as the formulation of criminal law is concerned, many legal systems still focus too much on the individual physical persons instead of shifting towards corporate criminal responsibility, which may (as was shown by Heine) be desirable in cases of oil pollution incidents. These and many other conclusions where just a few of the suggestions that followed from the contributions in this book.

8 POLICY

It is hoped that this book dealing with recent developments concerning the prevention and compensation of marine pollution damage in Europe, China and the US may have important policy implications and thus lead to an improvement in the prevention and compensation of marine pollution damage. In many contributions, specific suggestions were made for the improvement of the (international, European and national) legal regime.

It is further hoped and believed that the information provided in some of the (European and American) contributions may be useful for legal systems, like China's, that are developing a legal regime for the protection of the marine environment. In shaping new legislation, it is always extremely useful to mirror examples from abroad, even if it is only to learn from the mistakes that others have made.

It is also hoped that for a non-Chinese readership, it is interesting to become acquainted with the level of discussions in case law and legal doctrine in China and to discover how this system is in full evolution and to some extent heavily debated. For this non-Chinese readership, it may be interesting to discover how apparently similar problems arise in China as have arisen in Europe and the US, for example. For instance, the question whether the recovery of clean-up costs should be formulated as an administrative order or as a civil action has been heavily debated in other legal systems as well. In addition, in other respects the developments in China and in Europe are comparable. For instance, China is today contemplating creating a domestic oil pollution compensation fund following Europe's suggestion several years ago of launching a fund for the compensation of pollution damage to European waters. Thus, it is hoped that the contributions contained in this volume have provided scope for mutual learning and for improvement of current policy, leading to improved prevention of and compensation for oil pollution damage.

9 LOOK FORWARD

Since the fact that many contributions provide specific indications for policy reform is summarized here and since it was also indicated that in many legal systems (International, Europe, but also China) the legal regime is still in full evolution, it is of course hoped that the contributions contained in this volume may inspire policy makers when legislative changes are contemplated.

Therefore, it is also acknowledged that a similar comparative policy analysis can be of great use and could lead to high-quality improvement. It should, therefore, be undertaken more often. Thus, it is believed that this volume will certainly not constitute the end of the cooperation between the various contributors who worked on the project that constituted the basis for this book and it is hoped that when further policy changes and (hopefully positive) evolutions have taken place a follow up with further research will result.

COMPARATIVE ENVIRONMENTAL LAW AND POLICY SERIES

1. Environmental Contracts – Comparative Approaches to Regulatory Innovation in the United States and Europe, Eric W. Orts and Kurt Deketelaere (eds.). ISBN 90-411-9821-0
2. Environmental Regulation through Financial Organisations – Comparative Perspectives on the Industrialised Nations, Benjamin J. Richardson (ed.). ISBN 90-411-1735-0
3. Access to Justice in Environmental Matters in the EU/Accés à la justice en matiére d'environnement dans l'UE, Jonas Ebbesson (ed.). ISBN 90-411-1826-8
4. Civil Liability for Environmental Damage – A Comparative Analysis of Law and Policy in Europe and the United States, Mark Wilde (ed.). ISBN 90-411-1891-8
5. Public Environmental Law in the European Union and the United States – A Comparative Analysis, René J.G.H. Seerden, Michiel A. Heldeweg and Kurt R. Deketelaere (eds.). ISBN 90-411-1926-4
6. International, EC and US Environmental Law – A Comparative Selection of Basic Documents, Kurt Deketelaere and Jan Gekiere (eds.). ISBN 90-411-1938-8
7. Public Interest Environmental Litigation in India, Pakistan and Bangladesh, Jona Razzaque (ed.). ISBN 90-411-2214-1
8. Criminal Enforcement of Environmental Law in the European Union, Michael Faure and Günter Heine (eds.). ISBN 90-411-2337-7
9. Prevention and Compensation of Marine Pollution Damage. Recent Developments in Europe, China and the US, Michael Faure and John Hu (eds.). ISBN 90-411-2338-5